Emerging Topics in Ecotoxicology

Principles, Approaches and Perspectives

Volume 4

Series Editor

Lee R. Shugart
L.R. Shugart and Associates, Oak Ridge, TN, USA

For further volumes:
http://www.springer.com/series/7360

Bryan W. Brooks · Duane B. Huggett
Editors

Human Pharmaceuticals in the Environment

Current and Future Perspectives

 Springer

Editors
Bryan W. Brooks
Baylor University
Waco, Texas, USA

Duane B. Huggett
University of North Texas
Denton, Texas, USA

ISSN 1868-1344   ISSN 1868-1352 (electronic)
ISBN 978-1-4614-3419-1   ISBN 978-1-4614-3473-3 (eBook)
DOI 10.1007/978-1-4614-3473-3
Springer New York Heidelberg Dordrecht London

Library of Congress Control Number: 201293197

Printed on acid-free paper

Springer is part of Springer Science+Business Media (www.springer.com)

Contents

Perspectives on Human Pharmaceuticals in the Environment................... 1
Bryan W. Brooks, Jason P. Berninger, Alejandro J. Ramirez,
and Duane B. Huggett

**Environmental Risk Assessment for Human Pharmaceuticals:
The Current State of International Regulations**... 17
Jürg Oliver Straub and Thomas H. Hutchinson

Regulation of Pharmaceuticals in the Environment: The USA.................. 49
Emily A. McVey

Environmental Fate of Human Pharmaceuticals... 63
Alistair B.A. Boxall and Jon F. Ericson

**Environmental Comparative Pharmacology: Theory
and Application**.. 85
Lina Gunnarsson, Erik Kristiansson, and D.G. Joakim Larsson

**A Look Backwards at Environmental Risk Assessment:
An Approach to Reconstructing Ecological Exposures**............................. 109
David Lattier, James M. Lazorchak, Florence Fulk, and Mitchell Kostich

**Considerations and Criteria for the Incorporation of
Mechanistic Sublethal Endpoints into Environmental
Risk Assessment for Biologically Active Compounds**................................ 139
Richard A. Brain and Bryan W. Brooks

**Human Health Risk Assessment for Pharmaceuticals in the
Environment: Existing Practice, Uncertainty, and Future Directions**....... 167
E. Spencer Williams and Bryan W. Brooks

Wastewater and Drinking Water Treatment Technologies.......................... 225
Daniel Gerrity and Shane Snyder

Pharmaceutical Take Back Programs... 257
Kati I. Stoddard and Duane B. Huggett

Appendix A. Take Back Program Case Studies 287

Index.. 297

Contributors

Jason P. Berninger Department of Environmental Science, Center for Reservoir and Aquatic Systems Research, Institute of Biomedical Studies, Baylor University, Waco, TX 76798, USA

Office of Research and Development, National Health and Environmental Effects Research Laboratory, U.S. Environmental Protection Agency, Duluth, MN 55804, USA

Alistair B.A. Boxall Environment Department, University of York, Heslington, York YO10 5DD, UK

Richard A. Brain Ecological Risk Assessment, Syngenta Crop Protection LLC, Greensboro, NC 27409, USA

Bryan W. Brooks Department of Environmental Science, Center for Reservoir and Aquatic Systems Research, Institute of Biomedical Studies, Baylor University, Waco, TX 76798, USA

Jon F. Ericson Pfizer Global Research and Development, Worldwide PDM, Environmental Sciences, MS: 8118A-2026, Groton, CT 06340, USA

Florence Fulk National Exposure Research Laboratory, Ecological Exposure Research Division, US Environmental Protection Agency, Office of Research and Development, Cincinnati, OH 45268, USA

Daniel Gerrity Water Quality Research and Development Center, Southern Nevada Water Authority, River Mountain Water Treatment Facility, Henderson, NV 89015, USA

Lina Gunnarsson Department of Neuroscience and Physiology, Institute of Neuroscience and Physiology, The Sahlgrenska Academy, University of Gothenburg, 405 30 Göteborg, Sweden

Duane B. Huggett Department of Biological Sciences, University of North Texas, Denton, TX 76203, USA

Thomas H. Hutchinson   CEFAS Weymouth Laboratory, Centre for Environment, Fisheries and Aquaculture Sciences, Weymouth, Dorset DT4 8UB, UK

Mitchell Kostich   National Exposure Research Laboratory, Ecological Exposure Research Division, US Environmental Protection Agency, Office of Research and Development, Cincinnati, OH 45268, USA

Erik Kristiansson   Department of Neuroscience and Physiology, Institute of Neuroscience and Physiology, The Sahlgrenska Academy, University of Gothenburg, 405 30 Göteborg, Sweden

Department of Zoology, University of Gothenburg, 405 30 Göteborg, Sweden

D.G. Joakim Larsson   Department of Neuroscience and Physiology, Institute of Neuroscience and Physiology, The Sahlgrenska Academy, University of Gothenburg, 405 30 Göteborg, Sweden

David Lattier   National Exposure Research Laboratory, Ecological Exposure Research Division, US Environmental Protection Agency, Office of Research and Development, Cincinnati, OH 45268, USA

James M. Lazorchak   National Exposure Research Laboratory, Ecological Exposure Research Division, US Environmental Protection Agency, Office of Research and Development, Cincinnati, OH 45268, USA

Emily A. McVey   Office of Pharmaceutical Science, Center for Drug Evaluation and Research, U.S. Food and Drug Administration, Silver Spring, MD 20993, USA

WIL Research, 5203DL 's-Hertogenbosch, The Netherlands

Alejandro J. Ramirez   Mass Spectrometry Center, Mass Spectrometry Core Facility, Baylor University, Baylor Sciences Building, Waco, TX 76798, USA

Shane Snyder   Chemical and Environmental Engineering, University of Arizona, Tucson, AZ 85721, USA

Jürg Oliver Straub   F.Hoffmann-La Roche Ltd, Group SHE, LSM 49/2.033, Basle CH-4070, Switzerland

Kati I. Stoddard   Department of Biological Sciences, University of North Texas, Denton, TX 76203, USA

E. Spencer Williams   Department of Environmental Science, Institute of Biomedical Studies, Center for Reservoir and Aquatic Systems Research, Baylor University, Waco, TX 76798-7266, USA

Perspectives on Human Pharmaceuticals in the Environment

Bryan W. Brooks, Jason P. Berninger, Alejandro J. Ramirez, and Duane B. Huggett

Background

Human interaction with the environment remains one of the most pervasive facets of modern society. Whereas the anthropocene is characterized by rapid population growth, unprecedented global trade and digital communications, energy security, natural resource scarcities, climatic changes and environmental quality, emerging diseases and public health, biodiversity and habitat modifications are routinely touted by the popular press as they canvas global political agendas and scholarly endeavors. With a concentration of human populations in urban areas

B.W. Brooks (✉)
Department of Environmental Science, Center for Reservoir and Aquatic Systems Research,
Institute of Biomedical Studies, Baylor University, One Bear Place, #97266,
Waco, TX 76798, USA
e-mail: Bryan_Brooks@Baylor.edu

J.P. Berninger
Department of Environmental Science, Center for Reservoir and Aquatic Systems Research,
Institute of Biomedical Studies, Baylor University, One Bear Place, #97266,
Waco, TX 76798, USA

National Health and Environmental Effects Research Laboratory, National Research Council
Research Associates Program, Office of Research and Development, U.S. Environmental
Protection Agency, 6201 Congdon Boulevard, Duluth, MN 55804, USA
e-mail: Berninger.Jason@epamail.epa.gov

A.J. Ramirez
Mass Spectrometry Center, Mass Spectrometry Core Facility, Baylor University,
Baylor Sciences Building, One Bear Place, #97046, Waco, TX 76798, USA
e-mail: Alejandro_Ramirez@Baylor.edu

D.B. Huggett
Department of Biological Sciences, University of North Texas,
1155 Union Circle, #305220, Denton, TX 76203, USA
e-mail: dbhuggett@unt.edu

B.W. Brooks and D.B. Huggett (eds.), *Human Pharmaceuticals in the Environment:
Current and Future Perspectives*, Emerging Topics in Ecotoxicology 4,
DOI 10.1007/978-1-4614-3473-3_1, © Springer Science+Business Media, LLC 2012

unlike any other time in history, the coming decades will be defined by "A New Normal," as proposed by Postel [1], where the interplay among sustainable human activities and natural resource management will inherently determine the regional fates of human societies.

In recent years, few topics have captured the public's attention like the presence of human pharmaceuticals in environment. Fish on Prozac [2, 3]. Male fish becoming female [4, 5]? Drugs found in drinking water [6, 7]. India's drug problem [8]. Chances are you have seen these headlines or read related reports. Pharmaceuticals and trace levels of other contaminants (e.g., antibacterial agents, flame retardants, perfluorinated surfactants, harmful algal toxins) are increasingly reported in freshwater and coastal ecosystems. In the developed world, many of these chemicals are released at very low levels (e.g., parts per trillion) from wastewater effluent discharges to surface and groundwaters. But why were citizens so engaged by stories about fish on Prozac [3] and drugs in drinking water [7]? Because pharmacotherapy is now entrenched in everyday life, a realization that common drugs were found in the water we drink or the fish we eat likely produces a boomerang effect, where our daily reliance on well-accepted therapies was concretely linked in a new way with their potential consequences to the natural world. On an increasingly urban planet, pharmaceutical residues and traces of other contaminants of emerging concern represent signals of the rapidly urbanizing water cycle and harbingers of the "New Normal."

Over the past 2 decades the implications of endocrine disruption and modulation have permeated public consciousness, scientific inquiry, regulatory frameworks, and management decisions in the environmental and biomedical sciences. Publication of Colburn, Dumanoski, and Myers' "Our Stolen Future [9]," which is often referred to as the second coming of Rachel Carson's "Silent Spring [10]," stimulated the public, scientific, and regulatory attention given to endocrine disruptors and ultimately influenced the environmental studies of human pharmaceuticals [11]. For example, human reproductive developmental perturbations elicited by the estrogenic human pharmaceutical diethylstilbestrol and feminization of male fish exposed to municipal effluent discharges represent examples of causal relationships among endocrine active substances and biologically important adverse outcomes [12].

In the late 1990s, research in the area of endocrine disruption was taking off, particularly to identify constituents of effluents or other environmental matrices that were potentially responsible for endocrine perturbations in wildlife and humans. Because many xenoestrogens are present in effluent discharges, initial investigations in the UK employed toxicity identification evaluation studies to fractionate and identify causative components of the complex mixtures inherent with effluents [13]. At the same time in the USA, Arcand-Hoy et al. [14] highlighted the importance of considering human estrogen agonist and veterinary androgen agonist pharmaceuticals as potential causative toxicants from point and nonpoint source effluents. Also in 1998, two of the first review papers on pharmaceuticals in the environment, by Halling-Sorensen et al. [15] and Ternes [16], appeared in the literature. In 1999, another review paper, by Daughton and Ternes [17], considered

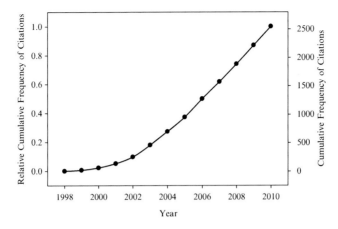

Fig. 1 Representative increase in peer-reviewed publications related to pharmaceuticals in the environmental through 2010, summarized by the cumulative and relative cumulative citation frequency of early review papers by Halling-Sorensen et al. [15], Ternes [16], and Daughton and Ternes [17]. Citation information from Web of Knowledge

Pharmaceuticals and Personal Care Products (PPCP) in the environment and by doing so coined the PPCP acronym, which remains pervasive. Subsequently, a precipitous number of workshops, symposia, special meetings, and publications related to pharmaceuticals in the environment have occurred. For example, Fig. 1 describes citation frequencies of just the Halling-Sorensen et al. [15], Ternes [16], and Daughton and Ternes [17] papers as surrogates for the trajectory of scientific inquiry in this important area of environmental science and public health.

Some of the most important developments related to pharmaceuticals in the environment included special issues of *Toxicology Letters* in 2002 and 2003, Pellston workshops by the Society of Environmental Toxicology and Chemistry (SETAC) on human pharmaceuticals (in 2003 [18]) and veterinary medicines (in 2007 [19]), formation of the SETAC Pharmaceuticals Advisory Group (in 2005; http://www.setac.org/node/34) and the Water Environment Federation's Microconstituents Community of Practice (http://www.wef.org), International Conferences on the Occurrence, Fate, Effects, and Analysis of Emerging Contaminants in the Environment (e.g., htpp://www.EmCon2011.com), the International Water Association's MicroPol conferences (e.g., htpp://www.micropol2011.org), and a special issue of *Environmental Toxicology and Chemistry* entitled "Pharmaceuticals and Personal Care Products in the Environment" in 2009. Following an editorial by Brooks et al. [20] entitled "Pharmaceuticals and Personal Care Products: Research Needs for the Next Decade," an international workshop entitled "Effects of Pharmaceuticals and Personal Care Products in the Environment: What are the Big Questions?" was held by Health Canada/SETAC in April 2011 [21]. In 2012, the SETAC Pharmaceutical Advisory Group is planning another Pellston conference on antimicrobial resistance, which represents a major threat to global public health. Though the information in this timely area continues to rapidly expand, it appears critically important to now consider the lessons learned from the study of human pharmaceuticals in the environment and formulate directions for future efforts.

Environmental Analysis and Exposure

To date, the majority of information for human pharmaceuticals in the environment is related to occurrence in various environmental matrices, which largely accounts for publication trends summarized in Fig. 1. Perhaps the most influential paper on occurrence was published by Kolpin et al. [22]. In 2002, this landmark article provided the first national reconnaissance study of a variety of contaminants of emerging concern, including a number of pharmaceuticals, in water [22] and promises to be the most heavily cited paper published in the history of the journal *Environmental Science & Technology*. In Table 1, we provide an overview of the representative literature related to the environmental analysis and occurrence of pharmaceuticals in the environment. Instead of performing an exhaustive survey and synthesis here, we instead relay some perspectives on environmental analysis and refer readers to the recent review of occurrence information for human pharmaceuticals by Monteiro and Boxall [23].

Table 1 Representative recent reviews on pharmaceutical analysis in various environmental matrices

Target analytes	Matrix	Type of review
Pharmaceuticals	Water	Analytical methods [64], multiresidue methods [65], LC–MS/MS methods [66], basic pharmaceuticals [67], antibiotics [68], anti-inflammatory drugs [69], recent advances [70]
	Solids[a]	LC–MS/MS [71], tetracycline antibiotics [72]
	Water, solids	Analytical methods [73], LC–MS/MS methods [74]
Conventional and/or contaminants of emerging concern, including pharmaceuticals	Water	Analytical methods [75, 76]
	Water, solids	LC–MS in environmental analysis [77]
	Various environmental matrices	Analytical methods [78, 79], methods applied to fate [80], environmental mass spectrometry [81], recent advances [82]
Pharmaceuticals and/or degradation products	Water	Advanced MS techniques [83], LC–MS methods [84], methods applied to fate and removal [85]
	Various environmental matrices	Mass spectrometry [86], analytical problems and sample preparation [87]
Other reviews related to pharmaceutical analysis and general occurrence information		Multivariate analysis [88, 89], sampling and/or extraction [90–94], chiral analysis [95], general occurrence [23], biological tissues [28, 29, 96]

[a]Sediment, biosolids and soil

Gas chromatography–mass spectrometry (GC–MS) was the primary analytical tool used to assess the environmental occurrence of PPCPs in initial studies (Table 1). The popularity of GC–MS in early work was due to its widespread availability and historical use in contract service laboratories for historical industrial chemical contaminants. The availability of electron-impact spectral libraries was initially important, as they increased confidence in analyte identification. Further, the distinctive nonpolar operating range of GC–MS was consistent with analysis of most personal care products (PCPs). In contrast, the use of GC–MS for analysis of pharmaceuticals, which are relatively polar compared to most PCPs, typically requires derivatization prior to analysis. For example, Brooks et al. [3] employed GC–MS with derivatization for initial identification of the antidepressants sertraline and fluoxetine in fish tissue. However, derivatization reactions are often unpredictable for complex samples and can limit the quality of quantitative data. Consequently, liquid chromatography–mass spectrometry (LC–MS) has become the technique of choice for analyzing pharmaceuticals in environmental samples.

Numerous studies have demonstrated the distinct advantages of LC–MS for analysis of pharmaceuticals (Table 1). LC–MS enables identification and quantification without derivatization and typically results in lower detection limits (below 1 ng/L and 1 ng/g for liquid and solid samples, respectively) and better precision than comparable GC–MS methodologies. In environmental applications, LC is typically combined with tandem MS (or MS/MS) to promote enhanced selectivity and sensitivity for target analytes. In a routine MS/MS analysis, a molecular ion is selected and subsequently fragmented to produce one or more distinctive product ions that enable both qualitative and quantitative monitoring. Recently introduced ultraperformance liquid chromatography (UPLC) provides a novel approach to chromatographic separation. UPLC differs from regular LC by the implementation of chromatographic columns with smaller particle diameters (i.e., sub-2-μm particles), which generates elevated back pressures and narrower chromatographic peaks. The overall effect is resolved peaks in shorter periods of time with increased sensitivity. UPLC requires fittings and pumps designed to support high back pressures, which increases the price of the LC system. An important feature of UPLC is the need of a fast detector to account for small peak widths (ca. 10 s). In other words to acquire enough data points through chromatographic peaks, selected mass spectrometer need to collect data points at high sampling rates. Q-TOF mass spectrometers are often coupled with UPLC systems due to their fast sampling rates. It is important to note, however, that LC–MS is not exempt from limitations. One of the limitations of LC–MS is that atmospheric pressure ionization (API) processes are influenced by coextracted matrix components. Matrix effects typically result in suppression or less frequent enhancement of analyte signal. There have been a number of methods proposed to compensate for matrix effects, including the method of standard addition, surrogate monitoring, and isotope dilution (Table 1). Although isotope dilution is the most highly recommended approach for analysis of human pharmaceuticals in environmental matrices, isotopically labeled standards are not always readily available for these target analytes. A further limitation is the paucity of available isotopically labeled standards for therapeutic metabolites. An alternative approach involves the use of an appropriate internal standard (i.e., a structurally similar compound expected to mimic the behavior of a target analyte(s)) with or without matrix-matched calibration. However, a given internal standard is typically effective over a limited retention time window. Accordingly, the use of more than one internal standard is recommended to compensate for matrix effects throughout the chromatographic run. Finally, it is important to point out that strategies to compensate for matrix effects should take into account the variability of matrix within each set of samples to be analyzed (e.g., surface water, effluent, sediment, fish tissue).

Due to potential regulatory implications of human pharmaceuticals in the environment, environmental analyses typically include rigorous quality assurance and quality control (QA/QC) metrics to confirm reliability of analytical data. Initial method validation provides essential performance parameters, such as method recoveries, precision, and limits of detection (LODs). Recurring analysis of quality control (QC) samples (e.g., method blanks, matrix spikes, laboratory control samples) is important to verify performance of the method over time, and to assess potential matrix effects. Considering the unpredictable nature of matrix interference in LC–MS analysis and the lack of effective strategies to deal with this difficulty, it has become imperative to use QA/QC data to document and qualify analytical results for human pharmaceuticals in environmental matrices. This is particularly important when reporting concentrations at or near the limit of detection for a given analytical method.

In this volume, an overview of global environmental regulatory activities relevant to human pharmaceuticals is provided in Chaps. 2 and 3. In Chap. 4, Boxall and Ericson examine important considerations for understanding the environmental fate of therapeutics. Below we provide some perspectives on bioaccumulation and effects of human pharmaceuticals in the environment.

Environmental Bioaccumulation and Effects

Though the potential for uptake of veterinary medicines by animals reared in aquaculture were understood for some time (see [24, 25]), Boxall et al.'s [26] study of the uptake of veterinary medicines from soils to plants highlighted the importance of considering potential accumulation of human medicines in terrestrial organisms because biosolids and effluents from wastewater treatment plants can be applied to agricultural fields. Such observations are particularly relevant for antibiotics. In fact, developing an understanding of the influences of human antibiotics and antimicrobial agents on antibiotic resistance was recently identified as critical areas of research need for environmental science and public health [21].

In aquatic systems, Larsson et al. [27] likely provided the first report of bioaccumulation of a human pharmaceutical, 17α-ethinylestradiol, in bile of fish exposed to Swedish effluent discharges. Brooks et al.'s [3] findings of the antidepressants fluoxetine and sertraline (and their primary metabolites) in brain, liver, and muscle tissues of three fish species from an effluent-dominated stream (a.k.a. fish on Prozac) appear to represent the second report in the literature of accumulation of human pharmaceuticals in wildlife and the first observation from North America. Such observations stimulated research related to the accumulation and effects of human pharmaceuticals in the environment and subsequently shaped the National Pilot Study of PPCPs in Fish Tissue by the US Environmental Protection Agency [28]. This study by Ramirez et al. [28] provided the first evidence of bioaccumulation of a number of human pharmaceuticals in fish collected across a broad geographic area. A summary of research on bioaccumulation of pharmaceuticals in aquatic organisms recently highlighted the need to understand thresholds of drug accumulation associated with adverse effects [29]. Unfortunately, an understanding of human pharmaceuticals accumulating in terrestrial wildlife is poorly understood [20] but has been recently identified as a major research question [21]. Several recent publications have started to further our understanding of the bioconcentration/bioaccumulation potential of pharmaceuticals in a laboratory setting, as well as publications aimed at understanding pharmaceutical metabolism in wildlife and its role in the accumulation of drugs [30–39]. Below we introduce important considerations for understanding relationships between pharmaco(toxico)kinetics and -dynamics of human medications in aquatic and terrestrial organisms. A more thorough examination of comparative pharmacological approaches for environmental applications is provided by Gunnarsson et al. in Chap. 5.

Understanding the environmental risks posed by historical contaminants has been challenged by the paucity of toxicity information available for most industrial chemicals [40]. In the case of human pharmaceuticals, however, intensive investigations occur prior to distribution, which yields a wealth of pharmacological and toxicological data compared to other industrial contaminants. To illustrate available data, Table 2 provides a summary of common characteristics for hundreds of pharmaceuticals. During the design of therapeutics, careful consideration is given to target-specific biomolecules (e.g., receptors, enzymes) and pathways to elicit beneficial outcomes. Because side effects are not desirable and large margins of safety (relationship between therapeutic and toxic doses) are ideal, pharmaceutical development often results in therapeutics with relative well-understood mechanisms/modes of actions (MOAs) and very low acute toxicity in mammals. For example, a recent study predicted that less than 8% of all pharmaceuticals are expected to be classified as highly acutely toxic to rodent models [41]. Similarly, Berninger and Brooks [41] predicted that less than 6% of all pharmaceuticals are acutely toxicity to fish below 1 mg/L.

As noted previously, concentrations of individual human pharmaceuticals in surface water of developed countries rarely exceed parts per billion levels; thus, limited acute toxicity is expected in surface waters of the developed world. Unfortunately, most studies to date have only examined acute toxicity in standard aquatic organisms [42]. However, chronic adverse responses resulting from therapeutic MOAs are more likely to be observed in the environment [41], particularly in systems with instream flows dominated by continuous release of effluent discharges [43] leading to longer effective exposure durations [11]. Early investigators

Table 2 A summary of the minimum and maximum values and 10th, 50th, and 90th centiles of common properties associated with pharmaceuticals

		MW	log P	LD_{50}	C_{max}	ATR	Cl	$T_{1/2}$	V_d	AqET
	Min	6.94	−9.4	0.00075	7.5×10^{-6}	1.6	0.0029	0.033	0.035	8.4×10^{-10}
Centiles	10th	164	−0.95	77	0.0017	97	0.49	0.77	0.15	2.9×10^{-5}
	50th	346	2.03	971	0.1300	8,127	3.71	5.01	1.03	0.0446
	90th	732	5.01	12,283	9.91	681,657	27.9	32.6	6.96	69.4
	Max	145,781	8.6	56,000	330	4.7×10^{8}	1,070	87,600	2,348	9.1×10^{9}
	n	1,042	797	1,035	832	741	936	979	944	831

MW molecular weight (g/mol); *log P* octanol–water partitioning coefficient; LD_{50} median oral lethal dose for rat model (mg/kg); C_{max} human peak plasma concentration (or therapeutic dose; µg/mL); *ATR* acute to therapeutic ratio margin of safety analog (LD_{50}/C_{max}; see Berninger and Brooks [41]); *Cl* clearance rate (mg/min/kg; $T_{1/2}$ half-life of elimination (hour); V_d apparent volume of distribution (L/kg); AqET is the aqueous effect threshold (mg/L) where fish plasma BCF/C_{max} = aquatic exposure concentration at the point in which C_{max} = fish plasma concentration and fish plasma BCF × exposure concentration = fish plasma concentration [29]

recognized the importance of leveraging mammalian pharmacological safety data to help understand various pharmaceutical effects in the environment, because many MOAs of human therapeutics appear to be evolutionarily conserved, particularly in vertebrates [14, 44–46].

In 2003, Huggett et al. [47] proposed a screening approach to identify pharmaceuticals in water that may result in fish plasma levels (or internal doses) ≥ human therapeutic levels (e.g., C_{max}). Huggett's plasma model was based on three core assumptions: (1) Evolutionary conservation of structure and function of drug targets among mammals and fish species; (2) Internal fish doses approaching mammalian C_{max} levels would result in similar therapeutic outcomes; and (3) A gill uptake model [48] for predicting rainbow trout plasma concentrations following waterborne exposure to nonionizable chemicals [48]. Subsequently, several recent studies have employed the Huggett et al. plasma model approach [49–51] or conceptually similar variations to account for ionization influences on bioavailability [29, 52, 53]. Of particular importance, Valenti et al. [53] recently provided an independent validation of the Huggett et al. [47] plasma model when ionization of the weak base sertraline [54] and an alternative gill uptake model [48] was considered. Valenti et al. [53] also employed an adverse outcome pathway (AOP) design [55], which included quantification of binding at the therapeutic target and anxiety-related behavioral responses stereotypical of the therapeutic efficacy of this model antidepressant. In the Valenti et al. [53] study, adult male fathead minnow were exposed via aqueous exposure to sertraline for 21 days. Fish plasma concentrations were accurately predicted from water exposures when pH influences on ionization and lipophilicity were considered [29, 52, 54]. When these plasma levels in fish exceeded the human therapeutic dose (C_{max}) of sertraline, binding to the serotonin reuptake transporter and antianxiety behavior were significantly affected [53]. The AOP approach was recently proposed by Ankley et al. [55] for linking molecular initiation events, such as those related to pharmaceutical interactions with a target site (e.g., a receptor), with cascading events leading to adverse outcomes at the individual and population level, which can be used as measures of effect in risk assessments. As demonstrated by Valenti et al. [53], linking predictions of uptake from surface waters to fish plasma with conceptual AOP models appear to represent a sound foundation from which potentially hazardous human pharmaceuticals may be identified.

Probabilistic hazard assessment approaches, which are commonly used to support environmental and public health decision making, can use existing mammalian pharmacological safety data to develop predictive models for various parameters [41]. These predictive tools can support prioritization activities for testing hypotheses regarding pharmacological parameters of various drug classes or chemical specific computational attributes that may result in hazards to wildlife [41]. For example, Table 2 presents the minimum and maximum values and 10th, 50th and 90th centiles of probabilistic pharmaceutical distributions (PPD) of molecular weight, logP, acute LD_{50}, C_{max}, acute to therapeutic ratio margin of safety analog (LD_{50}/C_{max}; see [41]), clearance rate, half-life of elimination, apparent volume of distribution (V_d), and the aqueous effect threshold (AqET; see [52]) based on data from hundreds of pharmaceuticals. PPD approaches can be used to predict the

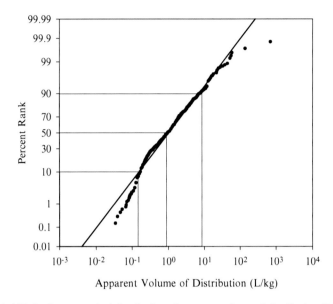

Apparent Volume of Distribution (L/kg)

Fig. 2 Probabilistic pharmaceutical distribution of apparent volume of distribution (L/kg) for 944 pharmaceuticals. Reference lines relate to the 10th, 50th and 90th centiles (Table 2), which correspond to 0.15, 1.03, and 6.96 L/kg, respectively. For example, apparent volume of distribution is predicted by this model to be at or above 6.96 L/kg for 10% of all pharmaceuticals

likelihood of encountering another therapeutic with attributes of interest. To illustrate the utility of PPD analyses, Fig. 2 depicts a PPD for V_d. Briefly, V_d data were ranked and converted to probability percentages then plotted against respective probability ranks on a log-probability scale; centiles were determined by regression (see [30] for a complete description of methods). Using this approach, we predict that 10% or less of all pharmaceuticals would have V_d values of 0.15 L/kg. In Fig. 3, we extend the PPD assessment to predict the likelihood of encountering a pharmaceutical in surface waters exceeding the AqET value, which is based here on the specific assumptions of Huggett et al.'s [47] plasma model. For example, 10% of all pharmaceuticals are predicted to result in internal fish plasma concentrations equaling the human C_{max} value at or below an environmentally relevant surface water concentration of 29 ng/L (Fig. 3, Table 2).

Based on the current state of the science, it appears critical to develop an advanced understanding of the risks associated with human pharmaceuticals in the environment. In Chaps. 6 and 7, Lattier et al. consider mechanistic characteristics of drugs for reconstructing environmental exposure scenarios and Brain and Brooks provide perspectives for incorporating non-standard endpoints in environmental risk assessments, respectively. In Chap. 8, Williams and Brooks examine human health risk assessment considerations for environmental exposures to therapeutics. When the outcome of an environmental risk assessment identifies unacceptable risks to wildlife or humans, risk management decisions and practices serve as interventions to protect public health and the environment. In the case of pharmaceuticals and other

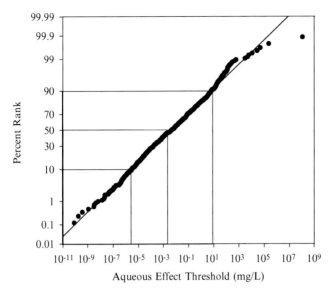

Fig. 3 Probabilistic pharmaceutical distribution of aqueous effect threshold (AqET; mg/L) for 831 pharmaceuticals. Reference lines relate to the 10th, 50th, and 90th centiles (Table 2), which correspond to 29 ng/L, 44.6 µg/L, and 66.4 mg/L, respectively. For example, an aquatic concentration leading to a plasma concentration in fish above the mammalian C_{max} value is predicted by the AqET model to be at or below 29 ng/L for 10% of all pharmaceuticals

contaminants in treated wastewater effluents, a number of treatment approaches, including appropriately designed and maintained constructed wetlands [56], appear viable for supporting risk management of indirect and direct potable water reuse. In this volume, Chaps. 9 and 10 examine timely issues related to environmental risk management. In Chap. 9, Gerrity and Snyder examine the available information related to the efficacy of various wastewater and drinking water treatment technologies for human pharmaceuticals. In Chap. 10, Stoddard and Huggett conclude this volume with an interesting perspective on pharmaceutical take back programs, which promise to divert unused medications from down the drain discharges and drug abuse by and poisonings of unintended users.

Lessons learned from human pharmaceuticals in the environment will continue to advance our understanding of the environmental risks of chemicals. For example, a number of organic contaminants are chiral, which remains an important environmental consideration because fate and effects often differ among enantiomers [57]. Herein, studies of chiral pharmaceuticals have advanced our understanding of risks posed by other chiral chemicals [58]. Similarly, many environmental contaminants, including metabolites and degradates, are weak acids and weak bases. Because site-specific pH influences environmental fate, uptake and toxicity, the study of ionizable therapeutics (~70% of all drugs are weak bases) has advanced our understandings of the impacts of climatic changes on bioaccumulation and toxicity of moderately polar and ionizable chemicals [59, 60]. Interestingly, lessons learned from the study and design of less-toxic pharmaceuticals, often described as benign by design [61], can be extended to advance green chemistry principles by developing sustainable molecular design guidelines for reducing the toxicity of other industrial contaminants [62, 63]. To the fields of aquatic toxicology and environmental risk assessment in particular, understanding the toxicity of human pharmaceuticals in the environment is beginning to advance our understanding of toxicity pathways. To date, relatively few toxicity pathways have been defined in ecological systems, but hundreds of pharmaceuticals targets are evolutionarily conserved across the various kingdoms. Developing an understanding of pharmaceutical MOAs and associated AOPs will improve prospective and retrospective diagnosis and management of environmental risks posed by industrial contaminants. Clearly a number of timely research questions remain unanswered [21].

References

1. Postel S (2010) Water: adapting to a new normal. In: Heinberg R, Lerch D (eds) The Post Carbon Reader: managing the 21st Century's Sustainability Crises. Watershed Media/ University of California Press, California
2. Fish on Prozac. http://news.sciencemag.org/sciencenow/2003/11/04-01.html
3. Brooks BW, Chambliss CK, Stanley JK, Ramirez AJ, Banks KE, Johnson RD, Lewis RJ (2005) Determination of select antidepressants in fish from an effluent-dominated stream. Environ Toxicol Chem 24:464–469
4. Male fish becoming female? Researchers worry about estrogen and pollutants in the water. http://www.msnbc.msn.com/id/6436617/ns/nightly_news/t/male-fish-becoming-female
5. Woodling JD, Lopez EM, Maldonado TA, Norris DO, Vajda AM (2006) Intersex and other reproductive disruption of fish in wastewater effluent dominated Colorado streams. Comp Biochem Phys Part C 144:10–15
6. Drugs found in drinking water. http://hosted.ap.org/specials/interactives/pharmawater_site
7. Benotti MJ, Trenholm RA, Vanderford BJ, Holady JC, Stanford BD, Snyder SA (2009) Pharmaceuticals and endocrine disrupting compounds in U.S. drinking water. Environ Sci Technol 43:597–603
8. India's drug problem. http://www.nature.com/news/2009/090204/full/457640a.html
9. Colburn T, Dumanoski D, Myers JP (1996) Our stolen future. Dutton, Peguin Books, New York
10. Carson R (1962) Silent spring. Houghton Mifflin Company, New York
11. Ankley GT, Brooks BW, Huggett DB, Sumpter JP (2007) Repeating history: pharmaceuticals in the environment. Environ Sci Technol 41:8211–8217
12. Hotchkiss AK, Rider CV, Blystone CR, Wilson VS, Hartig PC, Ankley GT, Foster PM, Gray CL, Gray LE (2008) Fifteen years after "Wingspread"—environmental endocrine disrupters and human and wildlife health: where we are today and where we need to go. Toxicol Sci 105:235–259
13. Desbrow C, Routledge E, Brighty G, Sumpter J, Waldock M (1998) Identification of estrogenic chemicals in STW effluent. 1. Chemical fractionation and in vitro biological screening. Environ Sci Technol 32:1549–1558
14. Arcand-Hoy L, Nimrod AC, Benson WH (1998) Endocrine-modulating substances in the environment: estrogenic effects of pharmaceutical products. Int J Toxicol 17:139–158
15. Halling-Sorensen B, Nielsen SN, Lanzky PF, Ingerslev F, Lutzhoft HCH, Jorgensen SE (1998) Occurrence, fate and effects of pharmaceutical substances in the environment—a review. Chemosphere 36:357–394
16. Ternes TA (1998) Occurrence of drugs in German sewage treatment plants and rivers. Water Res 32:3245–3260

17. Daughton CG, Ternes TA (1999) Pharmaceuticals and personal care products in the environment: agents of subtle change? Environ Health Perspect 107(suppl 6):907–938
18. Williams RT (ed) (2005) Human pharmaceuticals: assessing impacts on aquatic ecosystems. SETAC Press, Pensacola, Florida
19. Crane M, Barrett K, Boxall A (eds) (2008) Veterinary medicines in the environment. SETAC Press, Pensacola, Florida
20. Brooks BW, Huggett DB, Boxall ABA (2009) Pharmaceuticals and personal care products: research needs for the next decade. Environ Toxicol Chem 28:2469–2472
21. Boxall ABA, Rudd M, Brooks BW, Caldwell D, Choi K, Hickmann S, Innes E, Ostapyk K, Staveley J, Verslycke T, Ankley GT, Beazley K, Belanger S, Berninger JP, Carriquiriborde P, Coors A, DeLeo P, Dyer S, Ericson J, Gagne F, Giesy JP, Gouin T, Hallstrom L, Karlsson M, Larsson DGJ, Lazorchak J, Mastrocco F, McLaughlin A, McMaster M, Meyerhoff R, Moore R, Parrott J, Snape J, Murray-Smith R, Servos M, Sibley PK, Straub JO, Szabo N, Tetrault G, Topp E, Trudeau VL, van Der Kraak G (2012) Pharmaceuticals and personal care products in the environment: what are the big questions? Environ Health Perspect (in press)
22. Kolpin DW, Furlong ET, Meyer MT, Thurman EM, Zaugg SD, Barber LB, Buxton HT (2002) Pharmaceuticals, hormones, and other organic wastewater contaminants in U.S. streams, 1999–2000: a national reconnaissance. Environ Sci Technol 36:1202–1211
23. Monteiro SC, Boxall ABA (2010) Occurrence and fate of human pharmaceuticals in the environment. Rev Environ Contam Toxicol 202:53–154
24. Nordlander I, Johnsson H, Osterdahl B (1987) Oxytetracycline residues in rainbow trout analyzed by a rapid HPLC method. Food Addit Contam 4:291–296
25. Capone DG, Weston DP, Miller V, Shoemaker C (1996) Antibacterial residues in marine sediments and invertebrates following chemotherapy in aquaculture. Aquaculture 145:55–75
26. Boxall ABA, Johnson P, Smith EJ, Sinclair CJ, Stutt E, Levy LS (2006) Uptake of veterinary medicines from soils into plants. J Agric Food Chem 54:2288–2297
27. Larsson DGJ, Adolfsson-Erici M, Parkkonen J, Pettersson M, Berg AH, Olsson PE, Forlin L (1999) Ethinyloestradiol—an undesired fish contraceptive? Aquat Toxicol 45:91–97
28. Ramirez AJ, Brain RA, Usenko S, Mottaleb MA, O'Donnell JG, Stahl LL, Wathen JB, Snyder BD, Pitt JL, Perez-Hurtado P, Dobbins LL, Brooks BW, Chambliss CK (2009) Occurrence of pharmaceuticals and personal care products (PPCPs) in fish: results of a national pilot study in the U.S. Environ Toxicol Chem 28:2587–2597
29. Daughton CG, Brooks BW (2011) Active pharmaceuticals ingredients and aquatic organisms. In: Meador J, Beyer N (eds) Environmental contaminants in wildlife: interpreting tissue concentrations, 2nd edn. Taylor and Francis, Boca Raton, pp 281–341
30. Zhang X, Oakes KD, Cui S, Bragg L, Servos MR, Pawliszyn J (2010) Tissue-specific in vivo bioconcentration of pharmaceuticals in rainbow trout (Oncorhynchus mykiss) using space-resolved solid-phase microextraction. Environ Sci Technol 44:3417–3422
31. Paterson G, Metcalfe CD (2008) Uptake and depuration of the anti-depressant fluoxetine by the Japanese medaka (Oryzias latipes). Chemosphere 74:125–130
32. Nallani G, Paulos P, Vanables B, Constantine L, Huggett DB (2011) Bioconcentration of Ibuprofen in Fathead minnow (Pimephales promelas) and Channel catfish (Ictalurus punctatus). Chemosphere 84:1371–1377
33. Nallani G, Paulos P, Vanables B, Constantine L, Huggett DB (2011) Tissue specific uptake and bioconcentration of the oral contraceptive, Norethindrone, in two freshwater fishes. Arch Environ Contam Toxicol 62(2):306–313
34. Smith EM, Chu S, Paterson G, Metcalfe CD, Wilson JY (2010) Cross-species comparison of fluoxetine metabolism with fish liver microsomes. Chemosphere 79:26–32
35. Gomez C, Constantine L, Moen M, Vaz A, Huggett DB (2010) The influence of gill and liver metabolism on the predicted bioconcentration in fish. Chemosphere 81:1189–1195
36. Gomez CF, Constantine L, Moen M, Vaz A, Wang W, Huggett DB (2011) Ibuprofen metabolism in the liver and fill of rainbow trout, Oncorhynchus mykiss. Bull Environ Contam Toxicol 86:247–251
37. Schultz MM, Painter MM, Bartell SE, Logue A, Furlong ET, Werner SL, Shoenfuss HL (2011) Selective uptake and biological consequences of environmentally relevant antidepressant pharmaceutical exposures on male fathead minnows. Aquat Toxicol 104:38–47

38. Nakamura Y, Yamamoto H, Sekizawa J, Kondo T, Hirai N, Tatarako N (2008) The effects of pH on fluoxetine in Japanese medaka (*Oryzias latipes*): acute toxicity in fish larvae and bioaccumulation in juvenile fish. Chemosphere 70:865–873

39. Zhou SN, Oakes KD, Servos MR, Pawliszyn J (2008) Application of solid-phase microextraction for in vivo laboratory and field sampling of pharmaceuticals in fish. Environ Sci Technol 42:6073–6079

40. Environmental Defense Fund (1997) Toxic ignorance: the continuing absence of basic health testing for top-selling chemicals in the United States. Environmental Defense Fund, New York

41. Berninger JP, Brooks BW (2010) Leveraging mammalian pharmaceutical toxicology and pharmacology data to predict chronic fish responses to pharmaceuticals. Toxicol Lett 193:69–78

42. Brausch JM, Connors KA, Brooks BW, Rand GM (2012) Human pharmaceuticals in the aquatic environment: a critical review of recent toxicological studies and considerations for toxicity testing. Rev Environ Contam Toxicol 218:1–99

43. Brooks BW, Riley TM, Taylor RD (2006) Water quality of effluent-dominated stream ecosystems: ecotoxicological, hydrological, and management considerations. Hydrobiologia 556:365–379

44. Seiler JP (2002) Pharmacodynamic activity of drugs and ecotoxicology: can the two be connected? Toxicol Lett 131:105–115

45. Huggett DB, Brooks BW, Peterson B, Foran CM, Schlenk D (2002) Toxicity of select beta-adrenergic receptor blocking pharmaceuticals (β-blockers) on aquatic organisms. Arch Environ Contam Toxicol 42:229–235

46. Brooks BW, Foran CM, Richards S, Weston JJ, Turner PK, Stanley JK, Solomon K, Slattery M, La Point TW (2003) Aquatic ecotoxicology of fluoxetine. Toxicol Lett 142:169–183

47. Huggett DB, Cook JC, Ericson JF, Williams RT (2003) A theoretical model for utilizing mammalian pharmacology and safety data to prioritize potential impacts of human pharmaceuticals to fish. Hum Ecol Risk Assess 9:1789–1799

48. Fitzsimmons PN, Fernandez JD, Hoffman AD, Butterworth BC, Nichols JW (2001) Branchial elimination of superhydrophobic organic compounds by rainbow trout (*Oncorhynchus mykiss*). Aquat Toxicol 55:23–34

49. Brown JN, Paxeus N, Forlin L, Larsson DGJ (2007) Variations in bioconcentration of human pharmaceuticals from sewage effluents into fish blood plasma. Environ Toxicol Pharmacol 24:267–274

50. Fick J, Lindberg RH, Parkkonen J, Arvidsson B, Tysklind M, Larsson DGJ (2010) Therapeutic levels of levonorgestrel detected in blood plasma of fish: results from screening rainbow trout exposed to treated sewage effluents. Environ Sci Technol 44:2661–2666

51. Fick J, Lindberg RH, Tysklind M, Larsson DGJ (2010) Predicted critical environmental concentrations for 500 pharmaceuticals. Regul Toxicol Pharmacol 58:516–523

52. Berninger JP, Du B, Connors KA, Eytcheson SA, Kolkmeier MA, Prosser KN, Valenti TW, Chambliss CK, Brooks BW (2011) Effects of the antihistamine diphenhydramine to select aquatic organisms. Environ Toxicol Chem 30:2065–2072

53. Valenti TV, Gould GG, Berninger JP, Connors KA, Keele NB, Prosser KN, Brooks BW (2012) Human therapeutic plasma levels of the selective serotonin reuptake inhibitor (SSRI) sertraline decrease serotonin reuptake transporter binding and shelter seeking behavior in adult male fathead minnows. Environ Sci Technol 46:2427–2435

54. Valenti TW, Perez Hurtado P, Chambliss CK, Brooks BW (2009) Aquatic toxicity of sertraline to *Pimephales promelas* at environmentally relevant surface water pH. Environ Toxicol Chem 28:2685–2694

55. Ankley GT, Bennett RS, Erickson RJ, Hoff DJ, Hornung MW, Johnson RD, Mount DR, Nichols JW, Russom CL, Schmieder PK, Serrano JA, Tietge JE, Villeneuve DL (2010) Adverse outcome pathways: a conceptual framework to support ecotoxicology research and risk assessment. Environ Toxicol Chem 29:730–741

56. Mokry L, Brooks BW, Chambliss CK, Knight R, Keller C, Sedlak DL (2011) Evaluate wetland systems for treated wastewater performance to meet competing effluent quality goals. WateReuse Research Foundation, Alexandria, VA. 153 p

57. Garrison AW (2006) Probing the enantioselectivity of chiral pesticides. Environ Sci Technol 40:16–23

58. Stanley JK, Brooks BW (2009) Perspectives on ecological risk assessment of chiral compounds. Integr Environ Assess Manag 5:364–373

59. Valenti TW, Taylor JT, Back JA, King RS, Brooks BW (2011) Influence of drought and total phosphorus on diel pH in wadeable streams: implications for ecological risk assessment of ionizable contaminants. Integr Environ Assess Manag 7:636–647

60. Brooks BW, Valenti TW, Cook-Lindsay BA, Forbes MG, Scott JT, Stanley JK, Doyle RD (2011) Influence of Climate change on reservoir water quality assessment and management: effects of reduced inflows on diel pH and site-specific contaminant hazards. In: Linkov I, Bridges TS (eds) Climate: global change and local adaptation. NATO science for peace and security series C: environmental security. Springer, New York, pp 491–522

61. Kümmerer K (2007) Sustainable from the very beginning: rational design of molecules by life cycle engineering as an important approach for green pharmacy and green chemistry. Green Chem 9:899–907

62. Voutchkova AM, Kostal J, Steinfeld JB, Emerson JW, Brooks BW, Anastas P, Zimmerman JB (2011) Towards rational molecular design: derivation of property guidelines for reduced acute aquatic toxicity. Green Chem 13:2373–2379

63. Voutchkova AM, Kostal J, Connors KA, Brooks BW, Anastas P, Zimmerman JB (2012) Towards rational molecular design for reduced chronic aquatic toxicity. Green Chem 14:1001–1008

64. Fatta D, Nikolaou A, Achilleos A, Meric S (2007) Analytical methods for tracing pharmaceutical residues in water and wastewater. Trends Anal Chem 26:515–533

65. Gros M, Petrovic M, Barcelo D (2006) Multi-residue analytical methods using LC-tandem MS for the determination of pharmaceuticals in environmental and wastewater samples: a review. Anal Bioanal Chem 386:941–952

66. Hao C, Clement R, Yang P (2007) Liquid chromatography–tandem mass spectrometry of bioactive pharmaceutical compounds in the aquatic environment a decade's activities. Anal Bioanal Chem 387:1247–1257

67. Hernando MD, Gomez MJ, Aguera A, Fernandez-Alba AR (2007) LC-MS analysis of basic pharmaceuticals (beta-blockers and anti-ulcer agents) in wastewater and surface water. Trends Anal Chem 26:581–594

68. Hernandez J, Sancho JV, Ibañez M, Guerrero C (2007) Antibiotic residue determination in environmental waters by LC-MS. Trends Anal Chem 26:466–485

69. Wong CS, MacLeod SL (2009) JEM spotlight: recent advances in analysis of pharmaceuticals in the aquatic environment. J Environ Monit 11:923–936

70. Kim SC, Carlson K (2005) LC–MS² for quantifying trace amounts of pharmaceutical compounds in soil and sediment matrices. Trends Anal Chem 24:635–644

71. O'Connors S, Aga DS (2007) Analysis of tetracycline antibiotics in soil: advances in extraction, clean-up, and quantification. Trends Anal Chem 26:456–465

72. Buchberger WW (2007) Novel analytical procedures for screening of drug residues in water, waste water, sediment and sludge. Anal Chim Acta 593:129–139

73. Petrovic M, Hernando MD, Diaz-Cruz MS, Barcelo D (2005) Liquid chromatography–tandem mass spectrometry for the analysis of pharmaceutical residues in environmental samples: a review. J Chromatogr A 1067:1–14

74. Giger W (2009) Hydrophilic and amphiphilic water pollutants using advanced analytical methods for classic and emerging contaminants. Anal Bioanal Chem 393:37–44

75. Richardson S (2009) Water analysis: emerging contaminants and current issues. Anal Chem 81:4645–4677

76. Barcelo D, Petrovic M (2007) Challenges and achievements of LC-MS in environmental analysis: 25 years on. Trends Anal Chem 26:2–11

77. Hao C, Zhao X, Yang P (2007) GC-MS and HPLC-MS analysis of bioactive pharmaceuticals and personal-care products in environmental matrices. Trends Anal Chem 26:569–580
78. Morley MC, Snow DD, Cecrle C, Denning P, Miller L (2006) Emerging chemicals and analytical methods. Water Environ Res 78:1017–1053
79. Kot-Wasik A, Debska J, Namiesnik J (2007) Analytical techniques in studies of the environmental fate of pharmaceuticals and personal-care products. Trends Anal Chem 26:557–568
80. Richardson S (2008) Environmental mass spectrometry: emerging contaminants and current issues. Anal Chem 80:4373–4402
81. Rubio S, Perez-Bendito D (2009) Recent advances in environmental analysis. Anal Chem 81: 4601–4622
82. Perez S, Barcelo D (2007) Application of advanced MS techniques to analysis of human and microbial metabolites of pharmaceuticals in the aquatic environment. Trends Anal Chem 26: 494–514
83. Petrovic M, Barcelo D (2007) LC-MS for identifying photodegradation products of pharmaceuticals in the environment. Trends Anal Chem 26:486–493
84. Radjenovic J, Petrovic M, Barcelo D (2007) Advanced mass spectrometric methods applied to the study of fate and removal of pharmaceuticals in wastewater treatment. Trends Anal Chem 26:1132–1144
85. Kosjek T, Heath E, Petrovic M, Barcelo D (2007) Mass spectrometry for identifying pharmaceutical biotransformation products in the environment. Trends Anal Chem 26:1076–1085
86. Kostopoulou M, Nikolaou A (2008) Analytical problems and the need for sample preparation in the determination of pharmaceuticals and their metabolites in aqueous environmental matrices. Trends Anal Chem 27:1023–1035
87. Escandar GM, Faber NM, Goicochea HC, Muñoz de la Peña A, Olivieri AC, Poppi RJ (2007) Second- and third-order multivariate calibration data, algorithms and applications. Trends Anal Chem 26:752–765
88. Galera MM, Gil Garcia MD, Goicochea HC (2007) The Application to wastewaters of chemometric approaches to handling problems of highly complex matrices. Trends Anal Chem 26:1032–1042
89. Lambropoulou DA, Konstantinou IK, Albanis TA (2007) Recent developments in headspace microextraction techniques for the analysis of environmental contaminants in different matrices. J Chromatogr A 1152:70–96
90. Pavloivc DM, Babic S, Horvat AJM, Kastelan-Macan M (2007) Sample preparation in analysis of pharmaceuticals. Trends Anal Chem 26:1062–1075
91. Pichon V, Chapuis-Hugon F (2008) Role of molecularly imprinted polymers for selective determination of environmental pollutants—a review. Anal Chim Acta 622:48–61
92. Rodriguez-Mozaz S, Lopez de Alda MJ, Barcelo D (2007) Advantages and limitations of on-line solid phase extraction coupled to liquid chromatography–mass spectrometry technologies versus biosensors for monitoring of emerging contaminants in water. J Chromatogr A 1152: 97–115
93. Soderstrom H, Lindberg RH, Fick J (2009) Strategies for monitoring the emerging polar organic contaminants in water with emphasis on integrative passive sampling. J Chromatogr A 1216:623–630
94. Wardencki W, Curylo J, Namiesnik J (2007) Trends in solventless sample preparation techniques for environmental analysis. J Biochem Biophys Methods 70:275–288
95. Perez S, Barcelo D (2008) Applications of LC-MS to quantitation and evaluation of the environmental fate of chiral drugs and their metabolites. Trends Anal Chem 27:836–846
96. Ramirez AJ, Mottaleb MA, Brooks BW, Chambliss CK (2007) Analysis of pharmaceuticals in fish tissue using liquid chromatography—tandem mass spectrometry. Anal Chem 79: 3155–3163

Environmental Risk Assessment for Human Pharmaceuticals: The Current State of International Regulations

Jürg Oliver Straub and Thomas H. Hutchinson

Introduction

An overview is given on environmental risk assessment for pharmaceuticals (ERA), with a description of the current regulatory requirements for human pharmaceuticals ERA in Europe and the USA as well as developments worldwide. In addition, further developments on national levels concerning the environmental safety of pharmaceuticals are presented. Also, a short comparison with international veterinary pharmaceuticals guidelines and with biocides ERA is given.

As long as human population density is low and excreta are spread diffusely over a large area, no significant levels of PAS or metabolites are expected in the environment. But when population density increases, when excreta collect in sewage and the latter is discharged, after wastewater treatment or not, to receiving waters, measurable to significant concentrations in surface waters may be reached. With strong population growth in industrialised societies from the nineteenth century onward, with sewage collection systems in the growing cities and with the increase in the number of pharmaceutical companies and their biologically active products, a rise in environmental concentrations of at least certain PAS followed during the past century. A parallel development in analytical methods and power, expressed as constantly decreasing limits of detection and quantitation, inevitably led to determinations of PAS in environmental matrices.

J.O. Straub (✉)
F.Hoffmann-La Roche Ltd, Group SHE,
LSM 49/2.033, Basle CH-4070, Switzerland
e-mail: juerg.straub@roche.com

T.H. Hutchinson
CEFAS Weymouth Laboratory, Centre for Environment, Fisheries and Aquaculture Sciences,
The Nothe, Barrack Road, Weymouth, Dorset DT4 8UB, UK
e-mail: tom.hutchinson@cefas.co.uk

B.W. Brooks and D.B. Huggett (eds.), *Human Pharmaceuticals in the Environment:*   17
Current and Future Perspectives, Emerging Topics in Ecotoxicology 4,
DOI 10.1007/978-1-4614-3473-3_2, © Springer Science+Business Media, LLC 2012

The first analytical detections of PAS and metabolites in environmental media are reported from the USA in the 1970s [33, 37], where among others salicylic acid, the main metabolite of acetylsalicylic acid was detected in sewage works effluent. These initial detections initiated a rapidly growing list of similar publications and reviews covering sewage treatment effluent, surface, estuarine, marine, ground and tap water over the following decades (e.g. Richardson and Bowron [64], Aherne and Briggs [1], Ayscough et al. [4] and Thomas and Hilton [77] in the UK; Heberer et al. [35] and Ternes et al. [73] in Germany; Halling-Sørensen et al. [34] in Denmark; Buser et al. [10] and Tixier et al. [77] in Switzerland; Belfroid et al. [6] in the Netherlands; Stumpf et al. [72] in Brazil; Zuccato et al. [84] and Calamari et al. [11] in Italy; Farré et al. [28] and Fernández et al. [29] in Spain; Kolpin et al. [48] and Barnes et al. [5] in the USA; Metcalfe et al. [54] in Canada; Vieno et al. [81] in Finland; Nakada et al. [57] in Japan; Rabiet et al. [63] in France; Kim et al. [47] in South Korea). Note this is not meant to be a complete list but rather an illustration of the worldwide increase in publications in the 1990s and 2000s. Again, the scope of detections widened with massively refined analytical instruments and methods.

In parallel to these ubiquitous detections in environmental media, the question of possible adverse effects caused by PAS to environmental organisms and ecosystems also gained importance. Initial environmental risk assessments (ERAs), comparing environmental concentrations with known effects, began in the 1980s. The concerns about environmental safety of PAS, alone and in particular in combinations, strongly increased with accruing evidence for widespread endocrine disruption in wild fish [44], in particular downstream of sewage treatment works effluents and also with experimental adverse effects seen with a few PAS at very low concentrations (e.g. [19, 30, 43]), which in some cases were close to or within the range of measured environmental concentrations (MECs). In parallel, the use of PAS or similar substances has played an important role in other areas of aquatic research, including aquaculture [31, 40] and marine antifoulant paints [38, 50, 61].

In view of mounting evidence for widespread environmental exposure and potential or probable environmental effects of PAS, enquiries and investigations into environmental hazards and risks due to PAS began in the 1980s (e.g. [1, 18, 34, 36, 46, 65, 78]). In parallel to these often government-sponsored investigations, the necessity for and development of formal ERAs specifically for PAS (pharmaceuticals ERA or PERA) was recognised by regulators on both sides of the Atlantic, which led to legal requirements and, with some delay, to guidelines for such PERAs as part of the registration dossier from the 1990s onwards. Formal guidelines were developed and published in 1998 in the USA and in 2006 in the European Union (EU). In other countries, PERAs are requested (e.g. Australia) or formal own guidelines are in the making (Canada, Japan). In addition, Sweden led the way with a system for the ERA of "old" PAS already on the market. But even beyond the formal requirements for PERAs in the context of registration, PAS in the environment (PIE) may be the subject of other legislation than registration, which, however, may still require some kind of ERA. These developments and current states will be outlined in the following paragraphs.

Current State of PERA Regulation in Various Regions or Countries

PERA started in the USA and EU in the 1980s or early 1990s. Much of the methodology seems to derive from pesticides ERA, which came into focus and developed appropriate methodologies earlier than pharmaceuticals in general. All of the ERA procedures have in common a comparison between predicted (or measured) environmental concentrations (PECs or MECs) with predicted no effect concentrations (PNECs), both per environmental compartment under consideration. Such compartments may be wastewater treatment, surface waters, sediments, groundwaters, tidal and coastal/marine waters, soils (through landspreading of surplus sewage sludge, called biosolids in North American terminology) and, rarely, the atmosphere. PECs are derived from either predicted use or maximum daily use multiplied by a default use or penetration factor in the population, integrating human metabolism and depletion during sewage treatment or in the environment, sorption and distribution to other environmental compartments, dilution and advection (off-transport by the medium) in the receiving compartments. PNECs are mostly derived from either acute or chronic ecotoxicity tests, normally with standard organism groups representative for the compartment, by dividing by assessment factors (AFs) which are dependent on the character and number of ecotoxicity results available. In higher tiers of the ERA, the above deterministic procedure using AFs can be replaced by probabilistic methodology, where the distributional characteristics of a number of ecotoxicity test results (normally at least ten chronic datapoints) are used to derive a PNEC. PECs and PNECs are compared per compartment, in general through forming the PEC/PNEC ratio. If this ratio is <1, i.e. if the expected concentration is below the one predicted to cause no adverse effect, and there are no other concerns for all the compartments under consideration, there is no indication for significant risk and the ERA may be finalised. In case the PEC/PNEC ratio is ≥1, risk cannot be excluded and therefore the ERA must be refined by reappraisal of the PEC and/or PNEC through better, more in-depth methodology. An ERA may thereby progress from a relatively simple and crude assessment based on little data to a much more realistic assessment that, in turn, needs and incorporates far more experimental data and often also advanced models. However, even with a highly refined assessment there is never any guarantee that the outcome will be "no significant risk". A refined ERA can only characterise a possible risk better than a crude ERA, but it cannot make risks or concerns disappear—on the other hand, it certainly will identify compartments at potential risk, allowing the development of targeted risk management strategies if indicated.

PERA in the USA

Based on the 1969 US National Environmental Policy Act (NEPA) as amended, the Code of Federal Regulations (CFR) Title 21 Part 25 as amended details environmental assessments (EAs in US legal terminology) within the US Food and Drugs legislation (21 CFR 25; current version available at http://www.accessdata.fda.gov/scripts/cdrh/cfdocs/cfcfr/CFRSearch.cfm?CFRPart=25). By this, "all applications or petitions requesting Agency action must be accompanied by either an EA or a claim of categorical exclusion; failure to submit one or the other is sufficient grounds for refusing to file or approve the application" (cited from "Environmental Impact Review at CDER", http://www.fda.gov/AboutFDA/CentersOffices/CDER/ucm088969.html). In 1998 the US Center for Drug Evaluation and Research (CDER) and the Center for Biologics Evaluation and Research (CBER) within the US Food and Drug Administration published a "Guidance for Industry, Environmental Assessment of Human Drug and Biologics Applications", revision 1 [14], which is still current today.

The Guidance describes in which cases an EA can be waived and how to proceed with an EA in the remainder. Waivers, the so-called categorical exclusions, may be invoked in the following cases:

- If the application does not increase the use of active moiety (i.e. in case of extensions or additional applications by third parties for PAS already on the market).
- If the application may lead to increased use but the estimated concentration of the AS at the point of entry into the environment is less than 1 part per billion (ppb). This means that the entry into the environment concentration (EIC) of a particular PAS from US publicly owned treatment works (POTWs) must be below 1 µg/L, discounting all metabolism; calculating back from an EIC of 1 µg/L and the average annual total effluent of all POTWs results in a maximum annual amount of approximately 44 metric tonnes of PAS per year for the whole continental USA, based on daily POTW inflow data given in the Guidance ([14]; p 4). Hence, if the predicted annual use of a new PAS is below 44 tonnes/annum there is no need for an EA, except if the applicant has information to suggest that the use of even a lesser quantity may "significantly affect the quality of the human environment" ([14]; p 3).
- For biological PAS if their use will not lead to significant concentrations in the environment.
- For investigational new drugs still under development in clinical research.
- For specific biological products for blood or plasma transfusion.

In all other cases, the applicant needs to prepare an EA following a tiered, stepwise approach that follows the course of a PAS from human excretion into the environment. Hence, in a *first basic step*, if there is experimental evidence that a new PAS is rapidly depleted, e.g. through biodegradation in a POTW, and not inhibitory to microorganisms, the EA can be stopped and finalised with a Finding of No Significant Impact (FONSI). If the PAS is not rapidly depleted and if it is lipophilic (with an *n*-octanol/water distribution coefficient $\log D_{ow} \geq 3.5$ at a relevant environmental pH of approximately 7), suggesting bioaccumulation, the applicant should initiate chronic testing in tier 3; note the tier numbering is given according to the Guidance [14]. Further details as to depletion (degradation, hydrolysis or partitioning to other environmental compartments) and to interpretation of these fate processes are given.

In all other cases, the effects testing starts with one acute test in *tier* 1. If the ratio of the 50% effect or 50% lethal concentration (EC50 or LC50) in this test divided by the EIC or predicted (or expected in US terminology) environmental concentration (PEC or EEC), whichever is higher, is ≥1,000 and there were no adverse effects observed at the higher of EIC or EEC (termed maximum expected environmental concentration or MEEC), the EA can be stopped and finalised. This ratio corresponds to a margin of safety (MOS) in general ERA terminology. If there were effects at MEEC, the applicant should initiate chronic testing in tier 3.

If the tier 1 MOS is <1,000, acute base set testing in *tier* 2 is specified. For aquatic EA the base set consists of acute algal, aquatic invertebrate and fish tests, for terrestrial testing of plant growth, earthworm and soil microbial toxicity. The lowest EC50 or LC50 from the effects base set is again divided by the MEEC. If the obtained tier 2 MOS is ≥100 and there were no adverse effects observed at MEEC, the EA can be stopped and finalised. If there were such effects, the applicant should initiate chronic testing in tier 3.

In *tier* 3, an unspecified number and selection of aquatic or terrestrial species should be tested chronically; applicants are advised to contact CDER/CBER for test selection. If the obtained tier 3 MOS between the (lowest) chronic EC50 or LC50 and the MEEC is ≥10 and there were no adverse effects observed at MEEC, the EA can be stopped and finalised. If the MOS is <10 or if there were effects at MEEC, the applicant should contact CDER/CBER for further advice and strategy.

Overall, the US Guidance is characterised by a comparatively high threshold of 1 μg/L as the EIC (POTW effluent), respectively, as 0.1 μg/L as an average EEC, using the standard dilution factor of 10 ([14], p 19), which in turn translates to the above 44 tonnes/annum below which an EA can normally be waived. If this threshold or trigger is surpassed, the actual EA proceeds logically from excretion to sewage treatment and into further compartments along traditional methodology, comparing PECs and effect concentrations. Lower-tier PNECs are based on only one (tier 1) or a base set of three (tier 2) acute ecotoxicity tests. In case of only one acute ecotoxicity test in tier 1, the tier 1 MOS must be ≥1,000 for the EIC (POTW effluent), which corresponds to an implicit AF of 10,000 for surface waters, including the ten times default dilution from POTW effluent. For tier 2, with an acute ecotoxicity base set comprising three different groups of organisms, the surface water AF drops to 1,000, while for tier 3 with an unstated number of chronic ecotoxicity data the implicit AF is 100 for surface waters, based on chronic EC50s or LC50s, which is unusual and in contrast with other guidelines that use the chronic NOECs for PNEC derivation.

While the very first detections of single PAS in environmental media in the USA date from the 1970s [33, 37], it took a long time before a report of widespread detections in sewage works effluents, surface and groundwaters in the USA [48] brought the topic of pharmaceuticals in the environment (PIE) to scientific and regulatory, later also to public attention. A series of syndicated articles from Associated Press journalists in the late 2000s with a focus on PIE, specifically PAS in drinking water [3], attracted and widened public and political attention to the topic of PIE and tap water. Comparable reports continue being published from various States (e.g. [55], for Delaware drinking waters). Within half a year starting from the first AP report, according to the AP [3] site, a US Congressional Panel discussed monitoring and potential impacts of micropollutants including PAS in environmental waters, which are currently not regulated by the US Environmental Protection Agency (EPA) either as a group or as single substances in the USA. The discussion seemed to focus mainly on potential human risks from PIE through water abstraction, treatment and consumption as drinking water, but less on risks for environmental organisms or ecosystems. Also, questions on PIE and the safety of PAS in drinking waters were raised in the US Senate Committee on Environment and Public Works (http://epw.senate.gov/public/index.cfm, search for "pharmaceuticals" and "water"). Some investigations on potential human health risks from PIE via drinking water were published in the previous decade (e.g. [9, 12, 16, 17, 45, 67, 83, 84]), all of which have found no significant risks based on the available evidence.

In addition, on July 7, 2010, the Great Lakes Environmental Law Center and the Natural Resources Defense Council as petitioners submitted a "Citizen Petition" to the US Food and Drugs Administration Commissioner. A Citizen Petition in the US is a legal means to challenge existing regulations. In this Citizen Petition concerning an amendment to the current US PERA Guidance [14], the repealing of the categorical exclusion threshold of 1 ppt (1 µg/L, corresponding to approximately 44 metric tonnes of PAS per annum) EIC is requested, "because the current regulation does not reflect a safe standard supported by current scientific information". In case the threshold for a categorical exclusion is indeed repealed, this would mean that nearly all new human PAS would need an EA for registration.

It will remain to be seen whether the parliamentary discussions and legal motions in the USA will eventually have effects on US regulations, on PERA in general, on the US PERA Guideline, possibly also for "old" PAS already on the market, or for the regulation of water contaminants by the EPA.

PERA in the European Union

First requirements for PERA were laid down in EU Directive 93/39/EEC, which asked to "give indications of any potential risks presented by the medicinal product to the environment". The development of the PERA guideline in the EU took 13 years in all, with several draft guidelines published during that time [68, 69]. In 2006, the European Medicines Agency (EMA, London, UK; note that the former abbreviation EMEA for European Medicines Evaluation Agency is not being used any longer) published the first definitive Guideline for Environmental Risk Assessment of Human Medicines [26]. This guideline describes a tiered procedure, from categorical exclusion or direct referral, to a simple, worst-case exposure estimation of a pharmaceutical active substance to the investigation of fate and effects in sewage works and surface waters, up to a refined assessment for these or other environmental compartments.

A PERA is required for new registrations (Medicines Authorisation Application or MAA in EU terminology) and for all repeat registrations by the same applicant, termed "variations" in the EU, that may lead to significantly increased environmental exposure to the PAS; note that "significant" is not defined or quantified in this context. In the basic *Phase* 1 of the PERA, certain categories of PAS are excluded from PERA (amino acids, proteins, peptides, carbohydrates, lipids, electrolytes, vaccines and herbal medicines), while other PAS are directly referred to special ERA. Highly lipophilic PAS with a log K_{ow} > 4.5 are directly referred to a persistence, bioaccumulation and (high eco) toxicity (PBT) assessment, where these properties are to be tested and evaluated in that order, following the methodology of the EU Technical Guidance Document (TGD, [75]), now replaced by the REACH Technical Guidance Document [24]. As a second direct referral category, potential endocrine disrupters, viz. those PAS that "may affect the reproduction of vertebrate or lower animals at concentrations lower than 0.01 μg/L", should be assessed using a "tailored strategy that addresses the specific mode of action". Note that there is no technical guidance for assessing potential endocrine disrupters at present, the applicant should "justify all actions taken" and, to be on the safe side, would be well advised to contact the EMA Committee for Human Medicinal Products for scientific advice.

All remaining PAS in Phase 1 undergo a prescreening that involves a rigid worst-case PEC prediction which is compared with a threshold value or "action limit" in EMA terminology. The maximum daily dose of the PAS is multiplied with a default penetration factor (Fpen) of 0.01 or 1%, which was derived by probabilistic methods to model a reasonable-worst-case use of a medicine in the population [26], and divided by a default 200 L of wastewater per person per day and a default surface water dilution factor of 10, to give the Phase I surface water PEC. If this surface water PEC is <0.01 μg/L (i.e. <10 ng/L) and there are no other grounds for direct referral, the PERA can be finalised. Backcalculating with all the default values, a PAS would need to have a maximum daily dose of <2 mg for the surface water PEC to remain below 10 ng/L. If the PEC is ≥10 ng/L, the PERA has to go into Phase 2 Tier A for an initial ERA based on experimental data. Note that no metabolism, human or environmental, may be factored in the Phase 1 PEC. Further, the Fpen may only be changed in Phase 1 based on published epidemiology data for the medical indication(s) addressed by the PAS in question, but not by marketing predictions or other indicators. Both in case of a categorical exclusion and if the PEC is <0.01 μg/L, a justification letter for not producing an ERA Expert Report should be prepared and included with the registration dossier.

In *Phase* 2 *Tier A* a prescribed set of experimental environmental fate and effects data must be elaborated under GLP quality assurance. The results are then used to derive PEC/PNEC ratios for various compartments and for comparison with given threshold values, which will inform on the necessity or not of further evaluation of potential risks in certain environmental compartments in Phase 2 Tier B. Modelled data, e.g. by quantitative structure activity/property relationship algorithms (QSAR or QSPR) are not acceptable. The Phase 2 Tier A experimental data set consists of:

- n-Octanol/water partition coefficient (log K_{OW}, determined using OECD107, OECD117, OECD 123 or draft OECD122 technical guidelines)
- Adsorption constants to the organic carbon fraction in soils or activated sludges (K_{OC} and K_d; OECD106, OECD121 or OPPTS 835.1110)
- Ready biodegradability (OECD301) as a facultative test; if not readily biodegradable, a transformation test in aquatic sediment systems (OECD308) is mandatory
- An algal growth inhibition test (OECD201) with green algae, in case of antimicrobials with cyanobacteria
- A daphnid reproduction test (OECD211) with *Daphnia* sp. (meaning not with *Cerodaphnia dubia*, which has a shorter generation time)
- A fish early life stage toxicity test (OECD210)
- An activated sludge respiration inhibition test (OECD209)

These data are used for the following decision tree:

- If the substance is lipophilic with a K_{OW} > 1,000 (log K_{OW} >3), it is assumed that the PAS may bioaccumulate, which is why such a PAS is directed to an experimental bioaccumulation study in Phase 2 Tier B. Note that this provision is redundant to the Phase 1 direct referral for lipophilic PAS with a logK_{OW} > 4.5. The latter stems from concerns from the EU OSPAR panel (the Oslo-Paris Commission on PBT substances in the North Sea, http://www.ospar.org/), which uses a logK_{OW} threshold value of 4.5 for screening potential B substances. However, the more lipophilic a PAS is, the lower on the whole is its bioavailability (Roche own unpublished data); this entails an increase of daily dosage in order to attain pharmacologically active levels of the PAS. Thereby the action limit of 2 mg PAS per day will be breached and the substance will go into Phase 2 Tier A, with a sediment/water study (persistence), testing for bioaccumulation and the three base set chronic tests (toxicity). Hence, the whole PBT package is performed for all PAS with a logK_{OW} >3, anyway.
- If the substance adsorbs strongly with a K_{OC} >10,000 L/kg (logK_{OC} >4), it is assumed that it would be removed during wastewater treatment by adsorption to activated sludge, unless it proves to be readily biodegradable. As surplus sludge is often spread on arable land after treatment (dewatering, anaerobic digestion, etc.), strongly sorbing PAS are assumed to reach the terrestrial compartment, which is why such a PAS is directed to a terrestrial ERA in Phase 2 Tier B.
- If the substance is not readily biodegradable and the sediment/water environmental fate study shows >10% in the sediment at any time after 13 days, it is assumed that the PAS will partition to the sediment to a relevant degree, which is why such a PAS is directed to sediment toxicity testing and a sediment ERA in Phase 2 Tier B.
- The PNEC for surface waters is derived from the lowest chronic algal, daphnid or fish NOEC, dividing by an AF of 10. If the PEC/PNEC ratio for surface water

is below 1, the PAS is unlikely to represent a risk to the aquatic compartment. If the ratio is ≥ 1, a refinement preferably of the PEC should be made in Phase 2 Tier B. Note that there is no specific mention of further refining the PNEC, e.g. through probabilistic methods.

- The microorganism PNEC for wastewater treatment is derived from the NOEC of the activated sludge respiration inhibition test, dividing by an AF of 10. If the ratio of surface water PEC (which is extrapolated from the wastewater PEC with a dilution factor of 10) divided by the microorganism PNEC is <0.1 (i.e. if the implicit ratio of wastewater PEC divided by the microorganism PNEC is <1), the PAS is unlikely to represent a risk for wastewater treatment. If the ratio is ≥ 0.1, a refinement of the fate of the PAS in wastewater treatment or the effect on microorganisms should be made in Phase 2 Tier B.
- An initial groundwater assessment should be made, except for those PAS that are readily biodegradable or that have a 90% dissipation time (DT90) in the sediment/water study of <3 days or that have an average K_{oc} >10,000 L/kg, all of which would be largely removed during sewage treatment. The PNEC for groundwater is derived from the chronic daphnid NOEC (as the only potential higher organisms in groundwater are invertebrates, in contrast to green algae or fish) by dividing by an AF of 10. The groundwater PEC is approximated as surface water PEC $\times 0.25$. If the groundwater PEC/PNEC ratio is <1, the PAS is unlikely to represent a risk to the groundwater compartment. If the ratio is ≥ 1, a refinement preferably of the (surface water) PEC should be made in Phase 2 Tier B.

In *Phase 2 Tier B* of the EU PERA, referrals from Phase 1 or potential risks identified in Phase 2 Tier A should be investigated. Whereas in earlier phases Type II Variation For the Treatment of Granulomatosis With Polyangiitis (Wegener's) and Microscopic Polyangiitis only the parent compound was investigated based on a total residue approach (meaning that no demonstrable metabolism or degradation could be factored in), evidenced human metabolism may be used to refine PECs in Phase 2 Tier B, but then relevant (>10% of parent PAS) metabolites are to be assessed by PEC and PNEC as well:

- The surface water PEC may be further refined using sewage works modelling with the spreadsheet application SimpleTreat that is integrated in the EU substance assessment model EUSES (downloadable from http://ecb.jrc.ec.europa.eu/euses/). The sludge adsorption (K_{oc}) value from the Phase 2 Tier A adsorption test and ready biodegradability (if attained) must be entered into SimpleTreat, respectively, EUSES. In addition, a so-called local PEC can be calculated for refinement. Again, PEC/PNEC ratios as above are to be derived and the risk for the given compartments (wastewater treatment, surface waters, groundwater) characterised.
- For sediment ERA, a sediment PEC is to be calculated based on the TDG (2003) [75], respectively the REACH TGD [24] algorithms, based on surface water PEC, adsorption and default EU sediment parameters. The sediment PNEC is based on at least one chronic test with sediment-dwelling organisms (the crustacean *Hyalella*, the oligochaete worm *Lumbriculus* and the larvae of the insect *Chironomus* are specifically mentioned). In case of one chronic NOEC available,

the AF is 100; for two chronic NOECs the AF is 50 and for three the AF is 10, to derive the sediment PNEC.

- For refined wastewater treatment microorganism risk assessment, the PAS concentration in the aeration tank of a standard sewage works should be calculated using SimpleTreat. This should be compared with a PNEC refined based on microorganisms testing and AFs as set out in the TDG (2003) [75], respectively, REACH TGD [24]. If refinement does not result in a PEC/PNEC ratio <1, further PNEC refinement should be undertaken.

- Terrestrial assessment should be performed with a soil PEC calculated with a combination of SimpleTreat modelling to generate a sludge PEC and the derivation of the soil PEC from sludge spreading and the results, in particular the soil half-life, from an obligatory soil transformation test (OECD307), using the algorithm in the TGD [75], respectively, the REACH TGD [24]. The soil PNEC is derived from the lowest (no) effect value from the following obligatory terrestrial ecotoxicity tests soil microorganisms (nitrogen transformation test, OECD 216), terrestrial plants growth test (OECD 208), earthworm acute toxicity test (OECD 207), *Collembola* soil insects reproduction test (ISO 11267), again based on the above TGDs. Soil risk is then characterised with the PEC/PNEC ratio.

The above Phase 2 Tier B assessment concludes the EU PERA. The whole assessment is to be compiled in an Expert Report with all conclusions, with all references and test reports, and with the curriculum vitae and signature of the expert who produced the report. In case there remains residual risk in one or more compartments, this may not keep the medicine from the market, as patient benefit is given priority before environmental concerns [22]. However, to minimise environmental exposure from unused medicines, the following phrasing should be inserted in package/patient information leaflets: "Medicines should not be disposed of via wastewater or household waste. Ask your pharmacist how to dispose of medicines no longer required. These measures will help to protect the environment". Note that it is recommended that this phrase be included even for medicines that do not require special disposal measures. Also, in case of residual risk, the Expert Report should contain evaluation of precautionary and safety measures to be taken with a view to minimising environmental exposure both from disposal of unused medicines and from patient use; this information should also become part of the Specific Product Characteristics information.

The 2006 EMA PERA Guideline has a ten times lower threshold compared with the US EA guideline; moreover, if a PAS is directed to Phase 2 Tier A or B, much more experimental data must be elaborated for initial and in particular for refined assessment, notably water/sediment fate and chronic effects testing. Based on currently available knowledge, both aspects may be defended with good scientific reasons. However, there are still some shortcomings in the EMA approach.

In Phase 2 Tier A the data set for environmental fate seems somewhat imbalanced. On the one hand, only a facultative ready biodegradability study is listed, but if that does not meet the criteria for ready biodegradability, no additional higher-level biodegradation information is requested, even though wastewater treatment is by far the most important entry pathway of a PAS into the environment and removal in sewage works is often the most significant fate process. On the other hand, adsorption and sediment/water fate studies are requested, both of which (at least for OECD106 and 308) are exacting and expensive studies that normally use radio-labelled substance. But basically the results are only used for deciding on the necessity of a terrestrial ERA or of a sediment ERA, respectively, in Phase 2 Tier B. With the exception of PAS with either a $logK_{OC}$ >4 in the adsorption test or a systems half-life <3 days in the sediment/water study (in both of which cases the substance need not be assessed for groundwater risk), those two assays are not utilised any further. While half-lives must be stated in the Expert Report, they are not actually processed in a PEC refinement or used in the ERA. It is not easy to see why sophisticated sediment/water fate data should be determined if they are not really used; it is not easy to see, either, why the distribution to sediment cannot be read from the adsorption test, in particular as all sediment risk should be normalised to standard sediment parameters with a specified organic carbon content, anyway. Instead of the water/sediment fate test, which was developed to model a small ditch beside a field for pesticides ERA and never meant to be an assay for surface water fate, there is an OECD-validated alternative that really does test for surface water fate, the OECD309 surface water degradation test, where the biodegradation of a test substance in natural water with a small concentration of suspended natural sediment is investigated. As Richard Murray-Smith (pers. comm.) commented in several workshops and conferences, this test would give more realistic and useful information on surface water fate. Human PAS do not normally end up in small ditches close to fields, but they will show up in surface waters.

On the ecotoxicity side, the EMA [26] PERA guideline consequently addresses chronic effects. This is based on the realisation that (nearly) all PAS in surface waters, whether rapidly degradable or not, show a phenomenon termed "pseudopersistence", viz. relatively constant concentrations due to more or less continuous input or replenishment from human use (e.g. [18]). Hence, environmental organisms are exposed in a constant manner, which can only be scientifically evaluated using chronic-based PNECs. However, the EMA guideline does not give any guidance on how a chronic aquatic PNEC (normally based on a traditional deterministic approach) could be further refined. For example, a more refined approach may be useful in some cases through the use of probabilistic assessment methodologies originally developed to support pesticides risk assessment (e.g. [13]). Indeed, probabilistic approaches have been recently applied for PERA of a few "old" human PAS during the past years [70, 71].

In the EU, Guidelines are to be revisited and updated if necessary on a regular basis. The EMA [26] PERA guideline was only 4 years old at the time of writing, hence a revision may be somewhat premature. However, in view of some uncertainties in the guideline, the CHMP Safety Working Party of the EMA prepared and in March 2011 published a "Question and Answer Document" (Q&ADoc; [27]) to "provide clarification and harmonise the use of" the EMA [26] PERA guideline. This Q&ADoc has due to the time passed since originally writing the manuscript, this is now official become the official companion to the guideline. It gives pertinent information on how the regulators want to handle PERA in the EU in the next years. Only selected items deemed important will be shortly highlighted in the following paragraphs:

- Generics are not exempted from providing an ERA and crossreference to the ERA of the original applicant is not possible. Hence, a new applicant for a generic PAS, also in combination with a new PAS, must provide a full PERA following the EMA [26] guideline.
- "Significant increase" in environmental exposure due to a variation remains undefined.
- The sediment/water fate test remains compulsory (except in the case of a positive ready biodegradability test, as already stated). It may not be waived even if the applicant presents a sediment ERA under the assumption of all substance distributing to the sediment. However, the testing of fully anaerobic systems of the water/sediment fate test is not considered necessary in general, as even the aerobic systems will develop anaerobic parts. Similarly, for environmental fate testing in soil (OECD307), only aerobic systems are required.
- For combination products the ERA should be performed separately for each PAS.
- Metabolites are included up to Phase 2 Tier A in the total residue approach adopted in the EMA [26] guideline. They may be subtracted for refining the surface water PEC in Phase 2 Tier B, but then a full ERA is requested for all metabolites excreted as $\geq 10\%$ of the applied dose. PECs and PNECs are then calculated separately for all substances investigated, and all PEC/PNEC ratios are added together for the evaluation of the whole product.
- Many PAS are ionisable compounds and present as charged acids, bases or zwitterions at environmentally relevant pH range (commonly accepted as pH 5–9). Yet it still is the $\log K_{ow}$, measured for acids and bases at a nondissociating pH value, that decides on bioaccumulation testing in Phases 1 and 2 Tier A. In the Q&ADoc only the K_{oc} from an OECD106 adsorption test is recognised to be a possible function of the ionisability of a substance.
- In the sediment/water fate test, the so-called "bound residues" are commonly formed, which cannot be extracted even with appropriate solvents. However, the bound residue fraction may not be subtracted from the sediment PEC, i.e. bound residues are regarded as (ultimately) bioavailable.

The draft Q&ADoc does give more definition to the EMA [26] guideline, but it also maintains the same highly precautionary approach to PERA. With the publication of the 2006 guideline it was the regulators' clear statement that over the coming years they wanted to collect PERAs to analyse them also for the scientific content and usefulness of the guideline, and to review the scheme based thereon, but obviously this time has not come yet. The Precautionary Principle being a nondefined and very controversially handled concept [32], it would seem that for the time being the EMA considers it has not sufficient scientific information to include a weight of evidence analysis and therefore remains on the conservative side.

PERA in Switzerland

In Switzerland, which is not a member of the EU, the relevant Medicines Registration Ordinance (Arzneimittel-Zulassungsverordnung, AMZV; [2]) only requires information and documentation on ecotoxicity for human pharmaceuticals ([2], article 4,[2],d), while for veterinary pharmaceuticals both data on ecotoxicity and potential risks for the environment are required ([2], articles 9,[2],b and 9,[1],b, respectively). There is no specific guideline for PERA nor any detailed requirements for ecotoxicity basic data mentioned in the AMZV [2]. Swiss regulators accept EU PERAs following the EMA [26] guideline.

PERA Developments in Canada

For the time being, pharmaceuticals in Canada are regulated under the New Substances Notification Regulation (Chemicals and Polymers) [58], respectively, the New Substances Notification Regulation (Organisms) [59], based on the Canadian Environmental Protection Act [8, 15]. In the NSNR/C&P, all kinds of chemical substances or organisms imported into or manufactured in Canada that are not already on the Canadian Domestic Substances List (DSL; http://www.ec.gc.ca/subsnouvelles-newsubs/default.asp?lang=En&n=47F768FE-1) must be notified to the authorities. For substances, the substance-related information content of the notification package depends on the total amount brought to the Canadian market in one calendar year. For a chemical or biochemical substance (including PAS) not on the DSL, below a first threshold of 100 kg per annum, no assessment is necessary, while increasing, defined substance information base sets ("schedule X information") become necessary in higher tonnage bands (>1,000, >10,000, >50,000 kg/a). If the chemical or biochemical is already on the DSL, the first threshold (requiring no notification) is 1,000 kg/a, with the same schedule X information necessary in higher tonnage bands.

However, a proper PERA guideline is currently (2012) under development in Canada. For the time being there seems to be no official draft document available to the public. Based on an earlier, nonattributable crude sketch that circulated a couple of years ago (and which may not be relevant any longer), it is possible that certain experimental tests that are not among the lower-tier studies in US or EU PERA schemes might become standard first-tier studies in the Canadian scheme. It is expected that a final draft of the Canadian PERA scheme will be published for a short public discussion and comment phase in 2012/2013 and that a definitive version may be adopted in the same year.

PERA Developments in Japan

Japan has been developing a PERA guideline for some years, according to Yasuyoshi Azuma (pers. comm.) of AstraZeneca, who presented on these activities at an international conference on PERA in Barcelona in 2009. PIE have been a topic for public news and scientific investigations in Japan, with increasing concern about the environmental safety of PAS in the recent past. In view of ongoing developments and of the language barrier, only little information is available as to the probable contents of a draft guideline. Still, an intermediate report from a mixed PERA study group led by the Ministry of Health, Labor and Welfare in 2008 (cited by Dr Azuma) suggests that PERA will become mandatory for new PAS, that categorical exclusions will apply, that the actual PERA would be risk based (i.e. not only hazard based) and that a tiered approach was preferred. In early tiers, a simple PEC would be calculated, with the possibility of refinement in higher tiers, while effects characterisation would be through chronic testing. Also, it perspired that a negative outcome of the PERA would not be sufficient reason to deny registration and marketing approval.

Based on this information, assuming it is still current, Japan seems to be set on developing a PERA scheme generally in line with existing guidelines elsewhere. While no precise dates are known, a draft guideline for public comment and finalisation is generally expected by about the year 2012/2013.

PERA Requirements in Australia

Australia has a requirement for a PERA to be submitted with new medicines registration in Annex I to Module 1 of the Common Technical Document issued by the Australian Therapeutic Goods Administration [74]: "Applications to register prescription medicines for human use should include [...] an indication of any potential risks presented by the medicine for the environment. This requirement is particularly applicable to new active substances and live vaccines. Applications for new active substances may include [...] an indication of relevant environmental hazards, making reference to standard physicochemical tests and any appropriate testing they have conducted on biodegradability, including some testing in sensitive species. [...] The risk assessment overview should include an evaluation of possible risks to the environment from the point of view of use and/or disposal and make proposals for labelling provisions that would reduce this risk". ([74]; Annex I to Module 1). There is no specific Australian PERA guideline nor is there information about such a guideline being developed; however, the EU EMA [26] Guideline is linked on the TGA homepage (link: http://www.tga.gov.au/docs/pdf/euguide/swp/444700en.pdf, which directly opens the EMA Guideline). Based on own experience as an environmental risk assessor with an international research pharmaceuticals company, an EU PERA is acceptable to the Australian regulators.

Further PERA Requirements

Based on own experience, there are a few sporadic cases of further countries that have started requiring PERAs, e.g. in South America. These have so far accepted

Spanish translations of the respective EU PERAs. It is not known whether these requests were based on established national legislation or on a wish on the side of environmental regulators to receive more pertinent information on new PAS. It may be assumed that such requests will increase in numbers and that some countries will establish formal legal requirements for PERAs, in view of developments in other countries and regions.

Other PERA Initiatives: The Swedish Environmental Classification and Simplified ERA of "Old" PAS Already on the Market

At the EnvirPharma Conference on Human and Veterinary Pharmaceuticals in the Environment in Lyon, France, in 2003, Prof. Åke Wennmalm of the Stockholm County Council (SCC) presented his concept of Environmental Classification of Pharmaceuticals, by assessing the environmental hazards of PAS by three criteria, persistence, bioaccumulability and ecotoxicity (PBT). The SCC is one of the biggest healthcare providers (including pharmaceuticals distribution) in Sweden. In 2003 the SCC started assessing the hazards of "old" PAS that were already on the market, which are not covered by PERA guidelines in the EU or USA. This hazard assessment proceeds by assigning numerical values from 0 to 3 to indicators or substitutes for PBT properties (ready, inherent or nonbiodegradability; bioaccumulation or $\log K_{OW}$; ecotoxicity data); in case of no available data for a category, the maximum, worst-case of 3 points will be applied. This results in a total between 0 and 9 points per PAS. Updated results of this PAS hazard assessment are available as a printable booklet (current 2012 version at: http://www.janusinfo.se/Global/Miljo_och_lakemedel/miljobroschyr_engelsk_2012_uppslag.pdf).

The pharmaceutical industry, which has delivered the PAS basic data for the SCC classification since 1993, suggested to improve the classification by not only considering hazard but also relating hazard to exposure, i.e. extending the SCC hazard assessment to a simplified PERA for PAS already on the Swedish market. In this so-called Voluntary Environmental Drug Classification System [53], substance-relevant data on physicochemical properties, (bio)degradability, persistence, bioaccumulability and (acute or chronic) ecotoxicity are contrasted with surface water PECs for Sweden based on actual annual use of the respective PAS. The results are expressed in three different formulations, on level 1 as a simple phrasing of the risk for lay people, mainly patients (e.g. "use of the medicine has been considered to result in insignificant/low/moderate/high environmental risk", with the qualifiers as appropriate for the PAS in question). On a second level intended for prescribers of the medicines, the environmental risk is given as in level 1, with additional information as to environmental degradation/persistence, to bioaccumulability or to PBT characteristics, all as appropriate for the PAS in question. On level 3, the full information available is given for specialists to assess and judge themselves. However, all levels of information are open to the public.

All PAS on the Swedish market are going to be integrated into this scheme; moreover, existing assessments are updated with recent pharmaceutical use data and new substance-specific data if available, every 3 years. Available risk classifications can be searched at FASS.se (http://www.fass.se/LIF/miljo/miljoinfo.jsp; search by ATC code or substance name ("substans", e.g. "sulfamet"), select one single PAS (e.g. sulfametoxazol in Swedish), then one single product (e.g. Bactrim forte), click on "FASS" on top of the product window, scroll down to subheading "Miljöpåverkan", then click on "Läs mer>>" to see the detailed environmental information).

The Swedish hazard and PERA systems address, at least in part, questions about "old" PAS already on the market. Current US and EU PERA guidelines do not primarily address old PAS but mostly new ones. Hence, the Swedish classification is an important step towards a broader base for a risk overview for PIE. As a consequence, several other EU member states have shown interest in the Swedish classification as a model for themselves or for the whole of the EU for existing PAS. In particular the Nordic countries with Norway and Denmark (beside Sweden), but also Germany, the Netherlands and the UK are looking into the Swedish model, possibly also into elevating it with additions to EU level.

Other, Non-PERA Regulations that Still Have an Indirect Influence on PIE and PERA

Is PERA Beyond REACH?

Registration, Evaluation, Authorisation and Restriction (which is less often mentioned but still part of the full name) of Chemicals, the "new" chemicals management named REACH in the European Community [23], aims at regulating the production, marketing and use of all chemicals not covered by other pertinent legislation. REACH intends to improve chemicals safety throughout, by assessing hazard and risk in function of annual amounts put on the market on the one hand and of hazardous properties marking the chemical as a "substance of very high concern" (SVHC), viz. carcinogenic or mutagenic or reprotoxic (CMR) or persistent and bioaccumulative and highly ecotoxic (PBT) substances, on the other.

Registration under REACH is not necessary for non-hazardous substances used in amounts below one metric tonne per year; all other chemicals may only be used after they have been duly (pre-)registered. However, a notification of Classification and Labelling is required for all chemicals, irrespective of amounts. For low tonnages the dossier is comparatively simple, but with increasing amounts or in case of SVHCs, additional prescribed data sets become necessary, including physicochemical, toxicological and environmental substance basic data, defined use scenarios for the chemical in question and chemical safety reports based thereon (e.g. [66]). Human pharmaceuticals (also veterinary medicines, medical devices, cosmetics, food or feedstuffs) in general are exempt from REACH, as they are assessed under different legislation. But REACH is still highly relevant for the pharmaceutical industry, as all starting materials, intermediates and also ancillary compounds like solvents fall under the chemicals legislation. However, in some avowedly rare cases, even a PAS may fall under full REACH coverage. This is the case where a PAS is formulated as a prodrug, mostly for reasons of improved absorption and bioavailability, and is only metabolised back to the actual PAS in the body. If this PAS already occurs in the chemical synthesis, to be esterified or otherwise chemically converted to the prodrug as dispensed, the PAS technically is an intermediate and therefore falls under REACH.

On the other hand, according to the EMA [26] ERA guideline, in case of prodrugs, the actual PAS should be assessed in an ERA. This leads to a situation where the same PAS may be investigated and assessed following two different, non-congruent guidelines. In the EMA scheme, both a base set including a water-sediment fate test (OECD test guideline 308) and chronic ecotoxicity studies with algae, daphnia and fish are needed, while in the REACH scheme [25], up to a high tonnage, comparatively simple ready biodegradability and acute ecotoxicity studies suffice. On the other hand, under REACH chemicals may be restricted or even denied marketing (unless their necessity can substantiated) based on SVHC characteristics, which is not possible following the human medicines legislation [22], where patient benefit takes precedence before environmental concerns. Hence, there may be double legislative coverage, inconsistent ERA and contradictory regulatory options for some few PAS.

The European Water Framework Directive and PAS

The EU Water Framework Directive (WFD; [21]) aims at achieving "enhanced protection and improvement of the aquatic environment", which covers inland surface waters, ground, transitional and coastal marine waters, and (re-)establishing good ecological status of aquatic ecosystems. One express means of attaining these goals is through "specific measures for the progressive reduction [...] and the cessation or phasing-out of discharges, emissions and losses of the priority hazardous substances", thereby "ensur[ing] the progressive reduction of pollution of groundwater and prevent[ing] its further pollution". Besides these priority substances, there may well be additional Environmental Quality Standards (EQS), corresponding to legal limit values, for other pollutants. EQS are set based on a compilation and interpretation of basic substance data including in particular environmental fate and toxicity, mammalian toxicity and bioaccumulation information. PAS are not exempt from the WFD; on the contrary, several PAS were included in a first list of candidate EQS substances at a relatively late stage in development of that list, end of 2009. In a proposal for an update of the WFD dating to January, 2012, oestradiol, ethinyloestradiol and diclofenac were identified as candidate- for an official EQS within the scope of the WFD. Further PAS may be included in a proposed watchlist of additional substances that should be analysed in surface waters and some of them may also warrant the future development of EQS values.

However, the question may be, what would the regulators decide in case an EQS for PAS were regularly breached at one or more sampling sites? Theoretically, the WFD has the power to ban substances of concern from further use. But banning pharmaceuticals might not be that simple given the human health benefits in prophylactic and disease treatment contexts. Indeed, the current Human PAS EU Directive 2001/83/EC [24] specifically states that human PAS may not be kept from the market, even in case of a negative PERA (in contrast to veterinary PAS, biocides or pesticides, which may be restricted or even banned in such a case; [68]). Possibly, the best way forward would be a specific improvement of sewage treatment works with a view to increase the removal of PAS, through biological or physicochemical means, as already advocated years ago by O'Brien and Dietrich [60].

ERA for the Production of Pharmaceuticals?

Subsequent to a first Swedish publication [52] evidencing very high concentrations of PAS, particularly antibiotics, in the effluent of a wastewater treatment plant serving an industrial park near Hyderabad, India, the production of pharmaceuticals (both of PAS and of formulated products) came under further scrutiny. In a later publication from the same group [51] it was shown that many PAS contained in medicines marketed in Sweden were originally produced in India, most of them with the same insufficient control and treatment of production effluents. One of the consequences of these investigations was a proposal by the Swedish Medical Products Agency (MPA) that a "requirement for an environmental certification of the production facilities" be introduced into the international regulations on Good Manufacturing Practice and further that "the current EU legislation for the authorisation of medicinal products for humans should be changed so that an ERA [of the production of the PAS] is also included in the approval" [56]. If so, a production ERA of the PASs and of the finished medicines would also become part of the PERA. However, this would be in contradiction to both the EU REACH legislation [23], which already covers all intermediates of pharmaceuticals production, and to the EMA [26] PERA Guideline, which expressly excludes the production from the scope of the PERA.

What the Swedish initiative certainly does is to point the finger at preventable environmental exposures to PAS that have no therapeutical, palliative or preventative benefit. While some exposures from patients' excretions may prove to be difficult or even impossible to be prevented, these uses at least have important medical benefits, while exposure through insufficient retention or treatment of wastewaters have none. On the other hand, not all productions of PAS lead to inacceptable environmental exposure, as shown by Boegård et al. [7] and Hoerger et al. [39] for two production sites in Europe, but a recent publication does show high levels of PAS downstream of US formulation facilities [62]. It will remain to be seen whether the reports showing high receiving water concentrations of PIE due to production will entail changes in EU or US regulations for PAS.

A Short Comparison with PERA for Veterinary Pharmacueticals

The situation for veterinary medicinal products (VMP) ERA is formally different from that for human PAS. VMP ERA, which had been developed independently in several countries (similar to the current situation for human PERA), has been harmonised between the EU, Japan and USA in the so-called International Cooperation for the Harmonisation of technical requirements for the registration of Veterinary medicinal product (VICH) process. Canada, Australia and New Zealand have adopted (and further countries may accept) the VICH PERA guidelines, which are split in to two phases.

Based on the concept that VMPs with very limited use or a high rate of metabolisation will have limited environmental exposure and effects, *Phase I* [80] is mainly a decision tree for filtering out those VMPs where no ERA is needed. Such categorical exclusions comprise:

- VMPs that are formally exempt from ERA by legislation
- VMPs that are natural substances, the use of which will not alter the background concentrations in the environment
- VMPs for exclusively non-food animals (i.e. pets or "companion animals")
- VMPs for use in a minor species that is handled and treated similarly to a major species for which a VICH PERA already exists
- VMPs that are only used to treat a small number of animals in a herd
- VMPs that are extensively metabolised in the treated animal

Then the decision tree splits into two branches, depending on whether the treated species are aquatic or terrestrial. For VMPs used in aquaculture, the categorical exclusions comprise:

- VMPs that are not released into the aquatic compartment (e.g. aquarium species) by disposal of the aquatic waste matrix
- VMPs that are used in a confined facility *and* that are not endo- or ecto-parasiticides *and* where the entry into the EIC from the aquaculture facilities is <1 µg/L (or where this EIC is mitigated to <1 µg/L by installations or proven degradation mechanisms)

On the terrestrial side, the categorical exclusions comprise

- VMPs that are not released into the terrestrial compartment by disposal of the terrestrial waste matrix
- VMPs that are used in animals reared on pasture *and* that are not endo- or ecto-parasiticides *and* where the soil PEC is <100 µg/kg soil (or where this PEC is mitigated to <100 µg/kg by installations or proven degradation mechanisms)

- VMPs that are used in animals not reared on pasture *and* where the soil PEC from spreading manure is <100 µg/kg soil (or where this PEC is mitigated to <100 µg/kg by installations or proven degradation mechanisms)

For all the above exclusions, the VICH PERA stops with a Phase I report that discusses the basis for this decision. The Phase I document [79] also contains a lengthy Q&A part where explanations for certain questions are given. In case a VMP would normally go into Phase II but the PEC can be brought below the aquatic or terrestrial threshold, discussion with the regulators is necessary before deciding how to proceed.

All other VMPs must be assessed in *Phase II* [80], which is a two-tiered procedure. Tier A is a simpler, more conservative assessment; if a conclusion of no significant risk cannot be reached in Tier A, the assessment must progress into Tier B with more demanding data. In Phase I a total residue approach is taken, metabolites are not normally considered and PECs have to be calculated based on the applied dose; note that in Phase II Tier A, PNECs are calculated for every single group of test organisms.

For practical reasons, Phase II is divided into three major branches, aquaculture, intensively reared terrestrial animals and pasture animals, due to different entry pathways into the environment. Detailed guidance for the three branches is given regarding types of environmental exposure, experimental data for Tiers A and B and calculation and refinement of PECs for the respective compartments.

General data requirements for Phase II Tier A comprise:

- Water solubility (OECD105)
- Dissociation constants (OECD112)
- UV–visible absorption spectrum (OECD101)
- Melting point/range (OECD102)
- Vapour pressure (OECD104), normally by QSPR calculation, except if there is evidence that the vapour pressure is $>10^{-5}$ Pa at 20°C, in which case it should be determined experimentally
- *n*-Octanol/water partition coefficient (OECD107 or OECD117); note that for ionisable substances there is a cryptic footnote that "if appropriate, the $\log K_{ow}$ for such substances should be measured on the non-ionised form at environmentally relevant pHs"
- Soil adsorption/desorption (OECD106) reporting both K_d and K_{oc} values
- In case of primary exposure to soil, soil biodegradation (OECD307)
- In case of primary exposure to the aquatic compartment, degradation in water/sediment systems (OECD308); note that for marine applications, the possibility of doing the water/sediment study with seawater should be discussed with the regulators
- (Optional) photolysis in water (OECD316) for aquaculture (consider seawater photolysis test for marine applications) or on soil (OECD guideline in preparation) for terrestrial branches
- (Optional) hydrolysis (OECD111)

- Acute aquatic ecotoxicity base set, for both terrestrial and aquaculture applications (consider seawater tests for marine applications), with appropriate endpoint and AF to derive the group-specific PNEC

  - Freshwater algal growth inhibition (OECD201), EC50, AF = 100 or
  - Seawater algal growth inhibition (ISO10253), EC50, AF = 100
  - Freshwater *Daphnia* immobilisation (OECD202), EC50, AF = 1,000 or
  - Saltwater crustacean acute toxicity (ISO14669), EC50, AF = 1,000
  - Freshwater fish acute toxicity (OECD203), LC50, AF = 1,000 or
  - Seawater fish acute toxicity (seek guidance), LC50, AF = 1,000

- Terrestrial ecotoxicity base set, for terrestrial branches/soil exposures, with appropriate endpoint and AF to derive the group-specific PNEC

  - Microbial nitrogen transformation (OECD216), to be tested at 1× and 10× the soil PEC; note no AF, but "pass" if difference in nitrate formation is ≤25% compared with controls at any time before 28 days; else the study should be prolonged to 100 days in Phase II Tier B
  - Terrestrial plants seedling emergence and growth test (OECD208), EC50, AF = 100
  - Earthworm subacute toxicity (OECD220) or reproduction test (OECD222), NOEC, AF = 10

Specifically for endo- or ecto-parasiticides used in pasture treatments, the following tests on dung fauna are also recommended:

- Dung fly larvae acute toxicity test (OECD228), EC50, AF = 100
- Dung beetle larvae acute toxicity test (OECD test in preparation, seek guidance), EC50, AF = 100

For Phase II Tier A risk characterisation, the initial PEC for soil or the aquatic compartment from Phase I is to be compared with all appropriate PNECs derived as a PEC/PNEC risk quotient (RQ) in VICH terminology. If all RQs are <1 and there is no risk of accumulation of the VMP in the environment, based on persistence data, there is no significant risk and the VICH PERA can be concluded. If the RQ is ≥1 for any organism tested, the initial PEC should be refined with information on metabolism/excretion and on environmental degradation (OECD307, 308). If the RQ is still ≥1 for any organism tested, the PERA should progress to Phase II Tier B and chronic testing should be done on the organism concerned refine the PNEC. In the case of pasture applications, if the RQ is ≥1 for a dung organism, the initial dung PEC, which assumes the whole dose of VMP excreted in one day, should be refined based on realistic excretion patterns; if the refined RQ is ≥1, regulatory guidance should be sought. Further, if the $logK_{ow}$ is ≥4, bioaccumulation should be investigated in Phase II Tier B. If the aquatic invertebrate RQ is ≥1, an initial sediment assessment based on equilibrium partitioning is recommended; if the RQ is still ≥1, a refinement of the calculated sediment PNEC through a preferably chronic sediment toxicity study is recommended.

In Phase II Tier B, the following organism- or compartment-specific assessments are described:

- For bioaccumulation, a fish bioaccumulation study (OECD305) should be performed (normally with radio-labelled material). If the bioconcentration factor is >1,000, regulatory guidance should be sought.
- For chronic aquatic effects testing, the following are recommended for those organisms with a Tier A RQ ≥ 1

  - Algal growth inhibition (freshwater and marine), but use NOEC and an AF of 10 from the tests already performed in Tier A
  - Freshwater *Daphnia* reproduction test (OECD210), NOEC, AF = 10,
  - For seawater seek guidance as to test guideline, NOEC; AF = 10
  - For freshwater fish early life stage test (OECD211), NOEC, AF = 10
  - For saltwater seek guidance for a chronic fish test, NOEC, AF = 10
  - For freshwater sediment, invertebrate chronic toxicity (OECD219 if entry into the environment is through water, OECD218 if it is through sediment or adsorbed to soil in surface run-off), NOEC, AF = 10
  - For seawater sediment, invertebrate chronic toxicity (seek guidance), NOEC, AF = 10

- For terrestrial long-term effects, the following are recommended for those organisms with a Tier A RQ ≥ 1

  - Terrestrial plants seedling emergence and growth test (OECD208), repeat test with two additional species from the most sensitive group and the most sensitive species in the first Tier A test, NOEC, AF = 10
  - Earthworm test (neither test guideline nor endpoint, respectively, AF given, hence seek guidance)
  - Microbial nitrogen transformation (OECD216) prolonged to 100 days in Phase II Tier B; note no AF, but "pass" if difference in nitrate formation is $\leq 25\%$ compared with controls

If after Tier B testing any RQ, including dung fauna RQ, is still ≥ 1 or if the pass level was not reached in the soil nitrification test, regulatory guidance should be sought. Also, the specific guidance for Phase II approaches for the three branches helps to identify and deal with common problems. If there are still unresolved questions at the end of Phase II Tier B, the applicant should contact the regulators for discussing risk mitigation strategies (e.g. use only under specified conditions to minimise environmental exposure), which would result in a registration with restrictions on use and mandatory labelling of the product regarding the environmental hazards.

The VICH [79, 80] guidelines present a highly detailed scheme of investigating environmental risks from VMPs. Many applications are excluded in Phase I, where a non-significant environmental exposure is assumed to result. In view of the discontinuous administration, in contrast to many human PAS, Phase II Tier A relies on acute data for an initial RQ assessment. However, this assessment is rendered specific for the three branches, respectively, for the main (and secondary) environmental compartments exposed by the aquaculture, pasture and intensively reared applications. The VICH procedure differs from other PERA schemes in that additional and often highly important environmental degradation pathways, e.g. aquatic or soil surface photodegradation are specifically mentioned. Phase II Tier B then is more conventional on the effects side, with chronic-based PNECs. With restricted registration looming at the end of the process, VICH applicants have very good reasons to refine their PECs and PNECs to the maximum possible to attain RQs < 1.

Outlook and Conclusion

Through regional development, current international PERA guidelines (and probably the ones in preparation) are different in many details, agreeing only on very broad levels or methodologies. On the other side, the environment is not that different on both sides of the Atlantic or the Pacific Oceans, PAS are often exactly the same, entry pathways into and fate processes within that environment are highly comparable and so, basically, are environmental organisms and ecological functions within compartments.

Moreover, at higher tiers or levels in the PERA process, the published guidelines (including the VICH Phase II) are much better comparable than at lower tiers, where evident discrepancies exist. A comparison of the existing human and veterinary ERA guidelines, together with the EU Biocides [20] guideline as an outgroup representative, is given in Table 1. On the other hand, all stakeholders in the investigation of environmental risks from pharmaceuticals would be expected to have the same expectations: clear, transparent, scientifically sound and comparable (if not identical) schemes to assess substances, with confidence and decreased uncertainty, on a local, national and international level. This, for human PERA at least, is still missing.

A side glance to veterinary PERA regulations and the VICH process shows the way forward: Even though some of the human PERA guidelines are only being developed or finalised, all of them may already now be in need of revision with a view to international harmonisation. This would facilitate the work of both applicants and regulators and, most of all, it would render PERA much more transparent on an international level.

PERA promises to remain both scientifically interesting, economically and societally important but also at times somewhat politically confusing. The scientific lessons learned in PERA are also likely to provide positive benefits for other areas where chemicals are used to support diverse industries from antifoulant paints to food production.

[AU7]

Table 1 Comparison of existing guidelines (GLs) for the environmental risk assessment (ERA) of pharmaceuticals, with EU biocides ERA as an outgroup

	Human PAS ERA GLs				Veterinary medicinal products GLs		Outgroup GL
	US (1998)	EU EMA [26]	Draft GL Japan	"Old" PAS GL Sweden	VICH I [80]	VICH II [81]	EU biocides [20]
Internationally harmonised GL	No	No (EU member states)	NI	No	Yes (EU, USA, J, CAN, AUS, NZ)	Yes (EU, USA, J, CAN, AUS, NZ)	No (EU member states)
Categorical exclusions	Yes (chemical groups)	Yes (chemical groups)	Yes (chemical groups)	Yes (chemical groups)	Yes (regulation and application based)	NA (higher tier)	No
Categorical inclusions (special assessment)	(Yes) if the PAS may significantly affect the human environment	Yes. $\log K_{ow} > 4.5$ or potential endocrine disruptor/ reproductive effects at <10 ng/L	NI	No	Yes. Endo-/ ectoparasiticides	NA (higher tier)	(No) all biocides must de assessed
Tiered ERA	Yes	Yes	Yes	No	Yes	Yes	(Yes) partly, refinement possible
Threshold PEC/ trigger value	Yes. 1 µg/L EIC (sewage works effluent); 0.1 µg/L surface waters	Yes – 0.01 µg/L surface waters	NI	No	Yes. 1 µg/L EIC for surface waters or 100 µg/kg EIC for soil	NA (higher tier)	No

Technical guidance available	Yes. Straightforward flow scheme, some lack of guidance for highest tier 3	NI	Yes. Straightforward for phases 1 and 2 tier A, insufficient for phase 2 tier B and reproductive/endocrine assessment	(Yes) straightforward for PEC and acute-based PNEC, little guidance for refinements	Yes. straightforward flow scheme	Yes. Detailed for the three branches aquaculture, intensively reared and pasture animals	Yes. EU TGD [76] and TGD-based EUSES application
Defined base set	(No) data increasing with every tier, but no fixed genuine base set	NI	Yes. Phase 2 tier A base set and in part for phase 2 tier B	(Yes) insufficient data situation will be noted	(No) exposure threshold approach	Yes. Both for phase II tier A and compartment/organism specific in phase II tier B	Yes
Chronic testing compulsory	No. Only in highest tier	Yes	Yes. Phase 2	No	NA (First tier)	Yes (higher tier)	No. only in higher tier
Assessment factors for surface water PNEC	10,000 (one acute EC/LC50 tier 1), 1,000 (three acute EC/LC50s tier 2), 100 (unknown number of chronic EC/LC50s tier 3)	NI	10 (three chronic NOECs phase 2 tier A)	1,000 (three acute EC/LC50s), 100 (one, fish or daphnia, chronic NOEC), 50 (two chronic NOECs), 10 (three chronic NOECs)	NA (first tier)	10 (chronic NOEC), no overall PNEC for surface water or soil, but single-species risk quotients	1,000 (three acute EC/LC50s), lower for higher-tier chronic tests: 10 (three chronic NOECs)

(continued)

Table 1 (continued)

	Human PAS ERA GLs				Veterinary medicinal products GLs		Outgroup GL
	US (1998)	EU EMA [26]	Draft GL Japan	"Old" PAS GL Sweden	VICH I [80]	VICH II [81]	EU biocides [20]
Bioaccumulation testing/trigger	No. $\log D_{ow} \geq 3.5$ at pH of ~7 only requires chronic testing	Yes. $\log K_{ow} \geq 4.5$ in phase 1, $\log K_{ow} \geq 3$ in phase 2 tier A	NI	(No) but PAS are regarded as bioaccumulating if $\log K_{ow} \geq 3$	NA (first tier)	Yes. $\log K_{ow} \geq 4$ in phase II tier A	Yes. Part of the base set in Annex IIA/VII
Comprehensive ERA	Possibly tier 3	Phase 2 tier B	NI	No	NA (first tier)	Phase II tier B	(No) only acute testing required in annex IIA/VII
ERA applicable for new substances only	Yes	(Yes) except repeat registrations/variations leading to higher exposure, generics for new applicants	Yes	No specific programme for "old" PAS	(No)	NA (higher tier)	(No)
Registration dependent on result of ERA	No	No. Compulsory labelling and proposal for minimising exposure may apply	No	No	NA	(No) application restrictions and compulsory labelling may apply	Yes. Limited authorisation with use restrictions and compulsory labelling and requirement of a safety data sheet may apply

Based in part on table III in Straub [69]. As there is no information available on the draft Canadian PERA GL, this was not used in the comparison. For details and in case of uncertainties refer to the original guideline

Note (no) or (yes) in brackets means this question/criterion cannot be unambiguously decided/applied, but the tendency is indicated

EIC entry into the environment concentration (before dilution); *NA* not applicable; *NI* no information

Acknowledgements Our thanks to our colleagues in Europe and the USA (I Radtke; plus many others), Canada (L Jack, A Beck), Japan (Y Azuma, T Tosaka) and Australia (L Justice) for discussions, help and support.

References

1. Aherne GW, Briggs R (1989) The relevance of the presence of certain synthetic steroids in the aquatic environment. J Pharm Pharmacol 41:735–736
2. AMZV (2001) Verordnung des Schweizerischen Heilmittelinstituts vom 9. November 2001 über die Anforderungen an die Zulassung von Arzneimitteln (Arzneimittel-Zulassungsverordnung, AMZV), Stand am 12. September 2006. SR 812.212.22, The Swiss Confederation, Berne. http://www.admin.ch/ch/d/sr/c812_212_22.html. Accessed Aug 2010
3. AP (2009) An Associated Press Investigation: pharmaceuticals found in drinking water. http://hosted.ap.org/specials/interactives/pharmawater_site/index.html. Accessed July 2010
4. Ayscough NJ, Fawell J, Franklin G, Young W (2000) Review of human pharmaceuticals in the environment. R&D Technical Report P390. Environment Agency, Bristol, UK
5. Barnes KK, Kolpin DW, Focazio ET, Meyer MT, Zaugg SD, Haack SK, Barber LB, Thurman EM (2008) Water-quality data for pharmaceuticals and other organic wastewater contaminants in ground water and in untreated drinking water sources in the United States, 2000–01: US Geological Survey Open-File Report 2008–1293, 7 p. plus tables. http://pubs.usgs.gov/of/2008/1293/pdf/OFR2008-1293.pdf. Accessed July 2010
6. Belfroid AC, Van der Horst A, Vethaak AD, Schäfer AJ, Rijs BJ, Wegener J, Cofino WP (1999) Analysis and occurrence of estrogenic hormones and their glucuronides in surface water and wastewater in The Netherlands. Sci Total Environ 225:101–108
7. Boegård C, Coombe V, Holm G, Taylor D (1998) The concentrations and potential environmental impact of pharmaceuticals in the effluent from a major AstraZeneca manufacturing facility. Poster, KNAPPE Conference, Nîmes (France), February 19th–20th, 2008
8. Breton R, Boxall A (2003) Pharmaceuticals and personal care products in the environment: regulatory drivers and research needs. QSAR Comb Sci 22:399–409
9. Bruce GM, Pleus RC, Snyder SA (2010) Toxicological relevance of pharmaceuticals in drinking water. Environ Sci Technol 44:5619–5626
10. Buser HR, Müller MD, Theobald N (1998) Occurrence of the pharmaceutical drug clofibric acid and the herbicide mecoprop in various Swiss lakes and in the North Sea. Environ Sci Technol 32:188–192
11. Calamari D, Zuccato E, Castiglioni S, Bagnati R, Fanelli R (2003) Strategic survey of therapeutic drugs in the rivers Po and Lambro in Northern Italy. Environ Sci Technol 37:1241–1248
12. Caldwell DJ, Mastrocco F, Nowak E, Johnston J, Yekel H, PFeiffer D, Hoyt M, DuPlessie BM, Anderson PD (2010) An assessment of potential exposure and risk from estrogens in drinking water. Environ Health Perspect 118(3):338–344
13. Campbell PJ, Arnold DJS, Brock TCM, Grandy NJ, Heger W, Heimbach F, Maund SJ, Streloke M (1999) Guidance document on higher-tier aquatic risk assessment for pesticides (HARAP). SETAC Europe, Brussels (Belgium)
14. CDER (1998) Guidance for industry—environmental assessment of human drug and biologics applications. US Dept of Health and Human Services, Food and Drug Administration, US Center for Drug Evaluation and Research (CDER), Center for Biologics Evaluation and Research (CBER). July 1998, CMC 6, Revision 1. http://www.fda.gov/downloads/Drugs/GuidanceComplianceRegulatoryInformation/Guidances/ucm070561.pdf. Accessed July 2010
15. CEPA (1999) Canadian Environmental Protection Act, 1999. S.C.1999, c.33. Current version at http://laws-lois.justice.gc.ca/eng/C-15.31/20100726/page-0.html?rp2=HOME&rp3=SI&rp1=environmental%20protection%20act&rp4=exact&rp9=cs&rp10=L&rp13=50#idhit1. Accessed July 2010

16. Christensen FM (1998) Pharmaceuticals in the environment—a human risk? Regul Toxicol Pharmacol 28(3):212–221
17. Cunningham VL, Binks SP, Olson MJ (2009) Human health risk assessment from the presence of human pharmaceuticals in the aquatic environment. Regul Toxicol Pharmacol 53(1):39–45
18. Daughton CG, Ternes TA (1999) Pharmaceuticals and personal care products in the environment: agents of subtle change? Environ Health Perspect 107:907–938
19. De Lange HJ, Noordoven W, Murk AJ, Lürling M, Peeters ETHM (2006) Behavioural responses of *Gammarus pulex* (Crustacea, Amphipoda) to low concentrations of pharmaceuticals. Aquat Toxicol 78:209–216
20. EC (1998) Directive 98/8/EC of the European Parliament and of the Council of 16 February 1998 concerning the placing of biocidaly products on the market. Off J L123:1–63
21. EC (2000) Directive 2000/60/EC of the European Parliament and of the Council of 23 October 2000 establishing a framework for Community action in the field of water policy. Off J L327:1–72
22. EC (2001) Directive 2001/83/EC of the European Parliament and of the Council of 6 November 2001 on the Community code relating to medicinal products for human use. Off J L311:67–128. Current (end-2009) consolidated version available at http://eur-lex.europa.eu/LexUriServ/LexUriServ.do?uri=CONSLEG:2001L0083:20091005:EN:PDF. Accessed 13 Jan 2010
23. EC (2006) Regulation (EC) No 1907/2006 of the European Parliament and of the Council of 18 December 2006 concerning the Registration, Evaluation, Authorisation and Restriction of Chemicals (REACH), establishing a European Chemicals Agency, amending Directive 1999/45/EC and repealing Council Regulation (EEC) No 793/93 and Commission Regulation (EC) No 1488/94 as well as Council Directive 76/769/EEC and Commission Directives 91/155/EEC, 93/67/EEC, 93/105/EC and 2000/21/EC. Off J L396:1–849. Current (end-2009) consolidated version available at http://eur-lex.europa.eu/LexUriServ/LexUriServ.do?uri=CONSLEG:2006R1907:20090220:EN:PDF. Accessed 11 Jan 2010
24. ECHA (2008) European Chemicals Agency, Guidance for the implementation of REACH. Chapter R11, Guidance on information requirements and chemical safety assessment, Part C: PBT Assessment. http://guidance.echa.europa.eu/docs/guidance_document/information_requirements_part_c_en.pdf?vers=20_08_08. Accessed July 2010
25. ECHA (2010) European Chemicals Agency, REACH and Classification & Labelling Guidance. http://guidance.echa.europa.eu/guidance_en.html. Accessed 11 Jan 2010
26. EMA (2006) Note for guidance on environmental risk assessment of medicinal products for human use. CPMP/SWP/4447/00. European Medicines Agency, London. http://www.ema.europa.eu/pdfs/human/swp/444700en.pdf. Accessed 11 Jan 2010
27. EMA (2011) Questions and answers on "Guideline on the environmental risk assessment of medicinal products for human use", 17 March 2011. EMA/CHMP/SWP/44609/2010, EMA, London (UK). http://www.ema.europa.eu/docs/en_GB/document_library/Other/2011/04/WC500105107.pdf. Accessed April 2012
28. Farré M, Ferrer I, Ginebreda A, Figueras M, Olivella L, Tirapu L, Vilanova M, Barceló D (2001) Determination of drugs in surface water and wastewater samples by liquid chromatography—mass spectrometry: methods and preliminary results including toxicity studies with *Vibrio fischeri*. J Chromatogr A 938:187–197
29. Fernández C, González-Doncel M, Pro J, Carbonell G, tarazona JV (2010) Occurrence of pharmaceutically active compounds in surface waters of the Henares-Jarama-Tajo river system (Madrid, Spain) and a potential risk characterization. Sci Total Environ 408:543–551
30. Fong PP (1998) Zebra mussel spawning is induced in low concentrations of putative serotonin reuptake inhibitors. Biol Bull 194:143–149
31. Fong PP, Molnar N (2008) Norfluoxetine induces spawning and parturition in estuarine and freshwater bivalves. Bull Environ Contam Toxicol 81(6):535–538
32. Foster KR, Vecchia P, Repacholi MH (2000) Science and the precautionary principle. Science 288:979–980

33. Garrison AW, Pope JD, Allen FR (1976) GC/MS analysis of organic compounds in domestic wastewaters. In: Keith LH (ed) Identification & analysis of organic pollutants in water. Ann Arbor Science Publications, Ann Arbor, MI, pp 517–556

34. Halling-Sørensen B, Nors Nielsen S, Lanzky PF, Ingerslev F, Holten Lützhøft HC, Jørgensen SE (1998) Occurrence, fate and effects of pharmaceutical substances in the environment—a review. Chemosphere 36(2):357–393

35. Heberer T, Butz S, Stan H-J (1995) Analysis of phenoxycarboxylic acids and other acidic compounds in tap, ground, surface and sewage water at the low ng/l level. Int J Anal Chem 58:43–53

36. Henschel K-P, Wenzel A, Diedrich M, Fliedner A (1997) Environmental hazard assessment of pharmaceuticals. Regul Toxicol Pharmacol 25:220–225

37. Hignite C, Azarnoff DL (1977) Drugs and drug metabolites as environmental contaminants: chlorophenoxyisobutyrate and salicylic acid in sewage water effluent. Life Sci 20:337–342

38. Hilvarsson A, Hallsdorsson HP, Granmo A (2007) Medetomidine as a candidate antifoulant: sublethal effects on juvenile turbot (*Psetta maxima* L.). Aquat Toxicol 83(3):238–246

39. Hoerger CC, Dörr B, Schlienger C, Straub JO (2009) Environmental risk assessment for the galenical formulation of solid medicinal products at Roche Basle, Switzerland. Integr Environ Assess Manag 5(2):331–337

40. Honkoop PJC, Luttikhuizen PC, Piersma T (1999) Experimentally extending the spawning season of a marine bivalve using temperature change and fluoxetine as synergistic triggers. Mar Ecol Prog Ser 180:297–300

41. Huffman MA, Seifu M (1989) Observations on the illness and consumption of a possibly medicinal plant *Vernonia amygdalina* (Del.), by a wild chimpanzee in the Mahale Mountains National Park, Tanzania. Primates 30(1):51–63

42. Huffman MA, Wrangham RW (1996) Diversity of medicinal plant use by chimpanzees in the wild. In: Wrangham RW, McGrew WC, de Waal FBM, Heltne PG (eds) Chimpanzee cultures. Harvard University Press, Cambridge, MA, pp 129–148

43. Huggett DB, Brooks BW, Peterson B, Foran CM, Schlenk D (2002) Toxicity of select beta adrenergic receptor-blocking pharmaceuticals (b-blockers) on aquatic organisms. Arch Environ Contam Toxicol 43:229–235

44. Jobling S, Nolan M, Tyler CR, Brighty G, Sumpter JP (1998) Widespread sexual disruption in wild fish. Environ Sci Technol 32:2498–2506

45. Jones OA, Lester JN, Voulvoulis N (2005) Pharmaceuticals: a threat to drinking water? Trends Biotechnol 23(4):163–167

46. Kalbfus W, Kopf W (1998) Erste Ansaätze zur ökologischen Bewertung von Pharmaka in Oberflächengewässern. Münchner Beiträge zur Abwasser Fischerei und Flussbiologie 51: 628–652

47. Kim SD, Cho J, Kim IS, Vanderford BJ, Snyder SA (2007) Occurrence and removal of pharmaceuticals and endocrine disruptors in South Korean surface, drinking, and waste waters. Water Res 41:1013–1021

48. Kolpin DW, Furlong ET, Meyer MT, Thurman EM, Zaugg SD, Barber LB, Buxton HT (2002) Pharmaceuticals, hormones and other organic wastewater contaminants in US streams, 1999–2000: a national reconnaissance. Environ Sci Technol 36:1202–1211

49. Kümmerer K (ed) (2008) Pharmaceuticals in the environment: sources, fate, effects and risks, 3rd edn. Springer, Berlin, Heidelberg, p 521

50. Krang AS, Dahlstrom M (2006) Effects of a candidate antifoulant compound (medetomidine) on pheremone induced mate search in the amphipod *Corophium volutator*. Mar Pollut Bull 52(12):1776–1783

51. Larsson DGJ, Fick J (2009) Transparency throughout the production chain—a way to reduce pollution from the manufacturing of pharmaceuticals? Regul Toxicol Pharmacol 53:161–163

52. Larsson DGJ, de Pedro C, Paxéus N (2007) Effluent from drug manufacturers contains extremely high levels of pharmaceuticals. J Haz Mater 148:751–755

53. Mattson B, Näsman I, Ström J (2007) Sweden's voluntary environmental classification system. Raj Pharma 153–158. http://www.lif.se/cs/Medlemsservice/Sidinnehall/Nyheter/

Dokument/RAJ%202007%20Mar%20-%20Sweden%20Environment%20FINAL.pdf. Accessed July 2010

54. Metcalfe CD, Koenig BG, Bennie DT, Servos M, Ternes TA, Hirsch R (2003) Occurrence of neutral and acidic drugs in the effluents of Canadian sewage treatment plants. Environ Toxicol Chem 22(12):2872–2880

55. Montgomery J (2010) Delaware drinking water at risk: prescription drugs on tap from major suppliers. News J. Delaware Online, http://www.delawareonline.com/article/20100804/ NEWS02/8040343. Accessed 4 Aug 2010.

56. MPA (2009) Läkemedelsverket/Medical Products Agency. Redovisning av regeringsuppdrag gällande möjligheten att skärpa miljökrav vid tillverkning av läkemedel och aktiv substans. Report from the Swedish MPA (with an English abstract), 16th December 2009. http://www. lakemedelsverket.se/upload/nyheter/2009/2009-12-16_rapport_milj%C3%B6krav-1%C3%A4kemedel.pdf. Accessed July 2010

57. Nakada N, Tanishima T, Shinohara H, Kiri K, Takada H (2006) Pharmaceutical chemicals and endocrine disrupters in municipal wastewater in Tokyo and their removal during activated sludge treatment. Water Res 40:3297–3303

58. NSNR/C&P (2005) Canadian New Substances Notification Regulation (Chemicals and Polymers). SOR/2005–247. Current version available at http://laws.justice.gc.ca/PDF/ Regulation/S/SOR-2005-247.pdf. Accessed July 2010

59. NSNR/O (2005) Canadian New Substances Notification Regulation (Organisms). SOR/2005– 248. Current version available at http://laws.justice.gc.ca/PDF/Regulation/S/SOR-2005-248. pdf. Accessed July 2010

60. O'Brien E, Dietrich DR (2004) Hindsight rather than foresight: reality versus the EU draft guideline on pharmaceuticals in the environment. Trends Biotechnol 22(7):326–330

61. Peterson SM, Batley GE, Scammell MS (1993) Tetracycline in antifouling paints. Mar Pollut Bull 26(2):96–100

62. Phillips PJ, Smith SG, Kolpin DW, Zaugg SD, Buxton HAT, Furlong ET, Esposito K, Stinson B (2010) Pharmaceutical formulation facilities as sources of opioids and other pharmaceuticals to wastewater treatment plant effluent. Environ Sci Technol 44:4910–4916

63. Rabiet M, Togola A, Brissaud F, Seidel J-L, Budzinski H, Elbaz-Poulichet F (2006) Consequences of treated water recycling as regards pharmaceuticals and drugs in surface and ground waters of a medium-size Mediterranean catchment. Environ Sci Technol 40:5282–5288

64. Richardson ML, Bowron JM (1985) The fate of pharmaceutical chemicals in the aquatic environment. J Pharm Pharmacol 37:1–12

65. Römbke J, Knacker T, Stahlschmidt-Allner P (1996) Umweltprobleme durch Arzneimittel— Literaturstudie. UBA-Forschungsbericht 96–060. Umweltbundesamt, Berlin (Germany)

66. Rudén C, Hansson SO (2010) Registration, evaluation and authorization of chemicals (REACH) is but the first step—how far will it take us? Six further steps to improve the European chemicals legislation. Environ Health Perspect 118(1):6–10

67. Schwab BW, Hayes EP, Fiori JM, Mastrocco FJ, Roden NM, Cragin D, Meyerhoff RD, D'Aco VJ, Anderson PD (2005) Human pharmaceuticals in US surface waters: a human health risk assessment. Regul Toxicol Pharmacol 42:296–312

68. Straub JO (2002) Environmental risk assessment for new human pharmaceuticals in the European Union according to the draft guideline/discussion paper of January 2001. Toxicol Lett 135:231–237

69. Straub JO (2005) Environmental risk assessment of new human pharmaceuticals in the European Union according to the 2003 draft guideline. In: Dietrich DR, Webb SF, Petry T (eds) (2005) Hot spot pollutants; pharmaceuticals in the environment. Elsevier Academic Press, Burlington, MA, pp 299–317

70. Straub JO (2008) Deterministic and probabilistic environmental risk assessment for diazepam. In: Kümmerer K (ed) Pharmaceuticals in the environment; sources, fate, effects and risks, 3rd edn. Springer, Heidelberg, pp 343–383

71. Straub JO, Stewart KM (2007) Deterministic and probabilistic acute-based environmental risk assessment for naproxen for western Europe. Environ Toxicol Chem 26(4):795–806

72. Stumpf M, Ternes TA, Wilken RD, Rodrigues SV, Baumann W (1999) Polar drug residues in sewage and natural waters in the State of Rio de Janeiro, Brazil. Sci Total Environ 225:135–141

73. Ternes TA, Stumpf M, Schuppert B, Haberer K (1998) Simultaneous determination of antiseptics and acidic drugs in sewage and river water. Vom Wasser 90:295–309

74. TGA (2008) Australian Government, Department of Health and Aging, Therapeutic Goods Administration. Module 1, Administrative Information and Prescribing Information For Australia; Notice to Applicants; CTD-Module 1, TGA Edition November 2008. http://www.tga.gov.au/docs/pdf/euguide/tgamod1.pdf. Accessed August 2010

75. TGD (2003) Technical Guidance Document on Risk Assessment in support of Commission Directive 93/67/EEC on Risk Assessment for new notified substances, Commission Regulation (EC) No 1488/94 on Risk Assessment for existing substances, Directive 98/8/EC of the European Parliament and of the Council concerning the placing of biocidal products on the market; Part II, Environmental Risk Assessment. European Commission; 2003. http://ecb.jrc.ec.europa.eu/tgd/. Accessed 12 Jan 2010

76. Thomas KV, Hilton MJ (2004) The occurrence of selected human pharmaceutical compounds in UK estuaries. Mar Pollut Bull 49:436–444

77. Tixier C, Singer HP, Oeller S, Müller SR (2003) Occurrence and fate of carbamazepine, clofibric acid, diclofenac, ibuprofen, ketoprofen and naproxen in surface waters. Environ Sci Technol 37(6):1061–1068

78. van der Heide EF, Hueck-van der Plas EH (1984) Geneesmiddelen en milieu. Pharm Weekbl 119:946–947

79. VICH (2000) Environmental impact assessment (EIA) for veterinary medicinal products (VMPs)—phase I. VICH GL 6 (Ecotoxicity Phase I). http://www.vichsec.org/pdf/2000/Gl06_st7.pdf. Accessed July 2010

80. VICH (2004) Environmental impact assessment for veterinary medicinal products—phase II guidance. VICH GL 38 (Ecotoxicity Phase II). http://www.vichsec.org/pdf/10_2004/GL38_st7.pdf. Accessed July 2010

81. Vieno NM, Tuhkanen T, Kronberg L (2005) Seasonal variation in the occurrence ogf pharmaceuticals in effluents from a sewage treatment plant and in the recipient river. Environ Sci Technol 39:8220–8226

82. Watts C, Maycock D, Crane M, Fawell J, Goslan E (2007) Desk based review of current knowledge on pharmaceuticals in drinking water and estimation of potential levels. Final Report for the Drinking Water Inspectorate in DEFRA (UK Department of the Environment, Food and Rural Affairs) by WCA, contract number CSA7184/WT02046/DWI 70/2/213. http://randd.defra.gov.uk/Default.aspx?Menu=Menu&Module=More&Location=None&ProjectID=14717&FromSearch=Y&Publisher=1&SearchText=wt02046&SortString=ProjectCode&SortOrder=Asc&Paging=10. Accessed July 2010

83. Webb S, Ternes T, Gibert M, Olejniczak K (2003) Indirect human exposure to pharmaceuticals via drinking water. Toxicol Lett 142:157–167

84. Zuccato E, Calamari D, Natangela M, Fanelli R (2000) Presence of therapeutic drugs in the environment. Lancet 355:1789–1790

Regulation of Pharmaceuticals in the Environment: The USA

Emily A. McVey

Introduction

The fate of pharmaceuticals in the environment has been studied for more than 50 years, with the presence and potential effects acknowledged shortly thereafter [1–5]. It has gradually become apparent that risk assessments developed for the usual chemical contaminants cannot be applied carte blanche to pharmaceuticals, because they are developed to be highly active and specific in biological systems at low levels [6–9]. Therefore, when applying risk assessment models to the environmental assessment of pharmaceuticals, regulators must take into account not only the complexity of the entity to be protected (the ecosystem at large) but also the complexity of the regulated article (pharmaceuticals).

Regulation and policy are, at best, a merging of science, politics, social science, and stakeholder input. Environmental regulation is further complicated by the complex nature of the entity to be regulated. Environmental regulations and regulators have larger, long-term goals of "protection" of human health and ecosystems (large and small) from damage as a result of environmental exposures to contaminants. What this means in practice and in theory may depend on the interpretation of existing policies at any given time, but the most overarching goal for environmental regulators is promotion and protection of harmony (sustainability) within ecosystems. This is no small task, particularly when you consider how to define an "ecosystem." The traditional view of the ecosystem is the interacting organisms and biophysical components in a particular place, focusing on relationships and processes of the living and nonliving components [10]. This definition could be seen as

E.A. McVey (✉)
Office of Pharmaceutical Science, Center for Drug Evaluation and Research,
U.S. Food and Drug Administration, 10903 New Hampshire Avenue, Silver Spring,
MD 20993, USA

WIL Research, P.O. Box 3476, 5203DL 's-Hertogenbosch, The Netherlands
e-mail: emily.mcvey@gmail.com

B.W. Brooks and D.B. Huggett (eds.), *Human Pharmaceuticals in the Environment:*
Current and Future Perspectives, Emerging Topics in Ecotoxicology 4,
DOI 10.1007/978-1-4614-3473-3_3, © Springer Science+Business Media, LLC 2012

relatively distinct from typical public health goals; on its surface, however, ecosystem protection directly intersects and impacts with the protection of health, as defined by the World Health Organization (WHO) as the "state of complete physical, mental and social well-being" (rather than simply an absence of disease) [10, 11]. From this definition, a sustainable "ecosystem" encompasses all impacts on human health, including social and economic, as well as the health of the natural world, as was recently recognized in the 2003 Millennium Ecosystem Assessment [12]. Indeed, where ecosystems are concerned, sustainability means maintaining a wide variety of complex (and not fully understood) systems that support health and life [13].

Currently, the European Medicines Agency (EMA), United States Food and Drug Administration (FDA), and Health Canada oversee the assessment of environmental risk from pharmaceuticals at the time of registration in the European Union (EU), USA and Canada, respectively. For each of these, the assessment of hazard and exposure is used to come to a conclusion regarding the risk of a particular substance. A common problem with environmental regulation to protect ecosystem sustainability is the development of risk characterization (hazard assessment) testing schemes, since the majority of environmental exposures occur chronically, over long periods of time and (particularly in the case of pharmaceuticals) possibly at very low levels [7, 14–17]. In addition, exposure is simultaneously to a wide variety of entities, and a multitude of different vectors may be exposed in a multitude of ways [18–21]. Extrapolation from acute toxicity testing has been the most typical method utilized, as many acute toxicity tests are well established and validated using particular organisms as representative examples, they require less time and money, and a much larger amount of data for acute tests is already available [7, 22]. The use of acute toxicity studies for environmental risk assessment in general has been criticized, because of their focus on immediate endpoints such as lethality, which may not be appropriate when trying to assess risk, especially for highly potent and biologically specific contaminants such as pharmaceuticals [23–25]. Indeed, some studies have found high acute to chronic ratios for certain pharmaceuticals (mostly for estrogenic or hormonally active compounds) [26–28]. Recently, new testing schemes have been proposed and devised, and work is ongoing on developing and validating chronic toxicity tests or testing schemes to assess long-term risk of low-level environmental contaminants. However, whatever testing scheme is utilized, the outcome of the risk assessment is what really matters. What to do if an environmental impact *is* expected? In the case of human pharmaceuticals the risk/benefit analysis highly favors human health over any potential ecotoxicological effects, making it doubtful that registration, approval, or use of a drug would be limited based on ecological concerns.

Regulation in the USA

The regulation of pharmaceuticals in the environment falls under the purview of both the US Environmental Protection Agency (EPA), which has authority over most environmental media through the Clean Water Act (CWA), Clean Air Act (CAA), Toxic Substances Control Act (TSCA), and Resource Conservation and

Recovery Act (RCRA), among others, and the FDA. FDA has authority over pharmaceuticals through the Federal Food, Drug, and Cosmetics Act (FFDCA) and has a regulatory responsibility to investigate the environmental impact of pharmaceuticals through one of the oldest environmental statutes, the *National Environmental Policy Act of 1969 (NEPA)*.

NEPA requires any US federal governmental entity to assess the environmental impact of its actions. It is overseen by the White House's Council on Environmental Quality (CEQ) and the US EPA. It is important to note that NEPA itself does not give any federal agency extra authority—whatever the outcome of a NEPA process, the federal agency does not have any additional regulatory authorities. For this reason, it may be said that NEPA is a procedural statute, an interpretation which has been established by case law in the US Supreme Court (*Vermont Yankee Nuclear Power Corp. v. Natural Resources Defense Council*[1], *Kleppe v. Sierra Club*[2]). By this it is meant that NEPA places a statutory requirement upon federal agencies to perform certain procedural tasks, but does not provide them with any authority related to that task, other than those which may be applicable from their own authoritative statutes. When a federal entity is sued, it may be sued for not performing the NEPA *procedure* correctly (under the Administrative Procedures Act (APA)). (It is important to note that the Environmental Analysis under NEPA is directly tied to an *action* by the government.) Although NEPA is only required for actions by *federal* agencies, or actions that have significant federal involvement, in the USA there are a number of State Environmental Policy Acts (SEPAs) that require similar assessments for State Governmental Agencies.

The NEPA Process

Under NEPA, before performing any action, a federal agency must assess the environmental impact of that action by preparing an Environmental Assessment (EA), followed by a Finding of No Significant Impact (FONSI) or an Environmental Impact Statement (EIS). An EA is a "concise" (i.e., 10–12 pages) document that assesses the potential environmental impact of an action. From the EA, regulators then determine whether a more in-depth EIS is required or whether there can be a FONSI. If it is determined that an EIS is needed, a longer process begins to create a (in most cases) massive document which takes into account all possible environmental impacts of an action, including cultural and social impacts and those relating to environmental justice. The EIS process can be summed up by Fig. 1 and often takes years to complete (not to mention significant funds). An EIS is required for an action which is expected to have a "significant" impact or that is expected to impact a significant resource (such as an endangered species or the habitat of an endangered species). In practice, many governmental agencies produce more comprehensive

---

[1] 435 US 519, 558 (1978).

[2] 427 US 390 (1976).

Fig. 1 EIS process flowchart

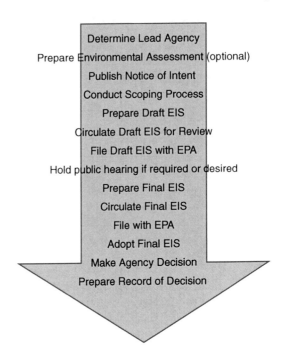

Determine Lead Agency
Prepare Environmental Assessment (optional)
Publish Notice of Intent
Conduct Scoping Process
Prepare Draft EIS
Circulate Draft EIS for Review
File Draft EIS with EPA
Hold public hearing if required or desired
Prepare Final EIS
Circulate Final EIS
File with EPA
Adopt Final EIS
Make Agency Decision
Prepare Record of Decision

EAs to assess potential environmental impacts fully before determining if an EIS is required. For this reason, most EAs nowadays are not particularly concise, though they are still dwarfed by the size of EISs.

Of course, if every federal entity had to perform even an EA for *every* action (payroll approvals, carpet changes, etc.) the entire federal government would grind to a halt. To circumvent this potential problem, NEPA and the CEQ provide a way for the respective agencies to exclude certain very common actions from having to perform NEPA analyses, through the use of agency-promulgated Categorical Exclusions (CEs or Cat Exes).

Agencies write implementing regulations which express their understanding of a particular statute, and how they plan to actually implement or enforce that statute. Therefore, while regulations are not law (but rather an Agency's interpretation of law), they are what govern an Agency's everyday actions. In the case of NEPA, each Agency's implementing regulations, including Categorical Exclusions they intend to use, are vetted by the CEQ and the US EPA and subject to public commentary, before they become common practice.

History of NEPA and the US FDA

To put the environmental regulation of pharmaceuticals by the FDA under NEPA into context, it is worthwhile to summarize the history of NEPA implementation by the FDA, since the advent of NEPA in 1969. Briefly, 3 years after NEPA passed, the

FDA performed an EIS on the use of plastic bottles for food and drugs. At this time, federal agencies were still trying to understand what NEPA meant for their "actions," particularly when they did not have statutory authority to address environmental issues. Following the plastic bottles EIS, the FDA promulgated a regulation which (very basically) said that FDA's legal interpretation of NEPA was that an adverse EIS does not permit the FDA to act if the adverse impact identified does not involve a threat to public health, adulteration, or misbranding or some other factor already identified by the FFDCA, and *therefore the FDA did not consider any of its actions to be under the purview of NEPA*. The interpretation was challenged (*EDF, Inc. v. Mathews*[3]) and from this case it was established that drug approvals and withdrawals are considered agency actions and are therefore subject to NEPA. The FDA therefore created implementing regulations to express how it intended to meet the NEPA burden to perform environmental analyses of Agency actions. In 1985, the FDA issued implementing regulations that detailed when the Agency would prepare or require an EA, when it would prepare an EIS, and what actions were categorically excluded from NEPA analysis [29].

In 1995, the President's National Performance Review issued a report "Reinventing Regulation of Drugs and Medical Devices," under the reinventing government (REGO) initiatives. One of the initiatives in the report was a proposal to reduce the number of environmental assessments (EAs) required to be submitted and, consequently, the number of reviews performed by the Agency. As a result, the FDA proposed to reduce the number of EAs by creating additional categorical exclusions from the EA requirements. The final rule for these new exclusions was published on July 29, 1997 and became effective August 28, 1997. The regulations that were promulgated at this time are the ones that are currently followed within the Agency and can be found in Title 21 of the Code of Federal Regulations, Part 25 (21 CFR 25).

Current Practice in Environmental Assessment at the US FDA

Since NEPA and CEQ allow an agency to contract the preparation of an EA, the US FDA requires sponsors and applicants to submit EAs as a part of the application package, similar to the requirement to submit safety and efficacy data. These EAs are then reviewed as part of the approval process for the pharmaceutical in question.

The document *Guidance for Industry: Environmental Assessment of Human Drug and Biologics Applications* (finalized in July 1998) summarizes the current practice for Environmental Assessment at the Center for Drug Evaluation and Research (CDER) and the Center for Biologics Evaluation and Research (CBER) at the US FDA [30]. A Guidance document describes the FDA's current thinking on a particular issue, to inform the regulated industry regarding the information the

---

[3] 410F. Supp. 336 (D.D.C. 1976).

Agency will be expecting and how they will be reviewing and interpreting it. As regulations are interpretations of statutes, so Guidances are interpretations of regulations.

Under the current regulations, an EA is required and received when the expected introductory concentration (EIC) into the environment will be greater than or equal to 1 µg/L (ppb), and/or when a noncultivated plant or animal is used in the production of the drug, and/or when the drug is expected to adversely impact the environment of any endangered species. An EA can also be required under the "extraordinary circumstances" provision. This provision allows the Agency to override its own categorical exclusions at times when it considers the weight of evidence to suggest that a "significant affect" on the environment might be expected (21 CFR 25.21). All applications to the FDA must have either an EA or a claim of categorical exclusion. If they do not, this is considered grounds for refusing to file or approve the application. An EA that is adequate for filing addresses relevant environmental issues with sufficient information to allow the FDA to determine whether the proposed action may affect the "quality of the human environment" [30].

In practice, both Investigational New Drug (IND) applications and Abbreviated New Drug Applications (ANDAs) typically claim and are granted exclusions: INDs because their EIC is under 1 ppb, and ANDAs because there is expected to be "no increased use" over the amount which was assessed in the EA for the originator's NDA for that product (if that amount was over 1 ppb to start with). These two are the major categorical exclusions that the majority of applications fall under.

"Increased use" of an active moiety occurs if the drug will be "administered at a higher dosage, for a longer duration, or for a different indication than previously, or if the drug is a new molecular entity" [30]. The Guidance provides lists of examples of actions that would or would not be considered to be increased use and suggests that if a sponsor has a question they can contact the Agency to find out whether they should provide an EA.

The equation used to calculate the concentration of a substance at the point of entry into the aquatic environment to determine qualification for the under 1 ppb categorical exclusion is calculated by multiplying the kg/year of the active moiety produced for direct use by 1/L water per day entering publicly owned treatment works. Conversion factors are included to convert to µg/L and from day to year. The number of L/day entering publicly owned treatment works (POTWs) is published in the *Needs Survey, Report to Congress* and can be found at http://www.epa.gov/owm. It is updated periodically and is at 1.321×10^{11} L/day currently.

The calculation assumes that all drug products produced in a year are used and enter the POTWs, that the drug product usage occurs throughout the USA in proportion to the population and the amount of waste generated, and that there is no metabolism. The estimate of the kg/year active moiety is based on the highest quantity of the active moiety expected to be produced for direct use in any of the next 5 years, excluding any quantity produced for inventory buildup or nonuse purposes. It includes the quantity of active moiety used in all dosage forms and strengths included in the application *and* the quantity used in an applicant's related applications. This calculation is an "out of pipe" calculation, meaning that the calculation is at the point of entry into the environment, and no dilution factor is added for dilution in surface or groundwater. If the applicants wish to factor in metabolism, depletion or dilution, they must document these in detail (see below) [30].

Once it has been established that the EA should be submitted, the Guidance provides specifics of what should be included: The first step is identification of the substance of interest, its chemical and physical characterization, and a discussion of the environmental fate of the substance. This section includes potential environmental depletion mechanisms. If the depletion mechanism is going to be used to reduce the expected introduction concentration (EIC) or eliminate effects testing,[4] a detailed analysis of the depletion mechanism is provided, otherwise a summary is acceptable.

If a moiety is expected to significantly partition into biosolids ($K_{oc} \geq 1,000$), a terrestrial EIC is calculated and terrestrial fate and effects testing undertaken. In the same manner, if a moiety is expected to have significant introduction into the atmospheric environment, this is considered and discussed in the EA. If no rapid and complete depletion mechanism is identified it is assumed that the substance will persist in the environment and toxicity testing will be required. The toxicity tests required for the EA follow a tiered system laid out in the Guidance document and summarized in Fig. 2.

Briefly, acute ecotoxicity testing (Tier 1) is performed on a minimum of one suitable test organism, and if the effect concentration for 50% (EC_{50}) or lethal concentration for 50% (LC_{50}) divided by the minimum expected effect concentration (MEEC) is greater than or equal to 1,000, no further testing is completed *unless sublethal effects are observed at MEEC.*[5] If the EC_{50} or LC_{50} divided by the MEEC is less than 1,000, Tier 2 testing is performed [30].

Tier 2 is acute ecotoxicity testing on the minimum base set of aquatic and/or terrestrial organisms (depending on where the active moiety or active metabolites are expected to accumulate), typically an acute fish toxicity test, an aquatic invertebrate acute toxicity test, and an algal species bioassay constitute the aquatic base testing; plant early growth tests, earthworm toxicity tests, and soil microbial toxicity tests constitute the terrestrial base tests. If the EC_{50} or LC_{50} for the most sensitive organism in the base set divided by the MEEC is greater than or equal to 100, no further testing is conducted *unless sublethal effects are observed at the MEEC.* If EC_{50} or LC_{50} divided by the MEEC is less than 100, Tier 3 testing should be performed.

Chronic toxicity testing (Tier 3) is considered if the compound has the potential to bioaccumulate or bioconcentrate, if indicated based on Tier 1 and/or 2 testing or if there are indications that the compound biotransforms into more toxic compounds. If the logarithm of the octanol–water partition coefficient (log K_{ow}) is greater than or

---

[4] If a rapid (defined in the Guidance document) and complete depletion mechanism is identified (with simple, polar by-products), no testing to determine environmental effects is necessary, except a microbial inhibition test (or other appropriate test to determine potential for effects on waste treatment processes).

[5] Sublethal effects at the MEEC indicate that chronic testing (Tier 3) should be performed.

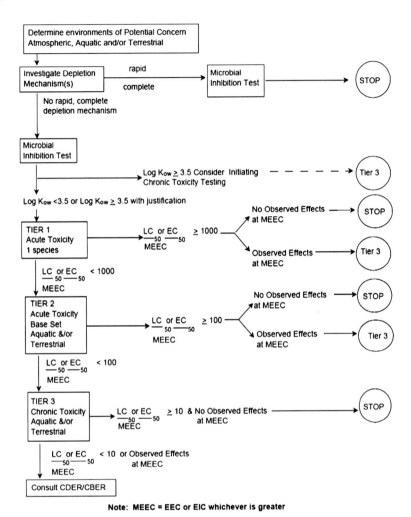

Fig. 2 CDER/CBER tiered approach to fate and effects testing. FDA CDER/CBER [30]

equal to 3.5 under relevant environmental conditions (e.g., neutral pH), chronic testing is required. If the EC_{50} or LC_{50} divided by the MEEC from the Tier 3 testing is greater than or equal to 10, no further testing is conducted unless sublethal effects are observed at the MEEC. The applicant consults with the Agency on how to proceed if the result is less than 10 or sublethal effects are observed at the Tier 3 MEEC.

While the information above summarizes the typical process for the submission and review of Environmental Assessments (EA)s for pharmaceutical applications at CDER and CBER, other Centers within the FDA have their own specific regulations for NEPA compliance. In particular, the Center for Food Safety and Nutrition (CFSAN) and the Center for Veterinary Medicine (CVM) have separate regulations and Guidances specific to the substances they are likely to encounter and the manner in which those substances may be introduced into the environment. The CVM participated in the Veterinary International Conference on Harmonization (VICH) for environmental assessment of animal pharmaceuticals to harmonize environmental assessment in animal medicines with requirements in the EU and Japan.

Regulation of Pharmaceuticals in the Environment in the European Union

In 2006, the European Medicines Agency/Committee for Medicinal Products for Human Use (EMA/CHMP) released a final "Guideline on the environmental risk assessment of medicinal products for human use." This guideline, like the FDA Guidance from a decade earlier, outlines when an environmental risk assessment (ERA) is required before marketing approval for a human drug, and if an assessment is required, what data should be submitted and reviewed.

An exposure assessment is the first step to determine if a certain "action limit" has been reached and an environmental assessment should be submitted. The action limit for the surface water predicted environmental concentration ($PEC_{surface\ water}$) for human pharmaceuticals in water was arrived at by dividing the 1 ppb threshold from the US FDA implementing regulations by 100, to take the precautionary principle into account and is therefore set at 10 ng/L [25].

The $PEC_{surface\ water}$ is calculated using the daily dose of the pharmaceutical in question, default values for wastewater production per capita, and estimated sales of the pharmaceutical product in question (market penetration factor), assuming no metabolism and no biodegradation or retention in the sewage treatment plant [31, 32]. In addition, the EMA specifies that highly lipophilic or potential endocrine-disrupting substances that may affect organisms below the threshold value must be evaluated via a risk assessment strategy [31]. A persistence, bioaccumulation, toxicity (PBT) assessment is required if log K_{ow} is ≥4.5 [31].

If any of the above situations exists, Phase II of the risk assessment is begun. Phase II is divided into Tier A and Tier B testing. In Tier A, a quantitative risk assessment is conducted for surface water, groundwater, and microorganisms in water (the predicted environmental concentration (PEC) value is compared with the respective predicted no effect concentrations (PNECs)). If one or more of the risk quotients suggest a potential for harm to the environment, Tier B studies are required. Tier B is known as the extended environmental fate and effects analysis, as the PEC can be refined according to specific sorption effects in sewage treatment plants (STPs), and water, sediment, microorganism, and terrestrial effects testing is performed. If there appears to be a potential for harm to the environment upon completion of the review of the ERA data, labeling, including proper disposal and highlighting of environmental risks can be employed [31, 32]. A potential risk to the environment is not grounds to block marketing approval.

Regulation and Environmental Risk Assessment

Beyond challenges of available data and appropriate testing for hazard identification, environmental risk assessment for regulatory purposes entails a number of other difficulties. For example, each regulatory body may define ecosystem protection differently, creating wide variety in regulatory goals. Therefore, harmonization of environmental protection goals and definitions, standards, and risk assessment models is essential for efficient and appropriate environmental assessment, whether on a local, national, or international level [33]. In systems with large variability (such as environmental and biological systems) it is important to identify and distinguish between variability and uncertainty in data used for the assessment. It is possible to reduce the level of uncertainty of data, by improvement of testing and collection of greater amounts of data, so this is often the focus of improved risk assessment, but it will always be necessary to account for variability within the risk assessment. Finally, a risk assessment must be science-based, with validated testing schemes and data, but must also be flexible enough to allow for a variety of new data and new models to be accommodated and considered, particularly with regard to the environmental compartments to be tested and addressed.

The FDA's approach for the 1998 CDER/CBER Guidance on EA was to achieve the goal of appropriate environmental assessment by requiring the information that was deemed necessary to identify potential hazards and from that perform an appropriate, science-based risk assessment. At the time of its creation, the Guidance was unique, and it continues to allow the data collection and flexibility that it was designed to achieve. The FDA continues to monitor regulations and developments elsewhere in the world and review current literature with eye toward ensuring that the EAs that the Agency receives and reviews are appropriate information for modern environmental risk assessment.

Disclaimer The views expressed in this chapter are those of the authors and should not be interpreted as the official opinion or policy of the US Food & Drug Administration, Department of Health and Human Services, or any other agency or component of the US government.

Acknowledgements The author would like to thank Dr. Nakissa Sadrieh for her thoughtful and extremely helpful comments and edits during the development of this manuscript.

Appendix: Table of abbreviations

FDA	United States Food and Drug Administration
EPA	United States Environmental Protection Agency
EMA	European Medicines Agency
WHO	World Health Organization
CWA	Clean Water Act
CAA	Clean Air Act

TSCA	Toxic Substances Control Act
RCRA	Resource Recover and Conservation Act
FFDCA	Federal Food, Drug, and Cosmetics Act
NEPA	National Environmental Policy Act
CEQ	Council on Environmental Quality
APA	Administrative Procedures Act
SEPA	State Environmental Policy Act
FONSI	Finding of No Significant Impact
EIS	Environmental Impact Statement
EA	Environmental Assessment
CFR	Code of Federal Regulations
CDER	Center for Drug Evaluation and Research
CBER	Center for Biologics Evaluation and Research
IND	Investigational New Drug
ANDA	Abbreviated New Drug Application
POTW	Publicly Owned Treatment Works
EIC	Expected Introductory Concentration
EC_{50}	Effect Concentration for 50% of population
LC_{50}	Lethal Concentration for 50% of population
MEEC	Minimum Expected Effect Concentration
PNEC	Predicted No Effect Concentration
CVM	Center for Veterinary Medicine
CFSAN	Center for Food Safety and Nutrition
VICH	Veterinary International Conference on Harmonization
EMEA/CHMP	European Medicines Agency/Committee for Medicinal Products for Human Use
ERA	Environmental Risk Assessment
PEC	Predicted Environmental Concentration
PBT	Persistence, Bioaccumulation, Toxicity
STP	Sewage Treatment Plant

References

1. Grimes DJ, Singleton FL, Colwell RR (1984) Allogenic succession of marine bacterial communities in response to pharmaceutical waste. J Appl Bacteriol 57(2):247–261
2. Lee WY, Arnold CR (1983) Chronic toxicity of ocean-dumped pharmaceutical wastes to the marine amphipod Amphithoe-valida. Mar Pollut Bull 14(4):150–153
3. Peele ER et al (1981) Effects of pharmaceutical wastes on microbial-populations in surface waters at the Puerto-Rico dump site in the Atlantic-Ocean. Appl Environ Microbiol 41(4):873–879
4. Soulides DA, Pinck LA, Allison FE (1962) Antibiotics in soils: V. Stability and release of soil-adsorbed antibiotics. Soil Sci 94(4):239–244
5. Voets JP et al (1976) Degradation of microbicides under different environmental-conditions. J Appl Bacteriol 40(1):67–72
6. Daughton CG, Ternes TA (1999) Pharmaceuticals and personal care products in the environment: agents of subtle change? Environ Health Perspect 107(suppl 6):907–938
7. Halling-Sorensen B et al (1998) Occurrence, fate and effects of pharmaceutical substances in the environment—a review. Chemosphere 36(2):357–394

8. Stan HJ, Heberer T (1997) Pharmaceuticals in the aquatic environment. Analusis 25(7):M20–M23
9. Sybesma W (1992) Do veterinary pharmaceuticals affect the environment. Tijdschrift Voor Diergeneeskunde 117(9):274–275
10. Marten GG (2001) Human ecology: basic concepts for sustainable development. Earthscan, London
11. WHO (1946) Preamble to the Constitution of the World Health Organization. In International Health Conference. Official Records of the World Health Organization, New York
12. Alcamo J, Bennet EM et al (2003) Ecosystems and human well-being: a framework for assessment. Millennium Ecosystem Assessment. Island Press, Washington DC
13. McMichael AJ (2006) Population health as the 'bottom line' of sustainability: a contemporary challenge for public health researchers. Eur J Public Health 16(6):579–581
14. Daughton CG (2000) Pharmaceuticals in the environment: overarching issues and concerns. Abs Papers Am Chem Soc 219:U622
15. Daughton CG (2001) Emerging pollutants, and communicating the science of environmental chemistry and mass spectrometry: pharmaceuticals in the environment. J Am Soc Mass Spectrom 12(10):1067–1076
16. Kolpin DW et al (2002) Pharmaceuticals, hormones, and other organic wastewater contaminants in US streams, 1999–2000: a national reconnaissance. Environ Sci Technol 36(6):1202–1211
17. Ternes TA, Joss A, Siegrist H (2004) Scrutinizing pharmaceuticals and personal care products in wastewater treatment. Environ Sci Technol 38(20):392a–399a
18. Daughton CG (2004) Non-regulated water contaminants: emerging research. Environ Impact Assess Rev 24(7–8):711–732
19. Lu GH et al (2010) Active biomonitoring of complex pollution in Taihu Lake with Carassius auratus. Chemosphere 79(5):588–594
20. Pomati F et al (2006) Effects of a complex mixture of therapeutic drugs at environmental levels on human embryonic cells. Environ Sci Technol 40(7):2442–2447
21. Inoue D et al (2010) Contamination with retinoic acid receptor agonists in two rivers in the Kinki region of Japan. Water Res 44(8):2409–2418
22. Floate KD et al (2005) Fecal residues of veterinary parasiticides: nontarget effects in the pasture environment. Annu Rev Entomol 50:153–179
23. Bound JP, Voulvoulis N (2004) Pharmaceuticals in the aquatic environment—a comparison of risk assessment strategies. Chemosphere 56(11):1143–1155
24. Montforts M (2005) The trigger values in the environmental risk assessment of pharmaceuticals in the European Union: a critical appraisal. Bilthoven, RIVM
25. Schmitt H, Boucard T, Garric J, Jensen J, Parrott J, Pery A, Rombke J, Straub JO, Hutchinson TH, Sanchez-Arguello P, Wennmalm A, Duis K (2010) Recommendations on the environmental risk assessment of pharmaceuticals: effect characterization. Integr Environ Assess Manag 6(S1):588–602
26. Ferrari B et al (2003) Ecotoxicological impact of pharmaceuticals found in treated wastewaters: study of carbamazepine, clofibric acid, and diclofenac. Ecotoxicol Environ Saf 55(3):359–370
27. Forbes VE, Calow P (2002) Extrapolation in ecological risk assessment: balancing pragmatism and precaution in chemical controls legislation. Bioscience 52(3):249–257
28. Webb SF (2001) A data-based perspective on the environmental risk assessment of human pharmaceuticals. In: Kummerer K (ed) Pharmaceuticals in the environment. Springer, New York, pp 203–219
29. Vincent PG (1993) Environmental assessment—us requirements in new drug applications. J Hazard Mater 35(2):211–216
30. CDER/CBER F (1998) Guidance for industry—environmental assessment of human drugs and biologics applications, U.D.o.H.a.H. Services, Editor. United States Food and Drug Administration, Center for Drug Evaluation and Research and Center for Biologics Evaluation and Research

31. EMA/CHMP (2006) Guideline on the environmental risk assessment of medicinal products for Human Use, E.M.A.C.f.M.P.f.H. Use, Editor. EMEA/CHMP/SWP/4447/00, London
32. Kuster A, Alder AC, Escher BI, Duis K, Fenner K, Garric J, Hutchinson TH, Lapen DR, Pery A, Rombke J, Snape J, Ternes T, Topp E, Wehrhan A, Knacker T (2010) Environmental risk assessment of human pharmaceuticals in the European Union: a case study with the beta-blocker atenolol. Integr Environ Assess Manag 6(S1):514–524
33. Montforts M (2001) Regulatory and methodological aspects governing the risk assessment for medicinal products; need for research. In: Kummerer K (ed) Pharmaceuticals in the environment. Sources, fate, effects and risks. Springer, Heidelberg, pp 439–462

Environmental Fate of Human Pharmaceuticals

Alistair B.A. Boxall and Jon F. Ericson

Introduction

Over the past 10 years, a wealth of data has been generated on the inputs, occurrence, transport, and fate of human pharmaceuticals in the environment. In addition, significant changes have occurred in the regulation of pharmaceuticals in the environment with guidelines being developed on which environmental fate studies need to be performed and on how to interpret these in terms of environmental risk. In this chapter, we attempt to draw upon the development in our scientific understanding of the fate and transport of pharmaceuticals in order to assess the suitability of current environmental risk assessment schemes for pharmaceuticals. We also attempt to provide some guidance on those factors that should be considered when interpreting existing regulatory fate studies as well as provide some ideas on alternative testing or modeling strategies for assessing environmental fate and exposure. The ultimate goal of the chapter is to contribute to the advancement of the understanding of environmental fate, and hopefully realign the context of regulatory guidance to relevant testing conditions, appropriate endpoints, and interpretation of depletion mechanism, as we have seen with ecotoxicity testing with adoption of chronic testing, and the use of relevant endpoints related to the mode of action of an API. The intent is not to provide a comprehensive review. The focus of subsequent chapters is on post consumer use rather than manufacturing operations.

A.B.A. Boxall
Environment Department, University of York, Heslington, York YO10 5DD, UK
e-mail: alistair.boxall@york.ac.uk

J.F. Ericson (✉)
Pfizer Global Research and Development, Worldwide PDM, Environmental Sciences,
MS: 8118A-2026, Eastern Point Road, Groton, CT 06340, USA
e-mail: jon.f.ericson@pfizer.com

B.W. Brooks and D.B. Huggett (eds.), *Human Pharmaceuticals in the Environment:
Current and Future Perspectives*, Emerging Topics in Ecotoxicology 4,
DOI 10.1007/978-1-4614-3473-3_4, © Springer Science+Business Media, LLC 2012

Physicochemical Characteristics

The physical–chemical properties of pharmaceuticals have been presented and discussed elsewhere in detail [1, 2] clearly demonstrating that they are large complex molecules (MW 300–1,000), typically with several functional/ionizable groups, highly ionic in nature (cations, anions, zwitterions) with relative low solubility (µg/L-mg/L range) with respect to other chemicals. From an organic chemistry perspective, pharmaceuticals are multifunctional diverse compounds composed individually or in combination of amines (primary, secondary, or tertiary), carboxylic acids, alcohols, polycyclic, aromatic/aliphatic, conjugated systems to only name a few. Most are solids and are designed as salts to enhance aqueous solubility (rather than volatility); and present in the environment either in a dissolved or sorbed state rather than as micelles or other biphasic solutions. Their partitioning into lipids, adsorption to environmental matrices, and bioavailability at trace levels are perhaps some of the more important fate and transport characteristics that determine their ultimate fate. Generally, the more we know about something, the better off we are to be able to predict and characterize its environmental behavior. But quite often the outcome is that we are overwhelmed with information without knowing what is pivotal, what properties we need to measure, and what is OK to estimate or model. The following section provides a perspective of what are the key chemical properties and why.

Nature and Extent of Ionization

The nature and extent of ionization is one of the more significant factors in determining a chemical's disposition in the environment and subsequent transport. It is also important in determining the bioavailability of a compound and is relevant to current discussions as to whether bound residues may be considered as "depleted." The following section provides a rationale around how ionization impacts ultimate environmental disposition of pharmaceuticals when present at trace levels in the environment.

Equilibrium Processes

Most pharmaceuticals are either an acid or base, present as a cation, anion, or a combination of both with respective charges of +, −, or a net charge of "0" [3]. The extent of the ionization is dependent on the type of functional group(s) present, and the pH of the surrounding environment. Very few pharmaceuticals are neutral or hydrophobic compounds, perhaps less than 5–10% all together. Generally, the more "neutral" a compound is, whether hydrophobic in nature or functionally neutral by carrying a net charge of "0," the greater the extent of partitioning into lipids found in the biomass (sludge) of wastewater treatment plants and/or in living organisms, such as fish. Conversely, the greater extent of ionization observed the greater extent of ionic complexation to minerals and clays typically found in suspended particulates, sediments, and soils as with cations; or greater dissolution in surface waters as with anions. Examples of ionic mechanisms of binding include ionic bonds, charge-transfer complexes, van der Waals forces, and H bonds to name a few [4].

Nonequilibrium Processes

This may potentially infer everything else that is not included in the above, such as what is entrapped in sediment and soil pores as part of the aging process. But of particular interest is what is found as "irreversibly bound" to soils and sediments. Perhaps not as significant for other chemicals, it is of interest for pharmaceuticals as it is one of the predominant end pathways as residues become either potentially depleted and/or inactivated as they become incorporated into the humic acid cycle. Though still debated, there is considerable evidence from other chemical sectors to suggest that such bound residues are not bioavailable [5, 6] and are essentially removed. Examples of pesticides and the link of covalent binding through amine functionalities [7] is of particular relevance for pharmaceuticals....and significance as most pharmaceuticals are cations containing amine functional groups.

Physical–chemical properties of pharmaceuticals are well characterized early on in development necessary to support active pharmaceutical ingredient (API) synthesis and formulation activities. Pharmaceutical APIs are usually: solids, available as a salt form, mp > 90°C with aqueous solubility in mg/L range. They have to be stable with an acceptable shelf life, even at elevated temperatures. When you think about it, they are bound by the conditions that are required to make an acceptable pharmaceutical product. Methods for determining physical–chemical properties are well established. Perhaps questions remain as to what information is really needed and what if any may be modeled. The following section is offered in the context of solid APIs which represent most pharmaceutical APIs.

Solubility and Melting Point

Both parameters are typically well characterized as part of drug development. Differential scanning calorimetry provides melting point endpoint and solubility is often determined for both aqueous and organic systems. Solubility information is typically more helpful in the handling and preparation of standard solutions used in environmental studies, than being a key determinant in its environmental fate.

Vapor Pressure and Density

Vapor pressure of drug substances and its potential impact to the atmospheric compartment is not a primary concern for most APIs found in the final product, except for rare cases where the API is a liquid or gas. Salt forms of APIs inherently raise the MP and increase aqueous solubility [8], thereby limiting vapor pressure and the ability to partition from water phase into the gas phase. Most vapor pressures are below the criteria established by the FDA that triggers an assessment of the risks to the atmosphere compartment (10^{-7} torr) as shown by several examples presented by Elder [9, 10]. When one also considers that decomposition upon melting is a common observation from differential scanning calorimetry [9], APIs as an atmospheric concern is not likely as a result of post consumer use. It is reasonable to assume vapor pressure determinations may be more relevant to manufacturing operations where drying operations at elevated temperatures and/or reduced pressure may make sublimation control a process control requirement. Estimations of vapor pressure may be made by several approaches [11, 12], including the EPIWIN predictive software, and these predictions may be used as a way of identifying circumstances where an experimental determination should be made.

Density falls into the same category as vapor pressure and also is not a significant parameter for consideration. Pharmaceuticals discharged as a result of post consumer use are at trace levels in the environment that are either full dissolved or partitioned onto solids and are not expected to be biphasic in nature or present as micelles. As part of a post consumer assessment, density is not an important parameter to measure.

Octanol–Water Partitioning

Octanol–water partition coefficient (K_{ow}) has been used historically in the general chemical industry to depict the distribution of a chemical between an oil and water phase and is useful in assessing how it may partition from the water into biomass of soils, sediments and sludge, or perhaps into the lipid of biota. This has been very useful for predicting behaviour of chemicals that are neutral in charge and applicable to many older pesticides, as well as many other general chemicals that don't ionize. Quite often one sees this expressed as $\log K_{ow}$ or $\log P$ and this has been a key parameter found in many regulations, especially those pertaining to bioaccumulation ($\log P > 3.0$). When one looks at pharmaceuticals, there is a need for a different expression. We believe that the $\log D$ is a more appropriate descriptor. Log D is defined as the distribution coefficient which accounts for the partitioning for all of the ionic species present in addition to the neutral form. When one considers how all of the ionic species for a given compound partitions from water into octanol, biomass, or lipids, one will see that $\log K_{ow}$ will often overpredict partitioning, as it does not account for the ionic species present and assume they all behave as the neutral form. The difference between these two values is directly related to the extent of ionization as predicted by the pH

Table 1 Impact of ionization on O/W partitioning; log P vs. log D; modeled vs. measured values log D (OECD 107)

Compound	Ionic specie	Modeled log P[a]	Modeled log D @ pH 7.0[a]	Measured log D @ pH 7.0[b]
Exemestane	Neutral	3.3	3.3	2.5
Amlodipine	Cation	3.7	2.0	2.15[c]
Pregabalin	Zwitterion	1.1	−1.38	−1.35

[a]ACD labs
[b]OECD 107
[c]pH 7.4

of the water and the pKa (s) of the compound. Table 1 compares the values of log P and log D for selected neutral, cationic, and zwitterionic compounds noting that they are essentially the same for neutral compound as the neutral species is the only species present; and considerable different for those that are ionic in nature as they have one or more ionic species present in addition to the neutral form. It is worth mentioning that while many of the fate parameters are difficult to model via a QSAR approach, log D is one parameter that has a fairly good history of modeling for pharmaceuticals with a range of software available including ALOGPS, Pallas Prolog D or ACD Labs software [13]. When determined experimentally, log D should be assessed at three different pHs in the range of 4.0–10. One pH around 7.0 should be tested as environmentally relevant, and one above and one below 7.0 to fully assess the impact of pH on extent of ionization. Actual values selected may depend on the pKa value(s) (dissociation constant) of the compound to assure that all potential ionic species are present when tested and preferably different from those at pH of 7.0.

Partitioning and Persistence in Environmental Systems

Partitioning Between Water and Air

The Henry's Law Constant (dimensionless) relates the concentration of a compound in the gas phase to that in the liquid phase ($H' = C_{sg}/C_{sl}$). It may also be expressed in terms of vapor pressure and solubility as $H = P_{vp}/S$ [12]. As noted in our earlier discussion on vapor pressure, most pharmaceuticals are salts to enhance solubility and bioavailability, thereby inherently diminishing its ability to partition into the gas phase. As a result, with a few exceptions (e.g., some of the pharmaceuticals used as local anesthetics) the Henry's Law Constants for pharmaceuticals are extremely low so contamination of the atmospheric environment is typically not a concern for most pharmaceuticals and will not be discussed further. It is not clear why vapor pressure data is requested in some environmental assessment guidance given the overall contribution it has on its overall environmental disposition. While many fugacity and other models require this data to run the model, quite often all that is needed is a default value indicating that no volatilization is likely ($V_p < 1 \times 10^{-7}$ mmHg).

Partitioning Between Water and Sludge, Sediment or Soil

The water-solid distribution coefficient (K_d) is key for understanding the mobility of a pharmaceutical through environmental systems and its availability for degradation and something where experimental data is required. K_d relates the amount of a chemical sorbed to a solid compared to that dissolved in solution at equilibrium. It is fundamental to understanding the overall disposition of a chemical once discharged into the environment. Pharmaceuticals display a wide range of sorption behavior and sorption of the same compound in different soil or sediment types can vary significantly (e.g., [14]). While quite often we see this normalized to the organic content expressed as K_{oc}, there is need to proceed with precaution when using this data. As with log D, ionization is a predominant factor that is also influential when it comes to mechanisms for sorption. For neutral compounds, K_{oc} may be used to normalize distribution to the amount of organic matter, and for estimating distribution from one matrix to another, such as from soil to sludge. In this specific case, partitioning into the organic matter is mainly driven by one mechanism, the compounds hydrophobicity or conversely its lipophilicity. But for all other pharmaceuticals that are ionizable, there is the potential for many mechanisms of sorption acting at any one time, including association with organic matter (OM), ion exchange, surface adsorption to mineral constituents, hydrogen bonding, and formation of complexes with ions such as Ca^{2+}, Mg^{2+}, Fe^{3+}, or Al^{3+}. For these compounds, normalizing data to the organic content will not necessarily compensate for the other mechanisms.

The sorption behavior is also influenced by the properties of the environmental system being studied, including pH, organic carbon content, metal oxide content, ionic strength, and cationic exchange capacity (e.g., [15–17]). The complexity of the sorbate–sorbent interactions means that modeling approaches developed for predicting the sorption of other groups of chemicals (e.g., pesticides and neutral organics) are inappropriate for use on pharmaceuticals in sludge, sediment, and soil. It is therefore dangerous to extrapolate sorption results across different matrices (e.g., sludge to soil) or extrapolating across the same matrix (e.g., one soil to another soil). The mismatch between sorption coefficients across matrices is demonstrated in Table 2 which compares sludge and soil K_{oc} data for three types of compounds. While some researchers have proposed models for understanding the sorption processes and estimating sorption behavior of pharmaceuticals in aquatic–solid systems, these are not yet in a state where they can be applied routinely in the environmental risk assessment process, as a consequence, sorption coefficient should be measured for sludge and for soils/sediments respectively with each matrix characterized for their properties.

For soils, the presence of biosolids may also alter the sorption behavior of pharmaceuticals. Recent studies have demonstrated that the addition of these matrices can either increase or reduce the sorption coefficients of pharmaceuticals in soils [18] Table 3. The reasons for the sludge effect are still unclear but similar work with veterinary medicines that have explored the effect of the animal manures on sorption, attribute the changes to changes in pH or the nature of dissolved organic carbon in the soil/manure system (e.g., [19, 20]).

Table 2 Impact of ionization on mechanism of sorption for neutral and cationic compounds; comparison of sludge vs. soil measured values OECD 106

Compound	Ionic specie	Sludge K_{oc}	Soil K_{oc}
Exemestane	Neutral	2,285	1,594–6,533; $n=5$
Azithromycin	Cation	40	22,800–59,600; $n=5$
Varenicline	Cation	62	6,500–15,000; $n=4$

Table 3 Effect of the presence of sludge on sorption coefficients (Kd) and persistence (DT50) for selected pharmaceuticals in soils (Monteiro and Boxall [18, 54])

	Soil only		Soil + sludge	
	Max	Min	Max	Min
Sorption (Kd)				
Carbamazepine	4.7	32.8	6.6	27.8
Naproxen	10.1	253	7.2	149
Fluoxetine	134	235	123	218
Sulfamethazine	1.7	98.2	5.0	44.9
Degradation (DT50)				
Naproxen	3.1	6.9	3.9	15.1

Persistence in Environmental Matrices

Since the first notable detection of pharmaceuticals in the environment, clofibric acid in the North Sea in the late 90s [21], and subsequent studies determining the occurrence of pharmaceuticals in the environment such as the work of Kolpin [22], questions remain around the disposition of pharmaceuticals and their ultimate biodegradability in the environment. What is clear is how they primarily enter into the environment from post consumer use with other minor sources from disposal and manufacture. The extent they enter into the environment is mitigated first by their metabolism in man and then secondarily by removal during wastewater treatment, both from sorption and biodegradation. From that point on, residues either enter the environment through land applied biosolids, or get dissolved in wastewater effluents. Concentrations of pharmaceuticals are generally in the ng/l range in river waters, variable based on overall volume of use, level of human metabolism, and the type of treatment plant and/or operating efficiencies of such [23–25].

Elimination Rates

Whether for the risk assessment or for classification purposes, elimination rates (half-lives) are needed to better characterize the fate of pharmaceuticals as they are transported through the environment and to better understand where persistence may be an issue. Historically, the fate of pharmaceuticals has been characterized by methods slated for general chemicals that are still found in current risk assessment guidance today. Many still follow ready and inherent methods that have been around since the late 1990s that are run at high test substance concentrations, low biomass concentrations and nonchemical specific endpoints such as DOC or mineralization. What is missing, however, are biodegradation methods that truly represent the trace conditions that pharmaceuticals are introduced into the environment, and methods that characterize the predominate mechanisms of biodegradation that some of these methods miss. It is only in recent times that we have begun to see some more methods specific for these conditions, as with the OECD 314 B [26]. Trace level of test substance, realistic biomass solid levels typical of wastewater treatment plants, with specific chemical and CO_2 analysis, are needed in sludge, soil, and sediment tests. For those cases where residues are transferred from one compartment to another, such as is with the land application of biosolids from wastewater treatment plants, there is a need to develop more rigorous protocols to specify how and when test substance is spiked to the soil. For example, to amend soils with biosolids, should the test substance be taken through the anaerobic digester process before the biosolids are amended to the soils; or is it sufficient to add the test material to biosolids at the start of the soil biodegradations study? Very little is known about the potential binding of pharmaceuticals during anaerobic digestion and its impact on its bioavailability to subsequent degradation in soil. Similarly, very little is known about the bioavailability of bound residues to sediments and whether the unextractable residues are truly "depleted" as they enter the humification process; or whether at some point in time they may be released. Also needed from these studies is more specific guidance on how to calculate elimination rates such when residues are highly bound. Simple, quick screening tests that assess the biodiversity present in the environment from the various compartments are also not available and require further development. A more detailed description of approaches for assessing elimination in different environmental matrices is given below.

Activated Sludge

The occurrence of pharmaceuticals in sludge and their removal during subsequent wastewater treatment is of interest and has been well studied. Historically, the biodegradability of compounds has been screened using a variety of ready and inherent tests. While many are economical and easy to perform, the outcome for many pharmaceuticals is that they are not readily biodegradable and require further testing. As shown in Table 4 , the results of the ready biodegradation test show the three compounds fail the ready biodegradation test and are assigned default rate constants of '0'. For the same three compounds, using the sludge die-away test (conducted at more realistic biomass and test concentrations) one is able to determine more meaningful elimination rate kinetics. As seen with exemestane and eplerenone, the ke values of 1.8 and 0.08 result in significant reduction of predicted surface water

Table 4 Comparison of ready biodegradation and sludge die-away endpoints and data output for assessing the biodegradation of pharmaceuticals

	OECD 301 ready biodegradation		OECD 314B sludge die-away		
Conditions	25 mg/L sludge solids, mg/L test material concentration,% CO_2		2,500 mg/L sludge solids, 10^{-03} mg/L test material concentration, loss of parent,% CO_2		
Endpoints	% CO_2 @ 28 days	K_e hr^{-1}	parent K_e hr^{-1}	% CO_2 @ 28 days	% Removed in effluent[a]
Exemestane	15.2	Default "0"	1.8	80.6	99.9
Eplerenone	−2.8	Default "0"	0.08	0.5	38
Varenicline	15.7	Default "0"	0.01	0.68	6

[a]K_e used as first order rate constant in modeling removal for 6 h hydraulic retention time

concentrations by 99.9% and 38% respectively. Such information may be very helpful in understanding the extent of removal during wastewater treatment and in refining the predicted environmental concentration (PEC) for the risk assessment. Data from these studies also provide a degradation profile illustrating the number of biotransformation products present and the sequence of formation of such over time. As discussed above, since biotransformation is a key pathway in its overall biodegradation, one can see how tests that provide this additional kinetic information are a good fit for risk assessment needs.

Water-Sediment or Soil

Occurrence of pharmaceuticals in surface waters is also very well studied [22] as with activated sludge. Their detection in rivers and surface waters, and in some cases in lakes and seas, would generally infer that their overall biodegradation rate in the water is somewhat slower than in other systems. It is reasonable that these observed elimination rates are not as great as activated sludge, or for that matter sediment just based on the low abundance of microorganisms found in surface waters [27]. Because of the microbial diversity encountered from one compart-ment to the next, it is also hard to translate elimination rates or type of transforma-tion products seen in the sludge compartment to that found in sediment or soil as noted with diclofenac [25, 27] for example. Occurrence of pharmaceuticals in sediments is somewhat less published than what is seen is surface waters, and even more so around lab studies investigating the fate in sediment systems [28, 29]. This is an area where more research is needed, especially when one considers the "sink" conditions that sediments offer for most pharmaceuticals, especially cat-ionic pharmaceuticals that are more likely to become highly bound. Table 5 shows the elimination rates for one pharmaceutical, exemestane in sludge, mixing zone and surface water determined in lab scale test systems (see Fig. 1 for structures of chemicals in Tables 1, 2, 4 and 5). One observes a general decrease in the elimina-tion rate with the overall amount of biomass present in each of these systems.

Table 5 Comparison of biodegradation potential in sludge, mixing zone, river water, and water-sediment systems: half-life for parent

Exemestane	Sludge die-away OECD 314B	Mixing zone OECD 314C	River water modified OECD 314C	Water-sediment OECD 308
DT-50 h	0.39	58	191	362
K_e hr^{-1}	1.8	0.012	0.0036	0.0019
Relative DT-50	1	149	490	923
Solids mg/L	2,500	15	12.3	–

Generic Name Trade Name CAS#	Structure
exemestane Aromasin® 107868-30-4	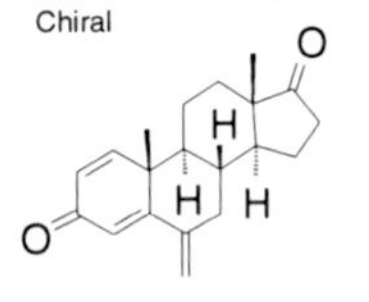 Chiral
eplerenone Inspra® 107724-20-9	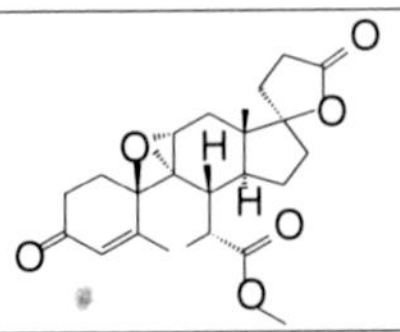
varenicline Champix® 375815-87-5	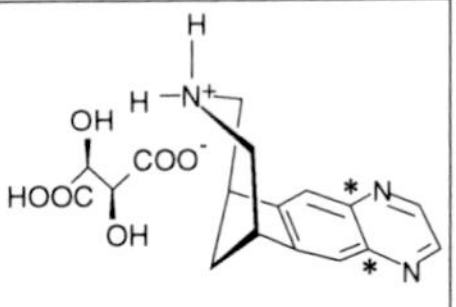
azithromycin Zithromax® 83905-01-5	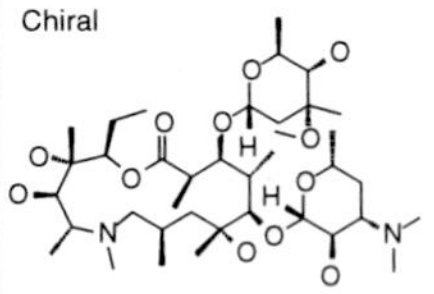 Chiral
pregabalin Lyrica® 148553-50-8	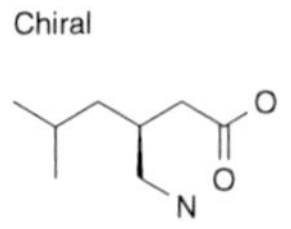 Chiral
amlodipine Norvasc® 88150-42-9	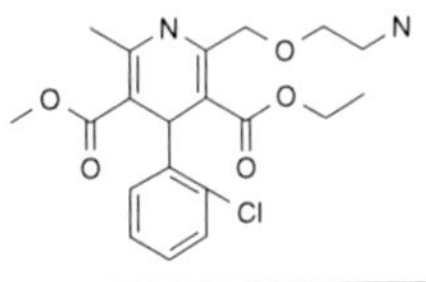

Fig. 1 Structures of compounds found in Tables 1, 2, 4 and 5

While this is oversimplistic, it is also noted that some of this is also likely due to changes in the overall microbial community structure and diversity found in these samples. When comparing the DT50 values from sludge and water to that of water-sediment system, we also see a progression to longer half lives. This may be explained in part by the role sediment sorption has on the overall dissipation from water and the resulting decrease in bioavailability in the water-sediment system. The interface between the overlying water, pore water, and the microorganisms found in sediment also plays a role in overall rates, suggesting that water flow may enhance transport of pharmaceuticals to the sediment environment where degradation may occur, thereby increasing the overall observed degradation rate [27]. It is not clear what role bound or unextractable residues have on the ultimate fate of pharmaceuticals, and how that may impact their overall bioavailability [29, 30]. Further work is needed in developing better test methods applicable to pharmaceuticals, their route of entry into the environment and subsequent release environment, and guidance as to how to apply such data in the risk assessment.

As biotransformation is a key process in the depletion of pharmaceuticals, one must be careful not to extrapolate its relative biodegradability from one compartment or matrix to the next. For each compartment (sludge, sediment, soil) the extent of biodiversity, the availability of other carbon sources and the extent of residue bioavailability will determine the overall rate of depletion and the types of potential metabolites formed. Consequently, there is a potential need to develop elimination rates for each of these matrices. Sludge elimination rate is pivotal to most risk assessments, as wastewater treatment plants are common to most discharge systems and provide some degree of removal relevant to the subsequent release environment. The need for water, sediment, and soil elimination rates is more contingent upon the specific target compartment of the analysis and the refinement of the risk assessment required, whether a worst case local assessment, a more general regional assessment or perhaps a more dynamic watershed analysis (High/Low/Mean Flow). A further discussion on how future needs and how these elimination rates should be used in risk assessment is found in "Future Needs" section.

Environmental factors such as soil type, temperature, and moisture are also very important in determining degradation rates of pharmaceuticals [31, 32]. Like, sorption, the presence of the biosolid matrix also seems to affect degradation rates compared to soil only (Table 3). For example, caffeine degradation rates in soils increased with addition of aerobically digested sewage sludge, whereas addition of anaerobically treated sewage sludge did not accelerate caffeine mineralization [33]. The degradation rate of naproxen was also reported to be increased by the addition of biosolids [31]. In only a few studies has the formation of metabolites been investigated [31, 32]. No detectable transformation products were found for naproxen or the hormones estrone and 17β-estradiol [31, 32].

Pharmaceuticals at trace levels are generally insufficient to support microbial growth as the sole carbon source. That is not to say that one does not see any mineralization, as quite often there are minor microbial communities capable of this. But for the most part, this is not the predominant acting mechanism, especially in carbon source rich environment such as is found in a wastewater treatment plant.

This is perhaps the most overlooked factor in assessing whether something is readily biodegradable or not, and whether something may mineralizes or not. Unless structurally similar to other carbon food sources, pharmaceuticals are unlikely to rapidly biodegrade and completely mineralize to CO_2 as is found with many other chemicals, such as detergents, where for example biodegradation is significant not only during wastewater treatment but also during transit to the wastewater facility and post discharge in the mixing zone. Without an higher exposure concentration to drive enzyme induction, rapid mineralization is unlikely to occur by microorganisms typically found in the wastewater treatment to plants that are rich in other carbon sources.

Transformation pathways in the environment, as with metabolic pathways in humans, are key to the degradation and elimination of pharmaceuticals. Xenobiotics are typically detoxified by the liver via P450 and other routes of metabolism and then excreted as more polar metabolites via the kidneys. Likewise, xenobiotics found in the environment are transformed by an abundant sources of P450 and other enzymes associated with microbial metabolism [34]. Many of the Phase I reactions [35] typical of human metabolism are are also observed in environmental transformations in environmental microorganisms. As a result, these transformations become a significant pathway in the overall depletion of pharmaceuticals in the environment [36–39]. The rate and extent of biotransformation are limited by its bioavailability, as well as its bioaccessibilty. Rather than being an issue of uptake from the gut, it is a question as to whether residues bound to suspended particulates, soils, and sediments are truly available to the microorganisms for subsequent biotransformation. Microorganisms unable to directly use pharmaceutical substrate as a carbon source quite often require several biotransformation steps to yield something more similar to other carbon sources before mineralization is observed.

Uptake into Organisms

Bioconcentration and bioaccumulation refer to the concentration of compound found in plasma and/or tissue of environmental organisms relative to the concentration found in the ambient water environment, either from direct exposure or, respectively, from direct exposure and other sources, such as food uptake. Values greater than 1 infers that the compound is bioconcentrating in the organism, and when values exceed 2,000, a compound is classified as "bioaccumulative" (B) or very bioaccumulative (vB) when exceeding 5,000 [40]. Historically both of these have been mainly discussed in the context of lipophilic compound and their ability to partition into the lipid fractions of biota from the surrounding water environment. Great examples are found with older chlorinated hydrocarbons such as DDT and polycyclic hydrocarbons [12] for example that have log K_{ow} values greater than 3.0 and some with values greater than 4.5. For pharmaceuticals, however, that are predominately ionic in nature, most do not have log K_{ow} values greater than 3.0 nor fall under the classification of "B" or "vB." No current pharmaceutical has exceeded a log K_{ow}

of 4.5 nor triggered a "PBT" assessment. What is interesting though is the focus on some pharmaceuticals such as fluoxetine, gemfibrozil, and diclofenac, for example, that are not classified as "B" but may have BCF values up to 500 [41, 42]. Most of these examples result in bioconcentration from other mechanisms than lipophilic partitioning, such as those mechanisms that either enhance uptake relative to the water, or diminish clearance. Compounds such as fluoxetine that have a pKa value close to environmental pH may appear to have an enhanced uptake of a drug as slight changes in environmental pH dramatically changes the extent of the neutral specie present [43, 44], thereby enhancing partitioning and uptake. Compounds that show cytological effects [45], such as with diclofenac, or alternatively enzyme inhibition [46] of diclofenac, gemfibrozil and some antidepressants, may result in decreased renal or hepatic clearance respectively and result in a slightly elevated BCF. It is not clear for any of these mechanisms how easy it would be to read-across to nonclinical drug safety studies as a means of predicting some of these more subtle BCF effects in environmental species.

Uptake of pharmaceuticals by soil organisms is also possible [47–49]. Antibiotics including florfenicol, trimethoprim, enrofloxacin, sulfamethazine, and chlorotetracycline have been shown to be taken up by plants from soils and sludge or manure-amended soils [48–52]. The occurrence of anthropogenic waste indicators, including the pharmaceutical trimethoprim, has also been reported in earthworm tissue. It is however difficult to develop a clear relationship between uptake and pharmaceutical properties, such as hydrophobicity, as some pharmaceuticals are taken up by some organisms and not by others and uptake into similar organisms in the different environments can vary. This is perhaps not surprising, as data for other environmental processes (e.g., sorption to soil) indicate that the behavior of pharmaceuticals in the environment is poorly related to hydrophobicity but is determined by a range of factors including H-bonding potential, cation exchange, cation bridging at clay surfaces, and complexation. Residues of fluoroquinolones have also recently been reported in the eggs of vultures and kites and associated with effects on the developing embryo (e.g., [53]). While the authors of this study indicated that the route of exposure was most likely from the consumption of carcasses of animals that have been treated with the drugs, there is a possibility that other environmental routes of exposure may be important. A more detailed understanding of the movement of pharmaceuticals through food chains would help to address this. Through controlled experimental studies it may be possible in the future to begin to understand those factors and processes affecting the uptake of veterinary medicines into plants and to develop modeling approaches for predicting uptake.

Transport of Pharmaceuticals Around the Environment

Pharmaceuticals released from sewage treatment works will typically be transported downstream and if not degraded will ultimately end up in marine systems. Depending on the hydrology of the system, pharmaceuticals may infiltrate into aquifers. The water may also be abstracted for use in drinking water or for irrigation of crops. A number of recent studies have detected pharmaceuticals in drinking waters (e.g., [54]). Our understanding of the fate of pharmaceuticals in common drinking water treatment processes is however not well developed, and this is an area where further research is required.

Following entry into the wastewater system, many pharmaceuticals will adsorb to the sludge phase and this may be subsequently be applied as a fertilizer to agricultural land [55, 56]. It is therefore not surprising that a plethora of pharmaceuticals and personal care products (including hormones and steroids, stimulants, antiepileptics, antidepressants, antibiotics, and musks) have been detected in biosolids (e.g., [55, 57–59]). These compounds will then be released to the soil environment when the biosolids are used as a fertilizer. Alternatively, pharmaceuticals may also enter soils from the irrigation of soils with contaminated wastewater [60]. In the current risk assessment process for pharmaceuticals, consideration of terrestrial risk is required if the sludge K_{oc} for the compound is 10,000 or greater. This approach however is unlikely to provide a realistic indication of which compounds are likely to be present in biosolids applied to land, the reason being that the approach does not consider information on the volume of the drug entering the sewage treatment plant. If we take two hypothetical chemicals, the first has a very high sludge K_{oc} (>10,000) and a low usage volume, and one with a medium sludge K_{oc} and a high usage volume, the first compound would require terrestrial assessment whereas the second compound would not even though the application rate to soil is in fact very similar for both substances. We would therefore advocate that instead of using a single K_{oc} trigger, an alternative trigger is developed that combines information on drug usage and sorption to sludge.

Contaminants applied to soil can be transported to aquatic systems via surface runoff, subsurface flow and drainflow. Most work to date on contaminant transport from agricultural fields has focused on pesticides, nutrients, and bacteria, but recently a number of studies have explored the fate and transport of some pharmaceuticals. For example, runoff of pharmaceuticals from soils amended with sewage sludge has been reported [61]. In a field work performed in Canada, sewage sludge was applied using two common practices: broadcast and injection application. In this study, it was concluded that the pharmaceuticals studied, such as carbamazepine, ibuprofen, acetaminophen, and naproxen, do run off with wet weather from a broadcast application [61]. Studies into the leaching behavior of antibiotics have shown that selected compounds have the potential to leach to groundwaters (e.g., [62]), and these data fit with groundwater monitoring campaigns that have detected a number of pharmaceuticals in groundwaters [63–65]. The extent of transport via any of the processes discussed above is determined by a range of factors, including the following: the solubility, sorption behavior, and persistence of the contaminant; the physical structure, pH, organic carbon content, and cation exchange capacity of the soil matrix; and climatic conditions such as temperature and rainfall volume and intensity.

The surface water exposure profile via these routes of exposure is likely to be very different from the exposure profile arising from releases from wastewater

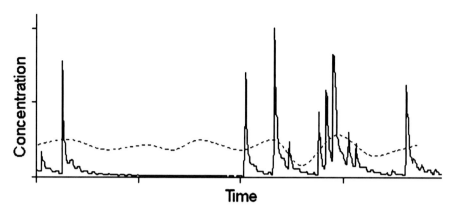

Fig. 2 Conceptual diagram of the aquatic exposure patterns for pharmaceuticals released to aquatic systems from wastewater discharges (*dashed line*) and via the soil environment (*solid line*)

treatment works. When substances move from the soil environment to surface waters, they will tend to enter in a series of pulses (corresponding to periods of rainfall), whereas emissions from wastewater treatment processes are more or less continuous, although with small variations in concentrations (Fig. 2). The modeling of aquatic exposure and subsequent risk assessment of pharmaceuticals that enter the soil environment therefore probably needs to be addressed differently from pharmaceutical releases from wastewater treatment plants.

Currently, the regulatory guidance for pharmaceuticals recommends that estimates of aquatic exposure, arising from releases from biosolid-amended soils, is estimated using a very simple algorithm from the soil sorption coefficient. More sophisticated models are available for predicting the movement of chemicals from soils to surface waters. These models generally originate from the pesticide risk assessment area. For example, the Forum for Coordination of Pesticide Fate Models and their Use (FOCUS) have established a suite of models for predicting the concentrations of pesticides in surface waters (PRZM, MACRO, and TOXSWA) and groundwaters (PEARL, PELMO) to support regulatory risk assessments (e.g., [66]). FOCUS has also developed a set of scenarios covering the different climatic, soil, and cropping characteristics encountered in the different member states. The FOCUS modeling framework may provide a basis for assessing the aquatic exposure arising from the application of pharmaceuticals to the terrestrial environment, although the climate/soil/crop scenarios may need to be adjusted to better reflect the characteristics of areas where biosolids are applied (e.g., [67]). Moreover, while the models have not been extensively evaluated against real monitoring data, the data that are available indicate that in some instances the models may greatly underestimate exposure (e.g., Fig. 3; [68]). One possible explanation for the mismatch between model outputs and measured concentrations is that key fate and transport processes that are important for pharmaceuticals are not covered in the model; one example of such a process would be facilitated transport of the pharmaceutical to

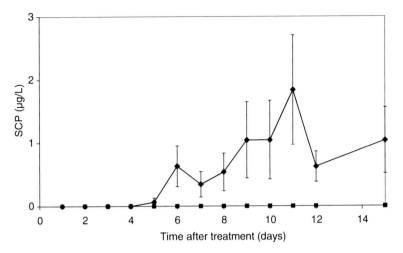

Fig. 3 Comparison of modeled (*square points*) and measured (*diamonds*) concentrations of the sulphonamide antibiotic, sulfachloropyiradzine (SCP), in a lysimeter system (adapted from Blackwell et al. [68])

groundwater or surface waters in biosolid particulate material or biosolid derived dissolved organic carbon. We would therefore advocate that much more work needs to be done to evaluate the suitability of these models and to adapt them to address some of these important fate and transport processes.

Future Needs

It is clear from the previous sections that the fate of pharmaceuticals in the environment is complex and highly dependent on the structure of the pharmaceutical, the nature of the receiving environment, and the mode of entry into the environment. However, current risk guidance for pharmaceuticals is often based on existing regulations for other chemical classes (e.g., industrial chemicals). Until we recognize that pharmaceuticals are fundamentally very different from many of these chemical classes, and begin to develop mechanisms explaining the environmental behavior of pharmaceuticals and refine the risk assessment process accordingly, we will not fully understand environmental risks of pharmaceutical products. There are a number of areas that we believe need attention:

1. A better understanding of those factors and processes affecting the persistence of pharmaceuticals. Ideally we should be able to predict microbial transformation (route and rate) of pharmaceuticals in environmental matrices (sludge, soil, surface waters, sediment) based on models or screening. In order to achieve this we need to

understand the key factors (species diversity, abundance, properties of the microenvironment, interactions with other chemicals, biochemical processes) modulating biotransformation. It would also be valuable to explore whether data from studies with mammals (e.g., ADME) can be used to inform our understanding of the rates and routes off degradation in the environment. We also urgently need to develop approaches to assess the environmental relevance, if any, of bound residues.

2. Understanding of fate in treatment processes. While we have a good understanding of the fate of pharmaceuticals in some treatment processes (e.g., activated sludge), our understanding of fate in other treatment processes is less well developed (e.g., anaerobic digestion, drinking water treatment processes). As we develop a better understanding of these processes, we should work to evaluate some of the more complex wastewater treatment models and drinking water treatment models (e.g., from engineering area) and begin to apply these in the environmental risk assessment processes.

3. Modeling of transformation product exposure. When assessing risks, we tend to focus on the parent pharmaceutical yet these may be transformed to other compounds in the different environmental compartments. In some instances, the fate of the transformation product (and hence the exposure) may be very different from that of the parent compound [69]. We should begin to develop an understanding of the formation and fate of transformation products and look at ways in which we can better incorporate transformation products into the risk assessment process. Potential approaches for doing this are described in [70].

4. Understanding of sorption mechanisms. Data from sorption studies show that existing models for prediction of sorption behavior or extrapolating to different matrices are probably not appropriate for pharmaceuticals. The reason being that sorption is determined by a range of mechanisms. Systematic studies are required to develop an understanding of the different sorption mechanisms as well as the effect of environmental properties for predicting sorption in different matrices. Ultimately, we should develop quantitative structure property relationships for estimating sorption from underlying chemical properties and environmental properties.

5. Exposure model evaluation and development. A range of exposure models are available for estimating concentrations of pharmaceuticals in different media. While some of these have been evaluated for some substances (e.g., [71]), the accuracy of many of the models for use on pharmaceuticals has yet to be established. We should work to use the wealth of monitoring data that are available for different matrices to evaluate the suitability of these models. Where models are found to fall down we should work to adapt them to cover key fate and transport processes not currently considered (e.g., DOC-facilitated transport).

6. Bioconcentration and transport through food webs. Very little work has been done to understand the uptake of pharmaceuticals into organisms and the potential for movement of pharmaceuticals through food chains. Work is required to explore the mechanisms of uptake (active and passive) of pharmaceuticals into plants, invertebrates and vertebrates as well as the potential movement through food webs, e.g., soil–plant–small mammal–red kite. In addition, we need to better understand

biotransformation in aquatic and terrestrial organisms and its potential role in mitigating bioconcentration and bioaccumulation. Like biodegradation, it is possible that information on the behavior of pharmaceuticals in mammalian systems (e.g., ADME data) may help to inform this issue.

7. Dealing with a changing landscape. It is important to recognize that the uses and risks of pharmaceuticals in the future could be very different from today due to changes in the environment as well as the application of new technologies. For example, climate change may well affect fate and transport processes and effects of pharmaceuticals in the environment as well as change drug usage patterns (e.g., [72]), meaning that risks are very different from today. The growth of new technologies, such as nanotechnology, also raise challenges for fate and exposure assessment.

Finally, to ensure the safe and sustainable use of pharmaceuticals, it is critical that regulatory fate testing guidelines are regularly updated in the light of new scientific knowledge.

References

1. Williams RT (ed) (2003) Human pharmaceuticals—assessing the impacts on aquatic ecosystems. SETAC Press, Pensacola
2. Kümmerer K (ed) (2008) Pharmaceuticals in the environment. Springer, New York
3. Kulshrestha PJ, Giese RF, Aga DS (2004) Investigating the molecular interactions of oxytetracycline in clay and organic matter: insights on factors affecting its mobility in soil. Environ Sci Technol 38(15):4097–4105
4. Gevao B, Semple KT, Jones KC (2000) Bound pesticide residues in soils: a review. Environ Pollut 108:3–14
5. Alexander M (2000) Aging, bioavailability, and overestimation of risk from environmental pollutants. Environ Sci Technol 34(20):4259–4265
6. ECPA (2000) ECPA position paper on soil non-extractable residues. European Crop Protection Association, Brussels
7. Northcott GL, Jones KC (2000) Experimental approaches and analytical techniques for determining organic compound bound residues in soil and sediment. Environ Pollut 108:19–43
8. Giron D (2003) Characterization of salts of drug substances. J Therm Anal Calorim 73:441–457
9. Elder JP (1997) Sublimation measurements of pharmaceutical compounds by isothermal thermogravimetry. J Therm Anal Calorim 49:897–905
10. Heng JYY et al (2005) The physiochemical and surface properties of solid state pharmaceutical solids. World Congress of Chemical Engineering, 7th Glowgow, UK, pp 1–10
11. ASC (1990) Handbook of chemical property estimation methods, 14-1—20. In: Lyman WJ, Reehl WF, Rosenblatt DH (eds) Vapor pressure. ASC, Washington
12. Boethling RS, Mackay D (eds) (2000) Handbook of property estimation methods for chemicals. Lewis Publishers, Boca Raton
13. Tetko IV, Poda GI (2004) Application of ALOGPS 2.1t predict log D distribution coefficient for Pfizer proprietary compounds. J Med Chem 47(23):5601–5604
14. Tolls J (2001) Sorption of veterinary pharmaceuticals in soils: a review. Environ Sci Technol 35:3397–3406
15. Ter Laak TL, Gebbink WA, Tolls J (2006) The effect of pH and ionic strength on the sorption of sulfachloropyridazine, tylosin and oxytetracycline to soil. Environ Toxicol Chem 25(4):904

16. Strock TJ, Sassman SA, Lee LS (2005) Sorption and related properties of the swine antibiotic carbadox and associated n-oxide reduced metabolites. Environ Sci Technol 39:3134

17. Sassman SA, Lee LS (2005) Sorption of three tetracyclines by several soils: assessing the role of pH and cation exchange. Environ Sci Technol 39:7452

18. Monteiro SC, Boxall ABA (2009) Factors affecting the degradation of pharmaceuticals in agricultural soils. Environ Toxicol Chem 28(12):2546–2554

19. Boxall ABA, Blackwell PA, Cavallo R, Kay P, Tolls J (2002) The sorption and transport of a sulphonamide antibiotic in soil systems. Toxicology Letters 131:19–28

20. Thiele-Bruhn S, Aust MO (2004) Effects of pig slurry on the sorption of sulfonamide antibiotics in soil. Arch Environ Contam Toxicol 47(1):31

21. Buser H et al (1998) Occurrence of the pharmaceutical drug clofibric acid and the herbicide mecoprop in various Swiss lakes and in the North Sea. Environ Sci Technol 32(1):188–192

22. Kolpin D et al (2002) Pharmaceuticals, hormones, and other organic wastewater contaminants in U.S. streams, 1999–2000: a national reconnaissance. Environ Sci Technol 36(6):1202–1211

23. Strenn B et al (2003) The elimination of selected pharmaceuticals in wastewater treatment-lab scale experiments with different sludge retention times. Water Resour Manage II 8:227–236

24. Strenn B et al (2004) Carbamazepine, diclofenac, ibuprofen and bezafibrate-investigations on the behavior of selected pharmaceuticals during wastewater treatment. Water Sci Technol 5:269–276

25. Swiener C et al (2003) Short-term tests with a pilot sewage plant and biofilm reactors for the biological degradation of the pharmaceutical compounds clofibric acid, ibuprofen and diclofenac. Sci Total Environ 309(1–3):201–211

26. OECD (2008) OECD guideline for the testing of chemicals. TG 314B biodegradation in activated sludge guideline. OECD, Paris

27. Loffler D et al (2005) Environmental fate of pharmaceuticals in water/sediment systems. Environ Sci Technol 39:5209–5218

28. Kunke U, Radke M (2008) Biodegradation of acidic pharmaceuticals in bed sediments: insight from a laboratory experiment. Environ Sci Technol 42:7273–7279

29. Ericson J (2007) An evaluation of the OECD 308 water/sediment system for investigating the biodegradation of pharmaceuticals. Environ Sci Technol 41(16):5803–5811

30. Diaz-Cruz MS et al (2003) Environmental behavior and analysis of veterinary and human drugs in soils, sediments and sludge. Trends Anal Chem 22(6):340–351

31. Topp E, Hendel J, Lapen D, Chapman R (2008) Fate of the non-steroidal anti-inflammatory drug naproxen in agricultural soil receiving liquid municipal biosolids. Environ Toxicol Chem 27(10):2005–2010

32. Collucci M, Bork H, Topp E (2001) Persistence of estrogenic hormones in agricultural soils (I—17-beta estradiol and estrone). J Environ Qual 30:2070–2076

33. Topp E, Hendel JG, Lu Z, Chapman R (2006) Biodegradation of caffeine in agricultural soils. Can J Soil Sci 86(2–3):533–544

34. Rosazza JP (ed) (1982) Microbial transformations of bioactive compounds, vol I & II. CRC Press, Boca Raton

35. Perez S, Barcelo D (2007) Application of advanced MS techniques to analysis and identification of human and microbial metabolites of pharmaceuticals in the aquatic environment. Trends Anal Chem 26(6):494–514

36. Schulz M, Loffler D, Wagner M, Ternes T (2008) Transformation of the X-ray contrast medium iopromide in soil and biological wastewater treatment. Environ Sci Technol 42(19):7207–7217

37. Perez S, Aga D, Barcelo D (2006) Biodegradation of pharmaceuticals in the environment. Top Issues Appl Microbiol Biotechnol 113–144

38. Farrre ML, Perez S, Kantiani L, Barcelo D (2008) Fate and toxicity of emerging polutants, their metabolites and transformation products in the aquatic environment. Trends Anal Chem 27(11):991–1007

39. Aga DS (2008) Fate of pharmaceuticals in the environment and in water treatment systems. Advances in the analysis of pharmaceuticals in the aquatic environment. CRC Press, Boca Raton, pp 53–80

40. European Chemicals Bureau (2003) Technical guidance document on risk assessment. Part II. European Communities, Luxembourg, p 164
41. Brown JN et al (2007) Variations in bioconcentration of human pharmaceuticals from sewage effluents into fish blood plasma. Environ Toxicol Pharmacol 24:267–274
42. Mimeault C et al (2005) The human lipid regulator, gemfibrozil bioconcentrates and reduces testosterone in the goldfish, *Carassisu auratus*. Aquat Toxicol 73:44–54
43. Zurita JOL et al (2007) Toxicological effects of the lipid regulator gemfibrozil in four aquatic systems. Aquat Toxicol 81:106–115
44. Nakamura Y et al (2008) The effects of pH on fluoxetine in Japanese medaka (*Oryzias latipes*): acute toxicity in fish larvae and bioaccumulation in juvenile fish. Chemosphere 70:865–873
45. Schwaiger J et al (2004) Toxic effects of the non-steroidal anti-inflammatory drug diclofenac Part I: histopathological alterations and bioaccumulation in rainbow trout. Aquat Toxicol 68:141–150
46. Thibaut R et al (2006) The interference of pharmaceuticals with endogenous and xenobiotic metabolizing enzymes in carp liver: an in-vitro study. Environ Sci Technol 40(16):5154–5160
47. Jjemba PK (2002) The potential impact of veterinary and human therapeutic agents in manure and biosolids on plants grown on arable land: a review. Agric Ecosyst Environ 93(1–3):267–308
48. Boxall ABA, Johnson P, Smith EJ, Sinclair CJ, Stutt E, Levy L (2006) Uptake of veterinary medicines from soils into plants. J Agric Food Chem 54(6):2288–2297
49. Kinney CA, Furlong ET, Kolpin DW, Burkhardt MR, Zaugg SD, Werner SL, Bossio JP, Benotti MJ (2008) Bioaccumulation of pharmaceuticals and other anthropogenic waste indicators in earthworms from agricultural soil amended with biosolid or swine manure. Environ Sci Technol 42(6):1863–1870
50. Migliore L, Cozzolino S, Fiori M (2003) Phytotoxicity to and uptake of enrofloxacin in crop plants. Chemosphere 52:1233–1244
51. Dolliver H, Kumar K, Gupta S (2007) Sulfamethazine uptake by plants from manure-amended soil. J Environ Qual 36:1224–1230
52. Kumar K, Gupta SC, Baidoo SK, Chander Y, Rosen CJ (2005) Antibiotic uptake by plants from soil fertilized with animal manure. J Environ Qual 34:2082–2085
53. Lemus JA, Blanco G, Arroyo B, Martinez F, Grande J (2009) Fatal embryo chondral damage associated with fluoroquinolone in eggs of threatened avian scavengers. Environ Pollut 157:2421–2427
54. Monteiro SC, Boxall ABA (2010) Occurrence and fate of human pharmaceuticals in the environment. Rev Environ Contam Toxicol 202:53–154
55. Golet EM, Xifra I, Siegrist H, Alder AC, Giger W (2003) Environmental exposure assessment of fluoroquinolone antibacterial agents from sewage to soil. Environ Sci Technol 37(15): 3243–3249
56. Oppel J, Broll G, Löffler D, Römbke J, Meller M, Ternes T (2004) Leaching behaviour of pharmaceuticals in soil-testing-systems: a part of an environmental risk assessment for groundwater protection. Sci Total Environ 328:265–273
57. Kinney CA, Furlong ET, Zaugg SD, Burkhardt MR, Werner SL, Cahill JD, Jorgensen GR (2006) Survey of organic wastewater contaminants in biosolids destined for land application. Environ Sci Technol 40(23):7207–7215
58. Lindberg RH, Wennberg P, Johansson MI, Tysklind M, Andersson BAV (2005) Screening of human antibiotic substances and determination of weekly mass flows in five sewage treatment plants in Sweden. Environ Sci Technol 39:3421–3429
59. Reddersen K, Heberer T, Dünnbier U (2002) Identification and significance of phenazone drugs and their metabolism in ground- and drinking water. Chemosphere 9:539–544
60. Ternes T, Bonerz M, Herrmann N, Teiser B, Andersen HR (2007) Irrigation of treated wastewater in Braunschweig, Germany: an option to remove pharmaceuticals and musk fragrances. Chemosphere 66:894–904
61. Topp E, Monteiro SC, Beck A, Coelho BB, Boxall ABA, Duenk PW, Kleywegt S, Lapen DR, Payne M, Sabourin L, Li H, Metcalfe CD (2008) Runoff of pharmaceuticals and personal care

products following application of biosolids to an agricultural field. Sci Total Environ 396:52–59

62. Blackwell PA, Kay P, Boxall ABA (2007) The dissipation and transport of veterinary antibiotics in a sandy loam soil. Chemosphere 62(2):292–299

63. Heberer T, Verstraeten IM, Meyer MT, Mechlinski A, Reddersen K (2001) Occurrence and fate of pharmaceuticals during bank filtration- preliminary results from investigations in Germany and the United States. Water Resour 120:4–17

64. Holm JV, Rügge K, Bjerg PL, Christensen TH (1995) Occurrence and distribution of pharmaceutical organic compounds in the groundwater downgradient of a landfill (Grinsted, Denmark). Environ Sci Technol 29(5):1415–1420

65. Ternes, T (2001) Pharmaceuticals and metabolites as contaminants of the aquatic environment. In: Daughton CG, Jones-Lepp T (eds) American Chemical Society, Symposium series 791. Washington, DC, pp 39–54

66. FOCUS (2000) FOCUS groundwater scenarios in the EU plant protection product review process. Report of the FOCUS Groundwater Scenarios Workgroup, EC document reference Sanco/321/2000. p 197

67. Schneider MK, Stamm C, Fenner K (2007) Selecting scenarios to assess exposure of surface waters to veterinary medicines in Europe. Environ Sci Technol 41:4669–4676

68. Blackwell PA, Kay P, Ashauer R, Boxall ABA (2009) Effects of agricultural conditions on the leaching behaviour of veterinary antibiotics in soils. Chemosphere 75(1):13–19

69. Boxall ABA, Sinclair CJ, Fenner K, Kolpin DW, Maund S (2004) When synthetic chemicals degrade in the environment. Environ Sci Technol 38(19):369A–375A

70. Boxall ABA (2009) Transformation products of synthetic chemicals in the environmental. Springer, Germany

71. Metcalfe C, Boxall A, Fenner K, Kolpin D, Servos M, Silberhorn E, Staveley J (2008) Exposure assessment of veterinary medicines in aquatic systems. CRC Press, Boca Raton

72. Boxall ABA, Hardy A, Beulke S, Boucard T, Burgin L, Falloon PD, Haygarth PM, Hutchinson T, Kovats RS, Leonardi G, Levy LS, Nichols G, Parsons SA, Potts L, Stone D, Topp E, Turley DB, Walsh K, Wellington EMH, Williams RJ (2009) Impacts of climate change on indirect human exposure to pathogens and chemicals from agriculture. Environ Health Perspect 117(4):508–514

Environmental Comparative Pharmacology: Theory and Application

Lina Gunnarsson, Erik Kristiansson, and D.G. Joakim Larsson

Introduction

Pharmaceuticals are intentionally selected or designed to interact with specific target proteins at relatively low doses. Similarly, their physicochemical characteristics often allow for their efficient uptake across biological membranes. As drugs most often are present in the environment at very low concentrations, high-affinity interactions are likely to mediate any adverse effects in wild life species. It can therefore be assumed that nontarget organisms with conserved drug targets have a higher risk of being affected by residual drugs, compared with species lacking conserved targets. Furthermore, the molecular mechanisms behind uptake, distribution, metabolism, excretion and pharmacological effects can be conserved between the organism that the drug is intended to affect (usually humans) and potential nontarget organisms in the environment. Accordingly, certain pharmaceuticals pose an environmental risk at very low concentrations [1–3].

The vast knowledge base of a new drug's molecular, pharmacokinetic, and pharmacodynamic properties in humans and other mammalian models, derived during its development, provides a basis for an expanded understanding of the potential action of residual pharmaceuticals in exposed nontarget species [4]. However, a comprehensive understanding of the physiology of the exposed wildlife species is also necessary in order to make well-founded predictions, and for the vast majority

L. Gunnarsson (✉) • D.G.J. Larsson
Department of Neuroscience and Physiology, Institute of Neuroscience and Physiology,
The Sahlgrenska Academy, University of Gothenburg, Box 434, 405 30 Göteborg, Sweden
e-mail: lina.gunnarsson@fysiologi.gu.se

E. Kristiansson
Department of Neuroscience and Physiology, Institute of Neuroscience and Physiology,
The Sahlgrenska Academy, University of Gothenburg, Box 434, 405 30 Göteborg, Sweden

Department of Zoology, University of Gothenburg, Box 463, 405 30 Göteborg, Sweden

B.W. Brooks and D.B. Huggett (eds.), *Human Pharmaceuticals in the Environment:*
Current and Future Perspectives, Emerging Topics in Ecotoxicology 4,
DOI 10.1007/978-1-4614-3473-3_5, © Springer Science+Business Media, LLC 2012

of species, this is currently a hampering factor. The recent advances in sequencing and characterization of genomes and transcriptomes have opened up new possibilities to advance the field of comparative pharmacology with an ecotoxicological focus. Information gained from deeper comparative efforts has the potential to aid in the prioritizing of drugs that need further attention for assessment of their environmental risks. Such data can also guide the selection of appropriate test species and methodologies (e.g. endpoints) [5–7]. Furthermore, genomic information could also indicate possible test-combinations of drugs and species which are not likely to be protective for others. In other words, toxicity data that are generated from a species that is lacking a human drug target ortholog might not be protective for a species with a conserved drug target. The copepod, *Nitocra spinipes*, does for example not have an ortholog to the estrogen receptor. Consequently, the NOEC value of 0.05 mg/L for chronic toxicity of EE_2 in the copepod [8] is not protective for fish, even if an assessment factor of 1000 was applied.

Much of the best ecological comparative pharmacology work today has been a result of examining the literature on drugs' target proteins, pathways, mechanisms, etc. in nontarget species. The European Centre for Ecotoxicology and Toxicology of Chemicals (ECETOC) has published a review on the use of intelligent test strategies in ecotoxicology [9]. They suggest and exemplify how information about the mode-of-action for specifically acting chemicals can be used in the environmental risk assessment. Many of the examples and case studies include pharmaceuticals. Other reviews, focusing on the comparative pharmacology in fish for selective serotonin reuptake inhibitors (SSRIs) and adrenoreceptorantagonists (beta-blockers), have also been published. Kreke and Dietrich (2008) summarize the current knowledge about the comparative pharmacology of SSRIs with an emphasis on possible physiological endpoints of potential SSRI interactions in fish, and conclude that the serotonergic system plays a modulatory role in several physiological processes in fish and that serotonin signaling transduction may be mediated by neuronal, endocrine and paracrine pathways. The influence of serotonin on different target tissues appears to be species-specific and may also depend on the gender and/or the development and reproductive status of the individual. Because of this complexity, it is difficult to assess the potential consequences of prolonged exposure of fish populations to SSRIs [10]. Owen et al. [11] describe the current knowledge about the comparative physiology, pharmacology and toxicology of β-blockers: to date, the full repertoire of β-adrenergic receptors has not been reported for any fish species, and even less is known about their expression and specificities for β-blockers. Another paper by Brain et al. [12] reviews the effects and risks of exposure to pharmaceuticals in aquatic plants. Plants provide a number of evolutionarily conserved target sites for antibiotic drugs, resulting from the bacterial ancestry of plastid organelles and conservation of certain metabolic pathways. The statin type of blood lipid regulators is a group of pharmaceuticals with a human target that also are conserved in plants. Indeed, measuring the downstream metabolites (sterols) of the target enzyme (HMG-CoA reductase) provided a specific biomarker in *Lemna gibba*, an aquatic plant, with a sensitivity 2 or 3 times lower than that of fresh weight. Apart from antibiotics and statins, there are few other classes of pharmaceuticals that are known to exert a strong toxicity in plants. Recently, Winter et al. [4] have published a review on the usage of drug development data in the environmental risk assessment of pharmaceuticals, discussing challenges associated with read across. Access to data from the drug development process and established strategies for how to perform comparative predictions is however lacking. Taken together, the different reviews conclude that future toxicological testing should encompass and reflect the known pharmacological effects of the substances studied, and should therefore focus more strongly on specific molecular targets. By identifying potentially affected pathways, it may be possible to identify sensitive endpoints.

In this chapter, we would like to start from a theoretical point-of-view and discuss ways to predict the conservation of proteins known to interact with drugs in the human body. Although such an approach has limitations, and of course must be followed up by empirical studies, it might enable predictions of both pharmacokinetic and pharmacodynamic properties for a large set of drugs in wildlife species with relatively limited effort. We therefore put some focus on how to predict proteins with a conserved function and the downstream pathways in nontarget organisms. Without any attempt at comprehensiveness, we also give some selected examples on both theoretically and empirically derived pharmacokinetics and pharmacodynamic data in nontarget species.

Information on Pharmacokinetics and Pharmacodynamics of Human Pharmaceuticals

The physiochemical, pharmacological and toxicological properties of an active pharmaceutical ingredient (API) are extensively studied during the development of a new drug. For most approved APIs, such information is easily accessible in different public databases, of which some of the most important are listed below.

- *ChEMBL* (www.ebi.ac.uk/ChEMBL) is a chemogenomic database for drug-like molecules that brings together chemical, bioactivity, and genomic data. To date, it contains more than 500 000 compounds.
- *DrugBank* database (http://www.drugbank.ca/) is an example of a bioinformatics and chemoinformatics resource combining detailed drug data with metabolizing enzyme and drug target information.
- *KEGG drug* (http://www.genome.jp/kegg/drug/) is an information resource for all approved drugs in Japan and the USA. Based on the drugs' chemical structure, but also contains information about drug targets and pathways.
- *Pharmacogenetics and Pharmacogenomics Knowledge base* (PharmGKB; http://www.pharmgkb.org/) contains rather few drugs and drug targets, but has information about the relationships between drugs, affected pathways and genes therein, diseases and genes, including their variations and gene products. It aims

to aid researchers in understanding how genetic variation among individuals contributes to differences in reactions to drugs.

- *RxList* (http://www.rxlist.com/script/main/hp.asp) is a comprehensive drug information database aiming to assist and support clinical decisions.

For ecologically relevant nontarget organisms, empirically derived pharmacokinetic and pharmacodynamic data are much more fragmented. Human drugs have been used for a long time, in order to gain insight into the physiology of different organisms. Thus, even literature with a very different purpose in mind, published before the environmental effects of pharmaceuticals became an issue of concern, can prove useful. During the past 10–15 years, large datasets, primarily on microalgae, daphnia, and to some extent fish, have been developed for the environmental risk assessments required for the approval process of new medical products in the EU [13] and USA [14]. The Swedish Association for Pharmaceutical Industries is responsible for the Web site www.fass.se, where they publish such data for products on the Swedish market. For most of these entries, there is, however, no pharmacokinetic information on nontarget species. Also, most of the data are based on results from short-term tests, and mainly cover gross endpoints such as lethality. In the EU, acute tests are no longer considered sufficient for environmental risk assessments of pharmaceuticals as short-term lethality (or inhibited growth) may not be reflective of the specific mode-of-action that could be expected to dominate at low, environmentally relevant concentrations [13]. Accordingly, moving to more chronic tests in general appears to have changed the species sensitivity distribution, with fish more often becoming the most sensitive organism [15]. Indeed, this is in agreement with a higher degree of conservation of drug targets in fish compared with daphnia and microalgae [6]. This also means that the value of the vast dataset on short-term toxicities for comparative pharmacology purposes may be relatively limited.

Predicting Conserved Function of Proteins

Translating pharmacological information from humans to other nontarget species is challenging and not without pitfalls. Such efforts are often based on inter-species comparisons of the human proteins affected by the drug, such as the primary target and the downstream pathway. If the biochemical properties of a protein in a nontarget species are similar to those of a human protein associated to a given drug, it is likely that the two proteins share similar pharmacological characteristics. Thus, the developmental history and evolution of proteins that are affected by APIs can provide information valuable in an ecotoxicological context.

The pharmaceutical industry has started to put more focus on the evolutionary aspects within the drug development. One reason for this is the rather low success rate in the pharmaceutical pipeline, which could partly be explained by the difficulties in successfully translating results from safety and efficacy studies in animal models

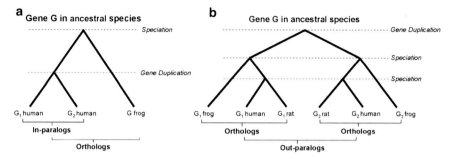

Fig. 1 Homologous proteins have a shared common ancestry. (**a**) An ancestral gene G undergoes one speciation event which is followed by a gene duplication event. The different variants of G appearing in different species are called orthologs (G: frog; G_1 or G_2: human). The different variants within each species are called paralogs (G_1 and G_2: human). Paralogs appearing as a result of a gene duplication event occurring after the latest speciation event are called recent or in-paralogs. (**b**) The ancestral gene G undergoes one gene duplication event followed by two speciation events. The different variants between the different species are still referred to as orthologs, but the paralogs in this case are called ancient or out-paralogs

to humans [16]. It is argued that the extrapolation of safety and efficacy could be improved if the evolutionary history, including information regarding functional similarities and discrepancies, is known for the target proteins of a drug. One apparent example is the choice of an appropriate animal model, which could be founded on information regarding the evolutionary conservation of the target pathway. Another example is minimizing the risk of unexpected off-target effects, which may be due to drug interaction with additional targets that have a shared ancestry with the primary drug target protein.

Homologs, Orthologs, and Paralogs

Functionally similar proteins between different species are generally identified based on sequence homology, a concept that is outlined in Fig. 1a, b. Two proteins are called homologous if they are evolutionarily related such that they stem from a common ancestral protein. The homologous proteins can be further classified as orthologs and/or paralogs, where orthologs are homologous proteins that exist in different species as a result of a former speciation event (Fig. 1), while paralogs, on the other hand, originate from gene duplication events within the genome of a single species. Although paralogous proteins often retain similar biochemical functions, they generally diverge after the duplication event. Paralogs can be divided into two groups, in-paralogs (Fig. 1a) and out-paralogs (Fig. 1b), depending on whether the duplication event occurred before or after the latest speciation event. These two groups of paralogs are therefore sometimes denoted as recent and ancient paralogs.

Since homologous proteins have a common evolutionary history, they often have similar biological functions. Even though orthology does not guarantee functional equivalence, the concept can be utilized to extrapolate the biochemical properties of drug target proteins from humans to nontarget species [17].

The identification of orthologous proteins between diverse species is in general a nontrivial undertaking, which is both conceptually and computationally challenging. In the eukaryotic kingdoms, this problem is particularly delicate since a substantial part of the genetic variation stems from recombination events giving rise to extensive functional redundancy. Reliable and accurate orthology predictions between different eukaryotic species are therefore dependent on the ability to detect both orthologs and paralogs [18]. Several different approaches are described in the literature, and these can be loosely dived into those that are phylogenetically based and those without an explicit tree structure. For reviews of existing approaches for predicting orthologs, please refer to [19, 20].

Orthology Predictions

Phylogenetics is the study of evolutionary relatedness based on molecular sequence data [21]. The phylogenetic topology, usually assumed to have the form of a tree, can be estimated from a group of related proteins. This structure encodes the causality of the evolutionary events, including speciation and gene duplication, and hence the orthologous proteins can be identified. Even though phylogenetically based approaches are generally considered to be relatively accurate and have a high resolution, they also exhibit number of weaknesses when applied to the en masse prediction of orthologs [22]. Inferring reliable phylogenetic trees is in general dependent on high-quality multiple sequence alignments, which in turn, depends on the a priori selection of relevant proteins [23]. Creating multiple alignments may also require manual intervention to achieve an optimal result, especially for environmental risk assessment purposes, where human orthologs are predicted in distantly related species. Furthermore, computing phylogenetic trees is also a computationally intensive task, which will limit the number of species that can be searched for orthologs. Popular methods for predicting orthologs based on phylogenetic trees are the EnsemblCompara Gene Trees [24] and PhiGs [25].

The identification of proteins with conserved functions can also be achieved without assuming an explicit evolutionary tree-like structure. A forthright approach is to use one-way comparisons, where a protein in one species is compared to all the proteins in a divergent species and those with a sequence similarity above a predefined threshold are defined as orthologs. For computational efficiency, these comparisons are usually performed using heuristic sequence comparison procedures, such as FASTA [26] or Basic Local Alignment and Search Tool (BLAST) [27], and the sequence similarity is usually based on a generated alignment score (e.g., E-value). Unfortunately, one-way sequence comparisons are often plagued by false positives. Since this approach does not take in-paralogs and out-paralogs into account, they are both usually identified as orthologs, a problem that is particularly severe in eukaryotic genomes exhibiting a large amount of paralogs [28]. Sequence comparisons from local alignments can also be too optimistic, such that divergent proteins sharing a single preserved functional domain can get a high sequence similarity score. The number of false positives from BLAST-based one-way comparisons was evaluated by Chen et al. [19], who estimated that 50% of the predicted orthologs between divergent eukaryotic species were false positives, while only 4% were false negatives. Nevertheless, the one-way comparison approach has still proven to be of some use, and was applied by Kostich and Lazorchak [7] to identify several conserved human drug targets in nontarget eukaryotic species.

It is possible to generalize from one-way comparisons, comparing one single protein against all proteins in a divergent species, to a symmetric approach, where all proteins in both species are compared against each other. In the most conspicuous setting, proteins from two species are called orthologous if they match each other, or in other words, they are each other's reciprocal best hit (RBH) [17, 29]. This approach has long been used to identify orthologs between prokaryotes and can easily be extended to include several genomes [30]. Since each pair of proteins needs to match each other, the RBH-approach is much more conservative than one-way comparisons, and the proportion of false positives was estimated by [19] to decrease to 8%, while that of false negatives increased to 30%. The reason for the high number of false negatives is the inability to recognize many-to-many and many-to-one ortholog relationships, i.e., in-paralogs and multiple orthologs.

There are several procedures that extend the basic RBH approach to also incorporate multiple orthologs and in-paralogs. The InParanoid algorithm applies clustering to identify orthologs between pairs of species [31]. One major drawback of the InParanoid algorithm is its inability to identify orthologs between several genomes simultaneously. Ortholog predictions from InParanoid are available at the InParanoid Web site (http://inparanoid.sbc.su.se/cgi-bin/index.cgi), which currently comprises 100 organisms with more than 1,600,000 proteins.

OrthoMCL is another algorithm for the identification of orthologs between eukaryotic genomes [18, 19]. In contrast to InParanoid, OrthoMCL can identify both multiple orthologs and in-paralogs in any number of species simultaneously. The algorithm works in a two-staged manner, where first all proteins are compared against each other using BLAST. These results are then interpreted as a graph where the nodes corresponds to proteins and the weighted edges their pair-wise sequence similarity. The graph is then partitioned into sub-graphs using a technique called Markov Clustering (MCL), a fast and efficient algorithm for clustering large graphs [32, 33]. The results are several clusters of proteins, each containing the orthologs and the in-paralogs for each species (Fig 2). The OrthoMCL algorithm was estimated to perform well (16% false positives and 7% false negatives) on a divergent set of eukaryotic species by Chen et al. [19], and the procedure has been used to predict orthologs for 1,318 human drug targets in 16 species of which many are relevant to ecotoxicological testing [6].

There are also several other considerations that can improve the power of orthology prediction. One example is secondary and tertiary protein structure, and another is

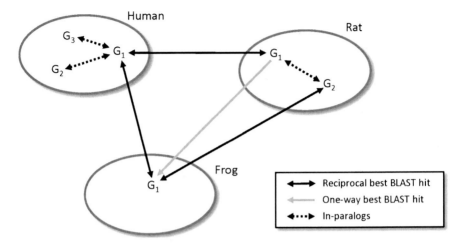

Fig. 2 Identification of both paralogs and orthlogs is necessary to understand the evolutionary history of a gene. The figure shows an example of an ancestral gene G, which appears as different variants in human, rat and frog. Since the reciprocal best BLAST hit (RBH) of G_1 in frog is G_1 in human and G_2 in rat, while the RBH of G_1 in human is G_1 in rat, algorithms solely using information from the RBHs would not identified orthlogs for G in all three species. However, the clustering approach utilized by OrthoMCL can detect all these nontrivial relationships and correctly assign all variants within a single cluster containing both the in-paralogs and orthologs

synteny, i.e., evolutionarily preserved chromosomal localization of genes. The latter has been combined with BLAST-based sequence similarity measures in the NCBI *Homologene* database (http://www.ncbi.nlm.nih.gov/homologene).

Comparative Pharmacokinetics

Pharmacokinetics describes how the body affects a specific drug after administration, and can be separated into absorption, distribution, metabolism, and excretion (ADME). This information is critical to understand how, for how long, and at what concentration an API gains access to its molecular target(s) in the organism.

Absorption and Distribution

The uptake of APIs in humans occurs most commonly through ingestion, while the uptake route can be very different in other species. A summary of the different uptake routes of pharmaceuticals in fish, invertebrates, plants and algae are available in [9]. For most pharmaceuticals, the major uptake route in fish and amphibians is likely to be through the gill/lung and skin. Randall et al. [34] showed that fish take up lipophilic xenobiotics (log K_{ow} >3) mainly across the gills, and the substances then enter the bloodstream directly. The amount of uptake via ingestion appears to be negligible for most pharmaceuticals in view of the normally low feeding rate and water intake of fish compared with their much larger breathing volume [34]. For invertebrates, the uptake route is heavily dependent on the life stage of the animal and the environment they are living in [35], while the uptake in plants and aquatic macrophytes depends on the type of plant tissue that is in contact with the drug [9].

Passive Diffusion and Carrier-Mediated Uptake

The uptake of pharmaceuticals is generally regulated by passage through the cell membrane in all organisms. A drug can permeate by passive diffusion or by an active or carrier-mediated uptake. Passive diffusion through the lipid bilayer of the cell is often considered to be the dominant process by which a drug is taken up by the cell. In drug development, Lipinski's "rule of five" is used to identify drug candidates with poor absorption, and it assumes that drugs are mainly taken up by passive diffusion. "The rule of five" predicts that poor absorption is more likely when there are more than five hydrogen bond donors, ten hydrogen bond acceptors, the molecular weight is greater than 500 and the calculated Log P is greater than five [36]. Huggett et al. [37] have similarly proposed a model, assuming passive diffusion, to predict the risk for fish to be affected by pharmaceuticals. The only inputs in this model are the water concentration of the drug, the LogP of the drug, and the human therapeutic plasma concentration. Empirical results suggest that a rather simplistic model like this could be valuable for identifying APIs with the potential to affect aquatic organisms at environmentally relevant concentrations [37–42]. Many pharmaceuticals are ionisable and the pH of the environment can affect the uptake and toxicity [43, 44]. Therefore, consideration of the ionisation of an API can be important for environmental risk assessments.

In many cases, drugs do not behave as expected based on passive diffusion alone, and other factors need to be considered in predicting the absorption and distribution of a drug. One factor may be carrier-mediated or active uptake of drugs. Dobson and Kell [45] argue that carrier-mediated and active uptake of pharmaceuticals may be more common than traditionally assumed. Indeed, many drugs are taken up by carriers in those specific cases where it has been studied (for a comprehensive list of examples see supplementary information S1 in [45]).

Carrier-mediated and active drug uptake might for example explain why certain drugs concentrate in specific tissues and also bioaccumulate in some aquatic organisms. Dobson and Kell [45] give examples of drug uptake by three of the most significant families of transporters. We used the NCBI Homologene database to predict the evolutionary conservation of these transporters. The solute carrier organic anion transporter family, member 1B1 (SCLO1B1), was the least conserved transporters, while four of the transporters were predicted to be conserved among all eukaryotes (Table 1). A conserved transporter protein might indicate a potential

Table 1 Evolutionary conservation of selected human transport proteins for pharmaceuticals based on the NCBI *Homologene* database

Pharmaceuticals	Transporter HUGO symbol	Description	Conserved according to Homologene
Amoxicillin, Cefaclor, Cefalexin, Bestatin, Amoxicillin, Ampicillin, Cefadroxil, Cefixime, Enalapril, Midodrine, Valacyclovir, Valganciclovir, Ceftibuten	SLC15A1	Oligopeptide transporter	Eukaryota
Amoxicillin, Cefaclor, Cefadroxil, Bestatin, Valganciclovir	SLC15A2	H[+]/peptide transporter	Eukaryota
Zidovudine, Acyclovir, Ganciclovir, Metformin, Cimetidine	SLC22A1	Organic cation transporter	Amniota[a]
Memantine, Metformin, Propranolol, Cimetidine, Zidovudine, Pancuronium, Quinine	SLC22A2	Organic cation transporter	Euteleostomi[b]
Cimetidine	SLC22A3	Extraneuronal monoamine transporter	Amniota[a]
Quinidine, Verapamil	SLC22A4	Organic cation transporter	Eukaryota
Quinidine, Verapamil, Valproate, Cephaloridine	SLC22A5	Organic cation transporter	Eukaryota
Adefovir, Acyclovir, Zalcitabine, Didanosine, Stavudine, Trifluridine, Ganciclovir, Lamivudine, Zidovudine, Methotrexate, Ketoprofen, Ibuprofen, Cimetidine, Tetracycline, Cephaloridine	SLC22A6	Organic anion transporter	Euteleostomi[b]
Zidovudine, Tetracycline, Salicylate, Methotrexate, Erythromycin, Theophyline	SLC22A7	Organic anion transporter	Euteleostomi[b]
Valacyclovir, Zidovudine, Methotrexate, Salicylate, Cimetidine, Cephaloridine	SLC22A8	Organic anion transporter	Eutheria[c]
Zidovudine, Cephaloridine	SLC22A11	Organic anion/ cation transporter	Eutheria[c]
Fexofenadine, Rocuronium, Enalapril, Temocaprilat, Rosuvastatin	SLCO1A2	Organic anion transporter	Amniota[a]
Benzylpenicillin, Pravastatin, Rifampicin, Atorvastatin, Capsofungin, Cerivastatin, Fexofenadine, Fluvastatin, Pitavastatin, Methotrexate	SLCO1B1	Organic anion transporter	Homo/Pan/Gorilla group
Digoxin, Rifampicin, Fexofenadine, Fluvastatin, Pitavastatin, Rosuvastatin, Methotrexate	SLCO1B3	Organic anion transporter	Amniota[a]

(continued)

Table 1 (continued)

Pharmaceuticals	Transporter HUGO symbol	Description	Conserved according to Homologene
Pravastatin, Glibenclamide, Atorvastatin, Benzylpenicillin, Fluvastatin, Rosuvastatin	SLCO2B1	Organic anion transporter	Amniota[a]
Methotrexate, Digoxin	SLCO4C1	Organic anion transporter	Bilateria[d]

The selection of drugs and transport proteins is based on Dobson and Kell [45]
[a]Mammals, reptiles, and birds
[b]Bony vertebrates
[c]Placental mammals
[d]All animals with bilateral symmetry

for bioaccumulation of the drugs taken up by the transporter. However, drugs could also concentrate in species that lack transporter orthologs and the significance of conserved carrier proteins needs to be evaluated. Data on uptake and distribution of pharmaceuticals, as well as molecular characterization of transporters in nontarget species, are needed.

Plasma-Protein Binding

Binding to proteins in the blood plasma is another factor that can affect the uptake and distribution of pharmaceuticals [46]. If a drug is bound to a plasma protein, it limits the drug's free motion, reduces its volume of distribution as well as its renal excretion, liver metabolism, and tissue penetration. Binding of plasma proteins can also increase the absorption and the half-life of the drug. Most drugs commonly bind to serum albumin (ALB) and orosomucoid (ORM1 and 2, alpha acid glycoprotein) in humans [47]. Neither serum albumin nor orosomucoid has orthologs in fish according to the NCBI *Homologene* database but an albumin-like protein has been described [48] although not predicted to be orthologous to human ALB. The plasma protein profile is also different in fish and the total protein levels are generally lower than in human plasma [49]. Thus, the characteristic binding of drugs to plasma proteins may not extrapolate to fish. For example, the antibiotic drug sulfadimethoxine and the antimicrobial ormetrophine are to a great extent associated to proteins in human plasma while the binding is very limited in trout [49]. Other drugs bind however in a similar manner. Sex hormone-binding globulin is conserved in euteleostomi (bony vertebrates) and it is the major transport protein for sex steroids in the blood both in humans and in fish [50]. It was shown that sex hormone-binding globulin controls the flux of sex steroids across fish gills and that its function can be hijacked, for example, by 17α-ethinylestradiol (EE_2) [51]. The synthetic progestin levonorgestrel bioconcentrates from water into blood plasma of trout considerably more than what is expected from its log P [40]. The high potency of this drug in fish

[3] is likely linked to a high bioconcentration factor facilitated by binding to sex hormone-binding globulin in the fish [40, 52]. Plasma binding proteins could thus be an important factor to consider both when uptake and bioavailability at the drug target should be predicted.

Metabolism and Excretion

Phase I and phase II drug metabolizing enzymes play a central role in the metabolism of drugs. Pharmaceuticals are often hydrophobic and need to be biotransformed to become more polar and water soluble so that they can be excreted. Biotransformation not only promotes drug elimination but can also change the overall biological properties of the drug, leading to the activation or inactivation of the pharmacological activity. Phase I metabolizing enzymes often catalyze oxidation, reduction, and hydrolysis reactions, while phase II enzymes catalyze conjugation reactions, which add polar functional groups to the drug. The metabolites generated by phase I and II reactions can be excreted from the body with the aid of membrane efflux pumps such as the multidrug resistance associated proteins [53]. Drug metabolism can occur in many diverse cell and organ systems, but the liver, intestine, kidney, and gill/lung play the largest role in vertebrates. In mammals, the kidney is by far the most important organ for excreting drugs. Short summaries of the metabolism and excretion of chemicals in fish, invertebrates, plants, and algae are available in [9].

The main phase I metabolizing enzymes are the cytochrome P450 protein superfamily, which catalyzes the incorporation of one oxygen atom from molecular O_2 into a substrate. This cytochrome reaction can be observed in virtually all living organisms, from bacteria to mammalian species [54]. In human liver, the most important drug metabolizing P450 enzymes are CYP1A2, CYP2B6, CYP2C9, CYP2C19, CYP2D6, and CYP3A4/CYP3A5, which metabolize about 95% of the drugs in clinical use [55, 56].

There are several problems with predictions of P450 mediated drug metabolism in nontarget species based on human data. It is very difficult to predict the orthology of P450s across distantly related species using sequence similarity based prediction methods. The great diversity of the P450 superfamily has arisen by extensive processes of gene duplication, conversions, genome duplications, gene loss and lateral transfers. This have created a large number of P450 paralogs, and many out-paralogs and in-paralogs are present in almost all eukaryotic genomes [54]. Mammalian CYP1A1 and CYP1A2 are believed to have diverged 250 million years ago by a duplication event [57]. Fish diverged from the mammalian line prior to that and does consequently only have one CYP1A gene [58]. The CYP2 gene family is the most diverse CYP gene family with 13 know CYP2 subfamilies in fish. None of the human CYP2 enzymes that are the most important for drug metabolism are believed to have orthologs in fish [58]. The ortholog relationships between various CYP3A enzymes are more unclear. To date, 13 teleost CYP3A genes have been identified, but the current nomenclature for the CYP3 gene family does not reflect orthologous relationship between organisms [58]. Despite the clear ortholog relationships and high conserved function of individual P450s in mammals, there are significant differences between species. For example, orthologous P450 enzymes in closely related species can have very different basal expression levels, different levels of induction by APIs, and also the enzyme substrate specificity may differ [59]. An example is omeprazole, a CYP1A inducer in humans that has little effect on CYP1A forms in rats, mice, and rabbits [55]. Another example is CYP1A mediated metabolism in fish and mammals. The fish CYP1A has similar substrate preferences, but the oxidation rate for some substrates can differ by orders of magnitude between fish and mammalian species [60]. Developmental differences in protein expression and activity of P450 enzymes are also important factors to consider. CYP1A is for example expressed in zebrafish 15h post fertilization but no protein expression or ethoxyresorufin O-deethylase activity could be detected until after hatching [61]. Taken together, we believe that high-throughput orthology predictions to extrapolate P450-mediated metabolism of drugs between distantly related species suffer from major limitations.

Nevertheless, in cases where the orthologous relations are clear there are also examples of functional similarities between P450 enzymes in, for example, mammals and fish. The fungicide ketoconazole inhibits CYP3A in both mammals [62] and fish [63], for example. A useful base for comparisons could be Lee et al. ([62]; Table IV) that summarizes drugs and xenobiotics that are inhibitors of one or more human cytochrome P450 enzymes involved in drug metabolism.

Phase II metabolizing enzymes often catalyze conjugation reactions, which add more polar functional groups to the drug. Sulfotransferases (SULTs), UDP-glucuronosyltransferases (UGTs), and glutathione S-transferases (GSTs) are examples of phase II enzymes that catalyze the addition of sulfate, glucuronate, and glutathione respectively. Methyltransferase, NAD(P)H:quinone oxidoreductase (DT-diaphorase), and acetyltransferase are other examples of phase II metabolizing enzymes [53]. The UGTs, SULTs, and GSTs are all superfamilies of proteins, with homologs present in almost all eukaryotic genomes. As many as 117 mammalian UGT genes and 56 distinct eukaryotic SULT isoforms and have been identified to date [64–66]. The UGTs are the most important phase II drug metabolism enzymes, and have therefore been extensively studied in humans. However the research on UGTs in lower vertebrates and invertebrates is much more limited [58].

Today, the lack of understanding of the detailed function and specificities of different phase I and II drug metabolism enzymes are two of several factors hampering the accurate prediction of the kinetics of drugs in nontarget species. There is a great need for empirical studies on ADME of APIs in wildlife species, both for generating and evaluating different predictive models. Predictions on the pharmacokinetics of drugs in different species may be a valuable component in identifying species at risk. For example, differences in pharmacokinetics of diclofenac between different vulture species may be an explanation behind the rather large difference in sensitivity between relatively closely related birds [67]. Computational models are frequently and successfully used in the drug discovery pipeline to predict ADME of a substance [68–70]. In the future, when we have better knowledge of the function of individual transporters and metabolizing enzymes, such *in silico* approaches might be possible to use in nontarget species as well.

Pharmacodynamics

Pharmacodynamics is the study of the biochemical and the physiological effects of drugs on the body. This includes the interaction between the drug and the target protein (i.e., activation or inhibition), the downstream mode-of-action and the affected physiological endpoints. Pharmacodynamics also encompasses drug interactions with off-targets that induce side effects and potentially toxic responses.

Drug Targets

A drug target can be defined as a molecular structure that undergoes a specific interaction with a pharmaceutical and the interaction has a connection with a clinical effect. The great majority of drug targets are proteins, and these can be classified into different broad functional groups, the most important of which are enzymes, receptors, ion channels and transporters. Drugs are selected or designed to induce their intended clinical effect and to cause a minimal amount of side effects at relatively low doses.

In Gunnarsson et al. [6], we performed an ortholog prediction of all human drug targets available in Drugbank [71, 72]. The study shows that roughly 80% of the human drug targets are conserved in the aquatic vertebrates *Xenopus tropicalis*, *Danio rerio*, and *Gasterosteus aculeatus,* roughly 60% are conserved in the invertebrates *Daphnia pulex*, *Drosophila melanogaster*, and *Caenorhabditis elegans*, and less than 40% are conserved in the plant *Arabidopsis thaliana* and the alga *Chlamydomonas reinhardtii* (Fig. 3). The protein sequence similarity of the orthologs show a similar pattern, with the vertebrate orthologs sharing about 60% sequence similarity with the human drug targets, whereas the invertebrates have about 40% sequence similarity (Fig. 3).

Another interesting pattern was revealed when different functional groups of the drug targets were analyzed. Receptors constitute an important group of validated pharmacological targets, and as much as 40% of all FDA approved drugs elicit their effects through receptors [73]. We show that the proportion of receptors decrease significantly while the proportion of enzymes significantly increase with the evolutionary distance to man (Fig. 4). Thus, the choice of environmental test species is particularly important for drugs that have receptors as targets, and effects on nonvertebrates would not be expected for most drugs from this class. Enzyme drug targets, on the other hand, were more ubiquitously present. Generally speaking, one could therefore expect drugs targeting enzymes to affect a wider range of species,

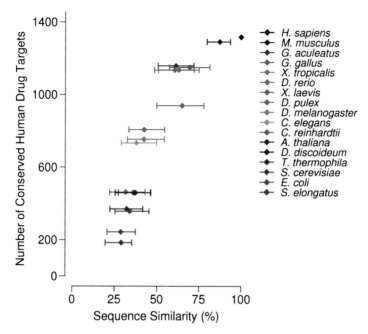

Fig. 3 The number of evolutionary conserved drug target proteins differs between species. The figure shows on the *x*-axis the median sequence similarity to the corresponding human drug targets and on the *y*-axis the number of conserved drug targets for 16 nontarget species. The boxes indicate 25 and 75% quantiles (Reprinted from Gunnarsson et al. [6] with permission from the American Chemical Society)

including invertebrates. The full list of ortholog predictions of 1,318 human drug targets in 16 species is available as supplementary information together with the published paper [6].

As already mentioned, the presence of a drug target ortholog in a nontarget species does not guarantee that a functional interaction with the drug can occur. However, in Gunnarsson et al. [6], we presented literature data supporting that an ortholog prediction often can indicate the ability of a conserved drug target protein to interact with the human drug. A more precise prediction of a potential drug target interaction might be possible with better knowledge about drug binding domains. Sakharkar et al. [74] summarized protein family (Pfam) domains related to drug-gable domains. However, many of these domains are too general to provide additional useful information to predict the ability of a protein to interact with a drug in an evolutionarily distant species. The database Supersite has recently been released. It contains 3D protein structures and ligand-binding site information for over 1,300 medically active compounds, as well as some evolutionary information [75]. Given that the structure of the proteins in nontargets species can be accurately modeled, this could improve current predictions.

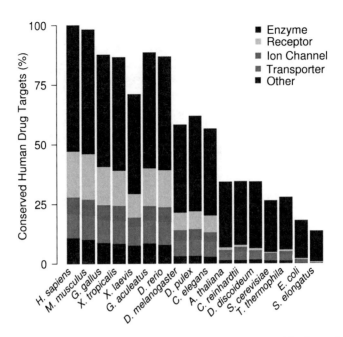

Fig. 4 The evolutionary conservation depends on the type of drug target. The figure shows evolutionarily conserved human drug targets in 16 nontarget species divided into functional categories (Reprinted from Gunnarsson et al. [6], with permission from the American Chemical Society)

Mode-of-Action

An interaction between a drug and its primary target generally leads to a number of subsequent reactions within the cell. This course of events can be described by one or several pathways, chains of causal biochemical reactions and interactions. Even though pathways, strictly speaking, are theoretical models, they can provide means for identification of affected physiological endpoints. The pathways affected by pharmaceuticals have not been as extensively studied as the drug targets themselves. An absolute identification of the therapeutic mode of action is not a requirement to market a new drug. Nevertheless, there is a considerable amount of human drug targets associated to pathways that at least partly describe the mode-of-action of the drug [76, 77].

The evolutionary process of a pathway is complex and not yet fully understood. Many of the pathways currently described are based on mammalian biology and can therefore be completely different or even nonexistent in nontarget species. Stimulation of the same target protein in two different species may indeed lead to very different outcomes depending on the physiology of the species. Knowledge about the evolutionary conservation of a pathway can therefore be helpful for identification of physiological endpoints affected by drugs in nontarget species.

Pathway Prediction

As for drug targets, predictions of conserved pathways are based on identification of orthologs, but for a pathway, information on all its members needs to be combined. The Reactome Project (www.reactome.org) uses OrthoMCL to predict the conserved counterparts of human pathways in 22 species, including mouse, rat and chicken, but also more distant species such as worm, fly and yeast. The information on pathways in nontraditional model organisms is however sparse.

The activation of a pathway often results, directly or indirectly, in a change in gene expression profiles. This is also true for pathways mediated by drug target proteins, and if the mode-of action is evolutionarily conserved between human and nontarget species, then it is likely that their transcriptional responses also have similarities. Large-scale gene expression analysis can thus provide information regarding the pharmacodynamics of drugs. These experiments are typically performed using DNA microarrays, which have been widely available for model species such as mouse, rat, and zebrafish. The introduction of more versatile platforms, such as Agilent or Nimblegen custom microarrays (www.aglient.com; www.nimblegen.com) and febit RT-analyzer (www.febit.de), has made the microarray technique more applicable to nonmodel organisms without a well-categorized genome [78, 79]. Even though the quality of data generated from microarrays was initially questioned [80, 81], the technique has matured and is today approaching the higher sensitivity and accuracy of methods such as quantitative PCR and high-throughput mRNA sequencing [82–84].

Microarray Analysis

Compared to more focused approaches, microarrays can be used in an explorative manner, i.e., without a clear hypothesis on how the nontarget species could be affected by the investigated drug. Some combinations of species and drugs will inevitably cause toxicity in unexpected ways via novel mechanisms, even at rather low concentrations. Thus, microarray analysis is an important complement of mode-of-action based tests relying on our ability to a priori identify relevant and sensitive endpoints.

Gene expression data generated by microarrays can be used to identify affected pathways. This can be used both to confirm hypotheses based on mammalian pharmacodynamics and to discover novel and unexpected effects. Heckmann et al. [85] showed, for example, that several of the genes involved in eicosanoid (for instance prostaglandin) metabolism were differentially expressed in *D. magna* exposed to the NSAID ibuprofen. In mammals, NSAIDs inhibit prostaglandin synthesis, and thus the mode-of-action for this drug is partly conserved. However, the physiological endpoints are different, as differences in eicosanoid metabolism may have direct consequences for reproduction in crustaceans.

Pathway analysis can thus be used to understand and interpret gene expression data in a pharmacological context. In general, this kind of analysis is performed by testing whether the genes from a given pathway have a higher tendency to be differentially expressed than other genes on the microarray [86]. The procedure is usually repeated for all pathways in a database, such as KEGG, and the models best describing the gene expression pattern can thus be identified. Several software applications performing pathway analysis of gene expression data have been developed, e.g., GenMAPP [87], Pathway Miner [88] and the SkyPainter tool at the Reactome Project [89], among which the latter allows for the identification of evolutionary conserved human pathways in microarray data from other species.

Even though pathway analysis has the ability to provide information regarding the molecular-level mechanistic of drug–target interactions, there are a number of potential issues that can make interpretations difficult. All proteins associated with a pathway affected by a certain drug will generally not be regulated at the transcriptional level. Indeed, later regulatory steps, such as post-transcriptional and post-translational modifications, may be more suitable in many cases, and such actions cannot be directly detected by microarray-based gene expression analysis. A pathway can also have one or a few specific bottlenecks, and hence increasing or decreasing the abundance of these proteins might be enough to change the activity of the entire pathway. Testing the entire pathway for upregulation or downregulation at the transcriptional level may thus not be the most appropriate solution. Another significantly hampering factor for pathway analysis in nonmodel organism is the lack of species-specific information about protein function, interaction, and thus pathways [89].

Transcription-Factor Proteins

Microarray data can also be used for identification of the activation or inhibition of transcription factor proteins. The synthetic estrogen used in contraceptives, EE_2, binds to the estrogen receptor, which is consequently relocated into the nucleus and initiates the transcription of hundreds of genes. This mechanism is conserved in all vertebrates [90], and the gene expression patterns produced also show similarities in many species [91]. Many pathways activate transcription of genes, which then can be measured using microarrays. The regulation of gene expression is generally performed by transcription factor proteins, which can initiate transcription by binding to short DNA sequences called *cis*-regulatory elements. Information regarding the *cis*-regulatory elements within the promoters of differentially expressed genes can therefore be used to untangle which transcription factors and biological process are responsible for the observed changes in gene expression [92]. Hence, if a *cis*-regulatory element is overrepresented among the differentially expressed genes, it is likely that the corresponding transcription factor regulates the transcription of these genes. However, the identification of *cis*-regulatory elements is dependent on information on the promoter regions, which is often lacking even for species with completely sequenced genomes [93]. The overrepresentation of *cis*-regulatory elements among differentially expressed genes can therefore yet only be performed reliably in a limited number of model species. The applicability of these approaches will, however, continue to improve as the number of sequenced species increases.

The evolutionary conservation of the mode-of-action of a specific drug can provide valuable information and facilitate the extrapolation of pharmacodynamics to nontarget species. Orthology predictions of drug targets and pathways are fundamentally connected to genomics data, and therefore continue to improve as more and more nontarget species become sequenced. Increased knowledge about protein function, interactions and the general physiology are equally important for reliable comparative pharmacodynamics. However, molecular-level responses identified by large scale comparative pharmacology approaches can aid in the identification of relevant and sensitive endpoints. Combining those endpoints with measurements of adverse effects can provide valuable information about possible chronic consequences of pharmaceutical exposure and for population health.

Conclusion

The currently available ecotoxicity data for most pharmaceuticals are insufficient, but the vast knowledge base derived in drug development could provide insights on possible effects of residual pharmaceuticals in exposed nontarget species. The rapid advances in genomics have opened up new possibilities to develop the field of comparative pharmacology in species of interest for ecological risk assessments. Accurate predictions of the conservation of proteins and pathways may provide important input to our understanding of pharmacokinetics and pharmacodynamics in nontarget species. However, a comprehensive understanding of the physiology of the exposed wildlife species is equally important to make well-founded predictions, and for the vast majority of species, this is currently a hampering factor. Nevertheless, large-scale comparisons of conserved proteins and pathways can still aid in the prioritization of which drugs need further assessment of their environmental risks, which organisms should be prioritized for testing, and what endpoints are most appropriate.

Acknowledgments The authors wish to thank the two anonymous reviewers for valuable comments and the Foundation for Strategic Environmental Research (MISTRA) and the Swedish Research Council (VR) for financial support.

References

1. Garric J, Vollat B, Duis K, Pery A, Junker T, Ramil M, Fink G, Ternes TA (2007) Effects of the parasiticide ivermectin on the cladoceran *Daphnia magna* and the green alga *Pseudokirchneriella subcapitata*. Chemosphere 69:903–910
2. Kidd KA, Blanchfield PJ, Mills KH, Palace VP, Evans RE, Lazorchak JM, Flick RW (2007) Collapse of a fish population after exposure to a synthetic estrogen. Proc Natl Acad Sci U S A 104:8897–8901

3. Zeilinger J, Steger-Hartmann T, Maser E, Goller S, Vonk R, Lange R (2009) Effects of synthetic gestagens on fish reproduction. Environ Toxicol Chem 28:2663–2670

4. Winter MJ, Owen SF, Murray-Smith RM, Panter GH, Hetheridge MJ, Kinter LB (2009) Using data from drug discovery and development to aid the aquatic environmental risk assessment of human pharmaceuticals: concepts, considerations and challenges. Integr Environ Assess Manag 6:38–51

5. Ankley GT, Brooks BW, Huggett DB, Sumpter JP (2007) Repeating history: pharmaceuticals in the environment. Environ Sci Technol 41:8211–8217

6. Gunnarsson L, Jauhiainen A, Kristiansson E, Nerman O, Larsson DGJ (2008) Evolutionary conservation of human drug targets in organisms used for environmental risk assessments. Environ Sci Technol 42:5807–5813

7. Kostich MS, Lazorchak JM (2008) Risks to aquatic organisms posed by human pharmaceutical use. Sci Total Environ 389:329–339

8. Breitholtz M, Bengtsson BE (2001) Oestrogens have no hormonal effect on the development and reproduction of the harpacticoid copepod Nitocra spinipes. Mar Pollut Bull 42:879–886

9. ECETOC (2007) Intelligent testing strategies in ecotoxicology: mode-of action approach for specifically acting chemicals. ECETOC Technical Report No.102. http://www.ecetoc.org/index.php?mact=MCSoap,cntnt01,details,0&cntnt01by_category=16&cntnt01template=displ ay_list_science&cntnt01document_id=281&cntnt01returnid=101. Accessed Sept 2008

10. Kreke N, Dietrich DR (2008) Physiological endpoints for potential SSRI interactions in fish. Crit Rev Toxicol 38:215–247

11. Owen SF, Giltrow E, Huggett DB, Hutchinson TH, Saye J, Winter MJ, Sumpter JP (2007) Comparative physiology, pharmacology and toxicology of beta-blockers: mammals versus fish. Aquat Toxicol 82:145–162

12. Brain RA, Hanson ML, Solomon KR, Brooks BW (2008) Aquatic plants exposed to pharmaceuticals: effects and risks. Rev Environ Contam Toxicol 192:67–115

13. EMEA (2006) Guideline on the environmental Risk Assessment of Medical Products for Human Use. The Committee for Medicinal Products for Human Use (CHMP), European Medicines Agency, CHMP/SWP/4447/00. http://www.tga.gov.au/docs/pdf/euguide/swp/444700en.pdf. Accessed Sept 2009

14. FDA (1998) Guidance for environmental assessment of human drug and biologics applications. Food and Drug Administration, Center for Drug Evaluation and Research, Center for Biologics Evaluation and Research. http://www.fda.gov/cder/guidance/1730fnl.pdf. Accessed Sept 2009

15. Maack G, Adler N, Bachmann J, Ebert I, Hickmann S, Küster A, B. R (2010) Environmental risk assessment of medicinal products for human use: does the risk assessment reflect the reality? SETAC Europe: 20th annual meeting. http://www.eventure-online.com/eventure/publicAbstractView.do;jsessionid=abcrY0Urm3inMfjVSloOs?id=115911&congressId=3358. Accessed June 2010

16. Holbrook JD, Sanseau P (2007) Drug discovery and computational evolutionary analysis. Drug Discov Today 12:826–832

17. Bork P, Dandekar T, Diaz-Lazcoz Y, Eisenhaber F, Huynen M, Yuan Y (1998) Predicting function: from genes to genomes and back. J Mol Biol 283:707–725

18. Li L, Stoeckert CJ Jr, Roos DS (2003) OrthoMCL: identification of ortholog groups for eukaryotic genomes. Genome Res 13:2178–2189

19. Chen F, Mackey AJ, Vermunt JK, Roos DS (2007) Assessing performance of orthology detection strategies applied to eukaryotic genomes. PLoS One 2:e383

20. Dolinski K, Botstein D (2007) Orthology and functional conservation in eukaryotes. Annu Rev Genet 41:465–507

21. Felsenstein J (1988) Phylogenies from molecular sequences: inference and reliability. Annu Rev Genet 22:521–565

22. Altenhoff AM, Dessimoz C (2009) Phylogenetic and functional assessment of orthologs inference projects and methods. PLoS Comput Biol 5:e1000262

23. Gabaldon T (2008) Large-scale assignment of orthology: back to phylogenetics? Genome Biol 9:235
24. Vilella AJ, Severin J, Ureta-Vidal A, Heng L, Durbin R, Birney E (2009) EnsemblCompara GeneTrees: complete, duplication-aware phylogenetic trees in vertebrates. Genome Res 19:327–335
25. Dehal PS, Boore JL (2006) A phylogenomic gene cluster resource: the Phylogenetically Inferred Groups (PhIGs) database. BMC Bioinformatics 7:201
26. Pearson W (2004) Finding protein and nucleotide similarities with FASTA. Curr Protoc Bioinformatics Chapter 3:Unit3.9
27. Altschul SF, Madden TL, Schaffer AA, Zhang J, Zhang Z, Miller W, Lipman DJ (1997) Gapped BLAST and PSI-BLAST: a new generation of protein database search programs. Nucleic Acids Res 25:3389–3402
28. Ding G, Sun Y, Li H, Wang Z, Fan H, Wang C, Yang D, Li Y (2008) EPGD: a comprehensive web resource for integrating and displaying eukaryotic paralog/paralogon information. Nucleic Acids Res 36:D255–D262
29. Tatusov RL, Koonin EV, Lipman DJ (1997) A genomic perspective on protein families. Science 278:631–637
30. Raymond J, Zhaxybayeva O, Gogarten JP, Gerdes SY, Blankenship RE (2002) Whole-genome analysis of photosynthetic prokaryotes. Science 298:1616–1620
21. Remm M, Storm CE, Sonnhammer EL (2001) Automatic clustering of orthologs and in-paralogs from pairwise species comparisons. J Mol Biol 314:1041–1052
32. van Dongen S (2000) Graph clustering by flow simulation. PhD thesis, University of Utrecht, Utrecht
33. Enright AJ, Van Dongen S, Ouzounis CA (2002) An efficient algorithm for large-scale detection of protein families. Nucleic Acids Res 30:1575–1584
34. Randall DJ, Connell DW, Yang R, Wu SS (1998) Concentrations of persistent lipophilic compounds in fish are determined by exchange across the gills, not through the food chain. Chemosphere 37:1263–1270
35. Barnes RSK, Calow P, Olive PJW, Golding DW, Spicer JI (2001) The invertebrates—a synthesis. Blackwell Science, Malden
36. Lipinski CA, Lombardo F, Dominy BW, Feeney PJ (1997) Experimental and computational approaches to estimate solubility and permeability in drug discovery and development settings. Adv Drug Deliv Rev 46:3–25
37. Huggett BD, Cook CJ, Ericson FJ, Williams RT (2003) A theoretical model for utilizing mammalian pharmacology and safety data to prioritize potential impacts of human pharmaceuticals to fish. Hum Ecol Risk Assess 9:1789–1799
38. Brown JN, Paxeus N, Forlin L, Larsson DGJ (2007) Variations in bioconcentration of human pharmaceuticals from sewage effluents into fish blood plasma. Environ Toxicol Pharmacol 24:267–274
39. Owen SF, Huggett DB, Hutchinson TH, Hetheridge MJ, Kinter LB, Ericson JF, Sumpter JP (2009) Uptake of propranolol, a cardiovascular pharmaceutical, from water into fish plasma and its effects on growth and organ biometry. Aquat Toxicol 93:217–224
40. Fick J, Lindberg RH, Parkkonen J, Arvidsson B, Tysklind M, Larsson DGJ (2010) Therapeutic levels of levonorgestrel detected in blood plasma of fish: results from screening rainbow trout exposed to treated sewage effluents. Environ Sci Technol 44:2661–2666
41. Fick J, Lindberg RH, Tysklind M, Larsson DGJ (2010) Predicted critical environmental concentrations for 500 pharmaceuticals. Regul Toxicol Pharmacol 58(3):516–523
42. Roos V, Gunnarsson L, Fick J, Larsson DGJ, Ruden C (2012) Prioritising pharmaceuticals for environmental risk assessment: Towards adequate and feasible first-tier selection. Science of the Total Environment 421–422:102–110
43. Rendal C, Kusk KO, Trapp S (2011) The effect of pH on the uptake and toxicity of the bivalent weak base chloroquine tested on Salix viminalis and Daphnia magna. Environ Toxicol Chem 30:354–359

44. Valenti TW, Perez-Hurtado P, Chambliss KC, Brooks BW (2009) Aquatic toxicity of sertraline to Pimephales promelas at environmentally relevant surface water pH. Environ Toxicol Chem 28:2685–94
45. Dobson PD, Kell DB (2008) Carrier-mediated cellular uptake of pharmaceutical drugs: an exception or the rule? Nat Rev Drug Discov 7:205–220
46. Benet LZ, Kroetz DL, Sheiner LB (1996) Pharmacokinetics. The dynamics of drug absorption distribution and elimination. McGraw Hill, New York.
47. Kratochwil NA, Huber W, Muller F, Kansy M, Gerber PR (2002) Predicting plasma protein binding of drugs: a new approach. Biochem Pharmacol 64:1355–1374
48. Gray JE, Doolittle RF (1992) Characterization, primary structure, and evolution of lamprey plasma albumin. Protein Sci 1:289–302
49. Kleinow KM, Nichols JW, Hayton WL, Mckim JM, Barron MG (2008) The toxicokinetics in fish. In: Di Gulio RT, Hinton DE (eds) The toxicology of fishes. CRC Press, Boca Raton
50. Bobe J, Guiguen Y, Fostier A (2010) Diversity and biological significance of sex hormone-binding globulin in fish, an evolutionary perspective. Mol Cell Endocrinol 316:66–78
51. Solange M-Q, Geoffrey LH (2008) Sex Hormone-Binding Globulin in Fish Gills is a Portal for Sex Steroids Breached by Xenobiotics. Endocrinology 149:4269–4275
52. Miguel-Queralt S, Hammond GL (2008) Sex hormone-binding globulin in fish gills is a portal for sex steroids breached by xenobiotics. Endocrinology 149:4269–4275
53. Iyanagi T (2007) Molecular mechanism of phase I and phase II drug-metabolizing enzymes: implications for detoxification. Int Rev Cytol 260:35–112
54. Werck-Reichhart D, Feyereisen R (2000) Cytochromes P450: a success story. Genome Biol 1(6):reviews3003
55. Graham MJ, Lake BG (2008) Induction of drug metabolism: species differences and toxicological relevance. Toxicology 254:184–191
56. Yang X, Zhang B, Molony C, Chudin E, Hao K, Zhu J, Gaedigk A, Suver C, Zhong H, Leeder JS et al (2010) Systematic genetic and genomic analysis of cytochrome P450 enzyme activities in human liver. Genome Res. doi:10.1101/gr.103341.109
57. Nebert DW, Gonzalez FJ (1987) P450 genes: structure, evolution, and regulation. Annu Rev Biochem 56:945–993
58. Schlenk D, Celander M, Gallagher EP, George S, James M, Kullman SW, van den Hurk P, Willett K (2008) Biotransformation in Fishes. In: Hinton DE (ed) Di Giulio DR. The Toxicology of Fishes CRC Press, Boca Raton (FL)
59. Fink-Gremmels J (2008) Implications of hepatic cytochrome P450-related biotransformation processes in veterinary sciences. Eur J Pharmacol 585:502–509
60. Prasad JC, Goldstone JV, Camacho CJ, Vajda S, Stegeman JJ (2007) Ensemble modeling of substrate binding to cytochromes P450: analysis of catalytic differences between CYP1A orthologs. Biochemistry 46:2640–2654
61. Mattingly CJ, Toscano WA (2001) Posttranscriptional silencing of cytochrome P4501A1 (CYP1A1) during zebrafish (*Danio rerio*) development. Dev Dyn 222:645–654
62. Lee MD, Ayanoglu E, Gong L (2006) Drug-induced changes in P450 enzyme expression at the gene expression level: a new dimension to the analysis of drug-drug interactions. Xenobiotica 36:1013–1080
63. Hegelund T, Ottosson K, Radinger M, Tomberg P, Celander MC (2004) Effects of the antifungal imidazole ketoconazole on CYP1A and CYP3A in rainbow trout and killifish. Environ Toxicol Chem 23:1326–1334
64. Blanchard RL, Freimuth RR, Buck J, Weinshilboum RM, Coughtrie MW (2004) A proposed nomenclature system for the cytosolic sulfotransferase (SULT) superfamily. Pharmacogenetics 14:199–211
65. Mackenzie PI, Bock KW, Burchell B, Guillemette C, Ikushiro S, Iyanagi T, Miners JO, Owens IS, Nebert DW (2005) Nomenclature update for the mammalian UDP glycosyltransferase (UGT) gene superfamily. Pharmacogenet Genomics 15:677–685
66. Blanchette B, Feng X, Singh BR (2007) Marine glutathione S-transferases. Mar Biotechnol (NY) 9:513–542

67. Rattner BA, Whitehead MA, Gasper G, Meteyer C, Link WA, Taggart MA, Meharg AA, Pattee OH, Pain DJ (2008) Apparent tolerance of turkey vultures (*Cathartes aura*) to the non-steroidal anti-inflammatory drug Diclofenac. Environ Toxicol Chem 27(11):2341–5

68. van de Waterbeemd H, Gifford E (2003) ADMET in silico modelling: towards prediction paradise? Nat Rev Drug Discov 2:192–204

69. Mohan CG, Gandhi T, Garg D, Shinde R (2007) Computer-assisted methods in chemical toxicity prediction. Mini Rev Med Chem 7:499–507

70. Chang C, Duignan DB, Johnson KD, Lee PH, Cowan GS, Gifford EM, Stankovic CJ, Lepsy CS, Stoner CL (2008) The development and validation of a computational model to predict rat liver microsomal clearance. J Pharm Sci 98(8):2857–67

71. Wishart DS, Knox C, Guo AC, Shrivastava S, Hassanali M, Stothard P, Chang Z, Woolsey J (2006) DrugBank: a comprehensive resource for in silico drug discovery and exploration. Nucleic Acids Res 34:D668–D672

72. Wishart DS, Knox C, Guo AC, Cheng D, Shrivastava S, Tzur D, Gautam B, Hassanali M (2008) DrugBank: a knowledgebase for drugs, drug actions and drug targets. Nucleic Acids Res 36:D901–D906

73. Overington JP, Al-Lazikani B, Hopkins AL (2006) How many drug targets are there? Nat Rev Drug Discov 5:993–996

74. Sakharkar MK, Sakharkar KR, Pervaiz S (2007) Druggability of human disease genes. Int J Biochem Cell Biol 39:1156–1164

75. Bauer RA, Gunther S, Jansen D, Heeger C, Thaben PF, Preissner R (2009) SuperSite: dictionary of metabolite and drug binding sites in proteins. Nucleic Acids Res 37:D195–D200

76. Gong L, Owen RP, Gor W, Altman RB, Klein TE (2008) PharmGKB: an integrated resource of pharmacogenomic data and knowledge. Curr Protoc Bioinformatics Chapter 14:Unit14 17

77. Frolkis A, Knox C, Lim E, Jewison T, Law V, Hau DD, Liu P, Gautam B, Ly S, Guo AC et al (2010) SMPDB: the Small Molecule Pathway Database. Nucleic Acids Res 38:D480–D487

78. Garcia-Reyero N, Adelman I, Liu L, Denslow N (2008) Gene expression profiles of fathead minnows exposed to surface waters above and below a sewage treatment plant in Minnesota. Mar Environ Res 66:134–136

79. Kristiansson E, Asker N, Forlin L, Larsson DGJ (2009) Characterization of the *Zoarces viviparus* liver transcriptome using massively parallel pyrosequencing. BMC Genomics 10:345

80. Woo Y, Affourtit J, Daigle S, Viale A, Johnson K, Naggert J, Churchill G (2004) A comparison of cDNA, oligonucleotide, and Affymetrix GeneChip gene expression microarray platforms. J Biomol Tech 15:276–284

81. Kristiansson E, Sjogren A, Rudemo M, Nerman O (2005) Weighted analysis of paired microarray experiments. Stat Appl Genet Mol Biol 4:Article30

82. Kuo WP, Liu F, Trimarchi J, Punzo C, Lombardi M, Sarang J, Whipple ME, Maysuria M, Serikawa K, Lee SY et al (2006) A sequence-oriented comparison of gene expression measurements across different hybridization-based technologies. Nat Biotechnol 24:832–840

83. Shi L, Reid LH, Jones WD, Shippy R, Warrington JA, Baker SC, Collins PJ, de Longueville F, Kawasaki ES, Lee KY et al (2006) The MicroArray Quality Control (MAQC) project shows inter- and intraplatform reproducibility of gene expression measurements. Nat Biotechnol 24:1151–1161

84. t Hoen PA, Ariyurek Y, Thygesen HH, Vreugdenhil E, Vossen RH, de Menezes RX, Boer JM, van Ommen GJ, den Dunnen JT (2008) Deep sequencing-based expression analysis shows major advances in robustness, resolution and inter-lab portability over five microarray platforms. Nucleic Acids Res 36:141

85. Heckmann LH, Sibly RM, Connon R, Hooper HL, Hutchinson TH, Maund SJ, Hill CJ, Bouetard A, Callaghan A (2008) Systems biology meets stress ecology: linking molecular and organismal stress responses in *Daphnia magna*. Genome Biol 9:R40

86. Curtis RK, Oresic M, Vidal-Puig A (2005) Pathways to the analysis of microarray data. Trends Biotechnol 23:429–435

87. Salomonis N, Hanspers K, Zambon AC, Vranizan K, Lawlor SC, Dahlquist KD, Doniger SW, Stuart J, Conklin BR, Pico AR (2007) GenMAPP 2: new features and resources for pathway analysis. BMC Bioinformatics 8:217
88. Pandey R, Guru RK, Mount DW (2004) Pathway Miner: extracting gene association networks from molecular pathways for predicting the biological significance of gene expression microarray data. Bioinformatics 20:2156–2158
89. Matthews L, Gopinath G, Gillespie M, Caudy M, Croft D, de Bono B, Garapati P, Hemish J, Hermjakob H, Jassal B et al (2009) Reactome knowledgebase of human biological pathways and processes. Nucleic Acids Res 37:D619–D622
90. Thornton JW, Need E, Crews D (2003) Resurrecting the ancestral steroid receptor: ancient origin of estrogen signaling. Science 301:1714–1717
91. Gunnarsson L, Kristiansson E, Forlin L, Nerman O, Larsson DG (2007) Sensitive and robust gene expression changes in fish exposed to estrogen—a microarray approach. BMC Genomics 8:149
92. Kristiansson E, Thorsen M, Tamas MJ, Nerman O (2009) Evolutionary forces act on promoter length: identification of enriched cis-regulatory elements. Mol Biol Evol 26:1299–1307
93. Schmid CD, Perier R, Praz V, Bucher P (2006) EPD in its twentieth year: towards complete promoter coverage of selected model organisms. Nucleic Acids Res 34:D82–D85

A Look Backwards at Environmental Risk Assessment: An Approach to Reconstructing Ecological Exposures

David Lattier, James M. Lazorchak, Florence Fulk, and Mitchell Kostich

Introduction

The primary goal for environmental protection is to eliminate or minimize the exposure of humans and ecosystems to potential contaminants. With the number of environmental contaminants increasing annually, more than 2,000 new chemicals are manufactured or imported each year for use in the USA, understanding the sources of contaminants, the movement of contaminants through environmental media, and the contact of contaminants with humans and ecosystems is critical to advancing environmental protection in the USA. A shift in emphasis from detection of chemical exposure to reconstruction of exposure scenarios will enhance the ability to assess the effectiveness of current environmental regulations and to improve environmental risk assessment for both humans and ecosystems. Exposure reconstruction is a concept that can guide this shift in research focus. Exposure reconstruction, as defined in this chapter, is the characterization of exposures, environmental concentrations, and/or sources from internal biological measurements that are used to inform environmental decision-making (Fig. 1).

This document has been reviewed in accordance with US Environmental Protection Agency policy and approved for publication. Approval does not signify that the contents necessarily reflect the views or policies of the Agency nor does mention of trade names or commercial products constitute endorsement or recommendation for use.

D. Lattier (✉) • J.M. Lazorchak • F. Fulk • M. Kostich
National Exposure Research Laboratory, Ecological Exposure Research Division,
US Environmental Protection Agency, Office of Research and Development,
26 W. Martin Luther King Drive, Cincinnati, OH 45268, USA
e-mail: lattier.david@epa.gov

B.W. Brooks and D.B. Huggett (eds.), *Human Pharmaceuticals in the Environment:* 109
Current and Future Perspectives, Emerging Topics in Ecotoxicology 4,
DOI 10.1007/978-1-4614-3473-3_6, © Springer Science+Business Media, LLC 2012

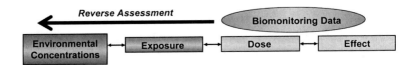

Fig. 1 Exposure reconstruction links an internal biological measurement to an external exposure and environmental concentration.

Background

Currently, information on the exposure to humans and ecosystems by environmental contaminants is primarily limited to biomonitoring studies which mainly collect data on the occurrence of a predetermined list of contaminants in environmental and biological samples (i.e., urine, blood, and tissues). Typically, the goals of national scale biomonitoring studies are to detect contaminant exposure and establish baseline measures, monitor exposure trends and identify geographic hotspots. Little to no information is collected that can contribute to elucidating where the contaminants originated, how they were transported through the environment, what pathways the contaminant took to reach living organisms (i.e., drinking water, food consumption, ambient air) or the routes of exposure by which people and other organisms came into contact with contaminants in question (i.e., ingestion, inhalation, adsorption). These studies are fundamentally unlike investigations of occupational exposure, wherein epidemiological events, usually industrial in nature [1], result in adverse health effects to specific members of a well-defined human population. Retrospective studies in human health, particularly relating to industrial exposures, presumes that the causative agent(s) are known and based on personal usage, proximity to exposures, historical presence and supportable interviews with affected individuals. Human health investigations in the realm of radiation dose metrics [2], and associated risk assessment such as inhalation studies [3, 4] lend support for efforts in *exposure reconstruction*. Epidemiological retrospectives, the underpinning of which is exposure reconstruction, are made possible by the limitless compilation of knowledge about a well characterized species. This cumulative assessment includes behavior and habits of individuals, psychology, medical histories, work histories, and the inestimable physiological, toxicological, cellular, and whole genome information. Ecological exposure reconstruction—illustrated in following pages—is predicated on identical specific aims; however, this undertaking is entirely deficient in the extensive knowledge bases readily accessible to human health investigators.

The Centers for Disease Control and Prevention (CDC) published the *First, Second, and Third Reports on Human Exposure to Environmental Chemicals* [5, 6]. The 2005 report provides exposure biomonitoring data for a representative sample of the US population and targets 148 environmental chemicals commonly found in the environment including lead, pesticides, herbicides, phthalates, polychlorinated biphenyls (PCBs), and polycyclic aromatic hydrocarbons (PAHs).

Of these 148 chemicals, only 25 have established reference values for safe levels of exposure. The current CDC survey gathers very limited exposure data and does not allow for exposure estimates by location or permit identification of sources of contaminants. The USEPA National Study of Chemical Residues in Lake Fish Tissue [7] conducted in 2005 faced similar challenges. One goal of this study was to develop national estimates of the mean levels of 268 persistent, bioaccumulative and toxic chemicals (PBTs) in fish, establishing a national baseline for tracking reductions in PBTs in freshwater fish as a result of pollution control activities. Information on the potential sources of the chemicals or the timing of the exposures of fish populations to PBTs is, by approach and design, absent from the study.

A major finding of a recent 2006 National Research Council (NRC) report on Human Biomonitoring for Environmental Chemicals [8] is the need for more extensive exposure information. The NRC report states that the collection of biomonitoring data and development of biomarkers has outpaced our ability to interpret the data with respect to both potential health effects and in retrospective source tracking. Biomonitoring is an important tool for understanding the linkages between external chemical exposures, internal doses, and potential health outcomes. However, biomarker data independently shows only that humans or organisms were exposed to a chemical at some point in time.

Exposure Reconstruction

Exposure reconstruction is the characterization of exposures, environmental concentrations, and/or sources from internal biological measurements to inform environmental decision-making (Fig. 1). The ability to reconstruct exposure scenarios requires a basic understanding of the relationship between an external exposure concentration and an internal biological measurement. The quantitative relationship between human and ecosystem exposures and biomarkers are estimated using a number of computational tools including physiologically based pharmacokinetic (PBPK) models and empirically based regression models. Exposure reconstruction as a research concept can guide the development of new biomarkers and the design of future biomonitoring studies and ultimately provide a critical component of environmental protection as well as the identification of important sources, pathways, and routes of exposure. There are two broad areas of research to support exposure reconstruction: (1) to leverage existing biomonitoring studies by collection of additional data and enhanced modeling techniques to aid the reconstruction of exposures and (2) to develop new biomarkers that can inform the what, when, where, and how much exposure.

A recent publication [9] provides an example of the first approach. The goal of this study was to evaluate plausible exposure scenarios of humans to chloroform from the activity of showering, consistent with measured concentrations of chloroform in human biomarkers. The authors approach included multiple steps: combined a PBPK model with an exposure model for showering to estimate the intake concentration of chloroform; evaluated the combined model using data from existing biomonitoring studies; developed potential exposure regimens based on typical levels of chloroform in residential water and accounting for multiple exposure routes (i.e., inhalation, dermal, ingestion); estimated distributions of exposure consistent with measured levels of chloroform in human biomarkers by a reverse dosimetry approach with the combined model. The foregoing study highlights the capability and the difficulty in reconstructing exposures from internal biological measurements. The authors were able to demonstrate that inhalation and dermal exposure substantially contributed to total chloroform exposure. However, sources of variability in model output from exposure conditions and from pharmacokinetics were significant.

The second research area that will further the capability to reconstruct exposure scenarios is the development of biomarkers of exposure. A biomarker of exposure, as defined in the NRC report, is the chemical or its metabolite or the product of an interaction between a chemical and some target molecule or cell that is measured in a compartment or an organism. Beyond the NRC definition, in order for an exposure biomarker to be useful in reconstructing exposures it must possess a number of characteristics: (1) specificity; the biomarker must identify a specific compound or class of compounds, if possible by way of mode of action (MOA), (2) sensitivity; the biomarker must be capable of measuring exposures above background levels and at concentrations that are environmentally relevant, (3) reproducibility; biomarker values are reproducible for both environmental sampling and laboratory analysis, (4) validated concentration response; the relationship between the biomarker and the external concentration is validated across a concentration gradient in single or multiple species, the relationship should be validated in both laboratory and field studies which account for physical conditions of the exposure, and (5) knowledge of the exposure kinetics; the biomarker or set of biomarkers are accompanied by an understanding of the kinetics of the exposure (i.e., biomarker level relative to the concentration and timing of exposure). These five characteristics establish a gold standard for biomarkers that will provide critical information to identify the contaminant, the exposure concentration and the timing of exposure. The NRC report recognized that "new technologies in biomonitoring have the potential to transform the nation's capacity to track exposure to pollutants and understand their impact on human (and ecosystem) health" [8]. Biomarkers of exposure developed using advances in genomics technology have the capability to meet the criteria for informing exposure reconstruction. In the area of ecosystem exposure, biomarkers for exposure of aquatic organisms to estrogenic endocrine disrupting chemicals (EDCs) were developed using genomic endpoints [10, 11]. In combination, an upregulated gene followed by production of a protein [12] allow for discrimination between an ongoing exposure (hours to days) from a recent exposure (days to weeks). Applied within the spatial context of a watershed, these biomarkers can inform the timing and the source of exposures to EDCs.

The Challenges of Reconstructing Ecological Exposures

An ecological exposure reconstruction program will aim to accurately identify conditions under which target or other indigenous species in aquatic ecosystems might have been receptors of geospatial point or nonpoint exposures from xenobiotic or natural stressors. This approach, which is not fundamentally dissimilar from that used in human health, presumes identification and quantification of specific stressors and exposure pathways (route and source, or the "what and how"), attempts to delineate exposure chronology and duration, employs strategies and models that will make possible recreation of acute and persistent exposures, integrates exposure and population information, and evaluates historical, current and prospective exposures.

Stepwise, the general approach in ecological exposure reconstruction in aquatic systems employs the same logic tree as do retrospective studies in human health.

- Identification of suspect xenotoxicants; subsequent to determining that an exposure has occurred, use available resources including analytical chemistry, histology, obvious physiological or behavioral aberrancies, ecological and community assemblages and toxicological data bases to limit the possible exposure agents—the assembled data conceivably leading to source.
- Identify the relevant ecological pathways and media of interest; if pathway not immediately obvious, such as known point source, invokes use of GIS and other spatial data including fate and transport models to reconstruct toxicant entry into watershed, such as the case of agrichemicals entering a watershed during a precipitation event. Presumption that application rates of agricultural chemicals comply with listed standards.
- Estimate external concentrations of toxicants in media at target sites and, if time-dependent bioindicators (molecular and cellular, histopathology, community assemblages) are available, approximate window of time(s), including duration of exposure.
- Establish routes of exposure; organisms in aquatic ecosystems, ingestion or absorption by way of physical contact.
- If indicated, by point source release or nonpoint source entry into water course (rain events, inadvertent spillage) develop cumulative exposure estimates during speculative time periods.
- Calculate internal dose for suspect xenotoxicant or, if determined, toxicant mixtures; given multifarious physical parameters, described presently, determination of internal dose presents the most rigorous scientific burden to the reconstructive process; however, using all available resources, and replicating physical and geochemical conditions in surrogate ecosystems, internal dose can be resolved using precise quantification of induced mRNA transcripts ("real-time" quantitative PCR) from a narrow set of mode of action-specific genes.

Molecular indicators (genomics, transcriptomics, proteomics and metabolomics) will play a crucial role in reconstruction of exposure events; however, these courses of action and associated technologies are incapable of functioning as the "stand alone" analytical scheme. The limiting factor for applied molecular biology in the realm aquatic ecosystems is the shortfall of knowledge with respect to genomes of nearly all species that inhabit inland surface waters; however, as cellular and molecular profiles for species of interest become more inclusive, use of molecular indicators for reconstructing ecological exposures will evolve into a pivotal asset. The nature of exposure reconstruction makes a case for a multifaceted scientific partnership, using analytical and descriptive assessments to accurately recreate an exposure episode. Meticulous reconstruction must first begin with analytical chemistry in order to exclude causal factors such as natural stressors or habitat demise. Included in the initial evaluation will also be watershed and ecological characterization such as analyses of community assemblages and Index of Biotic Integrity (IBI), remote imagery and GIS, and measures of genetic diversity within communities (biodiversity metrics, Fig. 2). Additionally, this undertaking must rely on available models (hydrogeologic, soil physics, fate and transport) for spatial depiction of contaminant deposition—particularly when considering exposure from agricultural chemicals and other possible nonpoint sources. In addition, the ER framework could be used to determine whether a chemical compound, detectable by analytical methods, contributes to observed physical and biologic effects in an aquatic ecosystem and, if so, the relative bioavailability of the suspect compound in a given physical or trophic state. As cited anecdotally in abundant environmental reviews, the average person is exposed, by way of dermal contact, diet, inhalation, etc., to about 10,000 discrete chemicals per day. The analogous argument could also be made regarding scores of organisms that inhabit aquatic ecosystems throughout the globe. One of the primary challenges in the area of ecosystems biomonitoring is to consider disparity in species sensitivities to myriad stressors and choose the appropriate organism with which to apply the correct suite of analyses for addressing a specific or suspected condition.

Fish populations have been steadily declining in second order Canyon River, but not in nearby Cottonwood River, also a second order stream located in the same watershed. Histology and biochemistry show signs of thyroid dysfunction in Canyon River fish, but not fish resident in Cottonwood River. Prolonged laboratory exposures using water from Canyon River and the standard aquatic toxicological model *Pimephales*, indicate thyroid histopathology and changes in thyroid hormone levels in fish. Concurrent laboratory exposures with water collected from Cottonwood River exhibit normal thyroid histology and function. Data suggests that xenobiotic stressor(s) in the water of Canyon River is damaging thyroid tissue in native fish populations and may contribute to the observed population declines. Chemical analysis of suspect water fails to reveal the presence of chemicals with known thyroid toxicity. Chemical scans show the presence of multiple minor peak differences between Canyon River and Cottonwood River, but the exact identity and potential toxicity of these peaks is not known. Exposure to primary cultures of fish thyroid cells of water from Canyon River, using a microarray platform and 2D gel protein analyses suggest induction of genes associated with cell death (apoptosis). Based on these first approximation results, narrow-spectrum molecular methods, such as

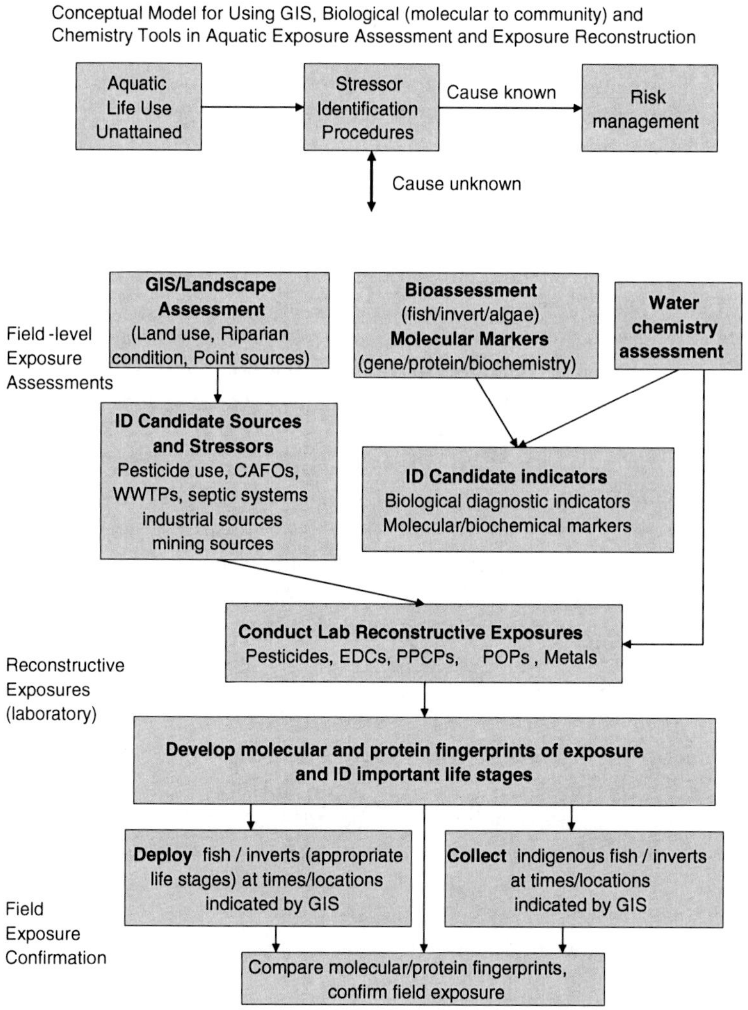

Conceptual Model for Using GIS, Biological (molecular to community) and
Chemistry Tools in Aquatic Exposure Assessment and Exposure Reconstruction

Fig. 2 Scenarios in exposure reconstruction that depict specific applications of molecular indicators in lotic environments

"real-time" PCR using transcription-specific synthetic oligonucleotides and apoptotic antibodies, are developed to track the newly identified biomarker genes in the thyroid cell assay. Successive chemical fractionation is performed on the water from Canyon River, with the focused biomolecular assays applied to monitor the active biologic fraction at each step, until individual active peaks are identified recovered. Analytical chemistry is used to identify the structure of the material in each active peak. The identity is confirmed by synthesizing the compound and demonstrating an identical profile using analytical chemistry, and replicating toxic effects at relevant concentrations, first in the inexpensive cell-based assay, followed by whole animal models. The suspect toxicant is added to water from Cottonwood River at concentrations originally identified in Canyon River, and the toxic potency of the spiked control is compared to the toxic potency of Canyon River water, first in cell-based assays, then in whole animal tests. The goal here is to determine what fraction of the total bioactivity related to Canyon River is attributable to the analytical concentrations of the suspect toxicant. Subsequent to establishment of the toxicity factor, source tracking—by way of detection chemistry—would comprise the next step, thereby paving the way for resource management teams to make informed decisions about hazard identification, mitigation or remediation. Note that, for this scenario, biomolecular methods are used to follow biologic activity, and not to determine chemical structure directly. Chemical analyses perform the complementary role of chemical structure determination, but have nothing to say about biologic activity or bioavailability. The two methods must be used in tandem to solve the problem. Molecular readouts are used in preference to whole animal readouts of biological activity because of (a) reduced sample quantities required—a critical factor during chemical fractionation, in which many fractions will be produced with limited volumes of active material—a volume vastly insufficient to reconstitute the liters of exposure medium needed for whole animal testing; cell-based methods require reconstitution of 1 mL or less (6^+ orders of magnitude less medium required); (b) reduced cost of a 24-well or 96-well based assay compared to a whole animal assay; (c) reduced whole animal testing—a worthy goal in itself, and the original objective of the United States Environmental Protection Agency Computational Toxicology initiative.

A similar scenario can be imagined where tumors are seen in fish from a particular river, and analytical chemistry fails to detect any of the "usual suspects" in the water collected from the site of biologic impairment. A similar fractionation/biomolecular readout approach can be used, except that this time the molecular readout will likely be different. Perhaps the comet assay, differential expression of one or more DNA repair enzymes, or induction of S-phase associated genes, will serve as a more appropriate indicator. The initial wide-spectrum molecular scans will facilitate selection of the indicator genes of interest, which then focal by employing low-cost, high-throughput assays for future tracking of relevant activity.

The foregoing two environmental monitoring scenarios aim to associate observed adverse outcomes to specific environmental exposures. When histopathology or any number of aberrant physiological traits—including behavior of individuals and populations—can be inexorably linked to a toxicant or xenobiotic stressor, then a point of *phenotypic anchoring* [13, 14] can be established as one component biomarker for exposure reconstruction in future monitoring activities. Environmentally induced temporal changes in gene expression in the context of conventional toxicology endpoints can make possible the anchoring of phenotypes as a consistent, valid biomarker in occurrences of ecological exposures.

A persistent, real world problem is posed by the estrogenic activity that are frequently a complement to posttreatment effluent released by waste water treatment plants (WWTPs), as well as other point and nonpoint sources. This is of concern at least in part because of the observation of declining populations, and histopathology suggestive of feminizing effects observed in fish downstream of some WWTPs compared to upstream organisms. Samples of WWTP effluent transported to laboratories have been shown to inhibit fish fertility in short-term reproductive assays. Additionally, as an initial screen for estrogenic potential in surface waters, we have developed the simple approach of thermally amplifying (PCR) estrogen induced transcripts of the vitellogenin gene in numerous species of freshwater teleosts [15, 16]. Because the vitellogenin gene is normally quiescent in males, detection of the gender-specific transcribed gene product (i.e., *vtg* mRNA[1]) provides a sensitive exposure marker for environmentally present estrogenic compounds. The gene for vitellogenin is categorically unique in expression and function, and discovery of other bioindicators with similar biologic profiles, such as gender-specific expression induced by a restricted mode of action, will likely be few and far between at best.

Chemistry analysis often indicates the presence of several known estrogenic substances in WWTP effluents, including the natural estrogens, estrone (E1), 17β-estradiol (E2), and the synthetic contraceptive estrogen, 17α-ethynylestradiol (EE2). These substances have been shown to result in adverse reproductive effects in laboratory validations using concentrations at which they occur in effluents. Does this mean we should embark on a strategy to reduce the levels of these three compounds in WWTP effluents? Will the substantial investment pay off in terms of improved wildlife health? One important question that needs to be answered is what fraction of the total estrogenic activity, released in effluent by the WWTP, is attributable to customary estrogenic substances. Several studies have suggested presence of other estrogenic materials in WWTP effluents, including nonylphenol and associated ethoxylates, congeners of PCB, certain metals, and dioxins. Additionally, estrogenic compounds, not amenable to analytic detection, might also be present in a given effluent. For the sake of all stakeholders involved, it is important to determine the fractional contribution of each conditional estrogen until the greater part of aggregate estrogenic activity in effluents has been clarified. Only then can an appropriate mitigation strategy can be implemented. Otherwise, a method may be chosen that addresses only a fraction of the problem (i.e., the structurally related steroidal estrogens E1, E2, and EE2, but not the structurally distant nonsteroidal estrogenic compounds), costing much, but producing minimal management benefit. A current approach to addressing the question of fractional analyses exploits estrogen-responsive cell lines transfected with reporter genes driven by estrogen responsive *cis*-active elements. This reporter system can effectively be used to measure the intrinsic estrogenicity of chemicals alone and in combination. The anticipated outcome is that in vitro systems can be used to characterize the fractional contribution of each known estrogenic agent, and also be used for isolation of previously uncharacterized estrogenic agents present in effluents. The main advantages offered by these

---

[1] Designation of macromolecular products: *vtg*, transcribed vitellogenin gene product (mRNA) and *Vtg*, circulating vitellogenin protein, follows the Zebrafish Nomenclature Guidelines, based on *Trends in Genetics* Genetic Nomenclature Guide (1998), found at http://zfin.org/zf_info/nomen. html#1.

cell-based molecular methods, when compared to whole animal studies, are low cost, speed of analysis, and limiting volumes of media required for analyses.

Besides this sort of retrospective exposure research, cell-lines may also prove useful for first pass high-throughput screening of new HPV industrial chemicals for estrogenic activity. Eventually, the development of batteries of molecular assays covering the most frequently encountered modes of toxicity seems like a sensible approach to preliminary screening of new chemicals (giving some hint of the relevance of predicted exposure levels). Such batteries of molecular assays can also help quickly and cheaply identify modes of toxicity and relevant molecular indicators for use during forensic and retrospective analyses intended to characterize the exposures causing observed population effects as noted in the previous three illustrations.

The argument for "scaled-down" and focused schemes of analyses would benefit greatly by use of embryos and early developmental stages (fish, amphibians, etc.) as a point of departure for developing exposure biomarkers. There are substantial numbers of similarities between fish and mammals in relation to developmental pathways, with approximately 75% of developmentally specific genes being homologous across metazoa. In addition to exploiting developmental plasticity for biomarker development and effects forecasting, physical exposures to early developmental stages require minimal experimental resources and reagents, resulting in significant cost savings. Additionally, this approach will comply with the charge of moving away from whole animal testing and use of adult animals as models for exposure.

Vitellogenin, the Answer in an Egg Shell; Once in a Genome Opportunity

Vitellogenin is an established and sensitive endpoint for analysis of exposure to estrogens, androgens and respective mimics in fish [10, 15, 17, 18]. There are several studies that have demonstrated links between high level induction of *vtg* and effects in fish [19–22]. One of the most popular test fish species for assessing chemical effects is the fathead minnow (*Pimephales promelas*, FHM), which is now used widely for studies into endocrine disruption [23, 24]. There is now unequivocal evidence showing that EDCs can have long-term effects on reproduction and subsequent population development in natural fish populations [21, 25].

Male fish downstream of some wastewater outfalls produce vitellogenin protein (Vtg), a protein normally synthesized by females during oocyte maturation, in addition to early stage eggs in their testes, and this feminization has been attributed to the presence of estrogenic substances such as natural estrogens (estrone or 17β-estradiol (E2)), the synthetic estrogen used in birth control pills (17α-ethynylestradiol (EE2)), or weaker estrogen mimics such as nonylphenol in the water. Despite widespread evidence that male fishes are being feminized, it is not known whether these low-level, chronic exposures adversely impact the sustainability of wild populations.

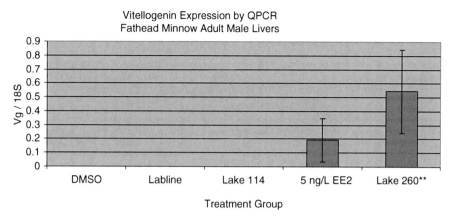

Fig. 3 Vitellogenin gene expression results from exposing male fathead minnows for 24 h to water collected from the Experimental Lake Area EE2 study, Lake 114 nondosed lake and Lake 260 EE2 dosed lake. Compared to a 5 ng L^{-1} (nominal) EE2 positive control **, $n=4$

A 7-year whole-lake experiment was conducted at the Experimental Lakes Area (ELA) in northwestern Ontario, Canada [21], which demonstrated that chronic exposure of fathead minnow (*P. promelas*) to low concentrations (5-6 ng L^{-1}) of the potent synthetic estrogen EE2, led to feminization of males through the production of *vtg* mRNA and protein, impacts on gonadal development as evidenced by intersex in males and altered oogenesis in females, and ultimately, a near extinction of this species from the lake. These observations demonstrate that the concentrations of estrogens and estrogenic mimics detected in freshwaters can impact the sustainability of wild fish populations.

There were several confirmation studies that were performed concurrent with the ELA study to demonstrate the utility of using *vtg* as a gene marker for exposure to estrogens in male fathead minnows. Figure 3 presents the *vtg* expression results of laboratory (USEPA Cincinnati Aquatic Facility) reared male fathead minnows exposed to water shipped from to the Cincinnati facility during the first year of lake dosing with EE2. Lake 114 was one of two control lakes where no EE2 was introduced and Lake 260 is the lake dosed with EE2. A positive control concentration of 5 ng L^{-1} EE2 was also tested. The target EE2 concentration for Lake 260 was 5 ng L^{-1} and measured concentrations during the first year were 6.1 ng L^{-1} (SD ± 2.8 ng L^{-1}). Males exposed to lab water, water with DMSO, and Lake 114 water showed no expression of the vitellogenin gene (Fig. 3). Males exposed to Lake 260 water showed extensive increase in levels of vitellogenin gene expression, even higher in some cases than males exposed to 5.0 ng L^{-1} EE2. Males had variable response to both the 5.0 ng L^{-1} water and the Lake 260 water. Two of the fish exposed to 5.0 ng L^{-1} EE2 showed no increase in expression of vitellogenin. One male exposed to water from Lake 260 showed no expression, while two had expression levels comparable to that of fish exposed to 5.0 ng L^{-1} EE2. Gene expression results were found to be very similar in male fathead minnows exposed to both pure EE2 at 5.0 ng L^{-1} and water from Lake 260 at a concentration of 6.18 ng L^{-1} EE2.

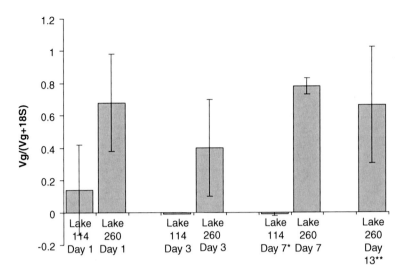

Fig. 4 Results of deploying indigenous fathead minnows from Lake 114* (only 7 fish left) non dosed lake into Lake 260 EE2 dosed Lake for 13 days** (only 4 fish left)

A second experiment was conducted in which indigenous fathead minnows were collected using minnow traps in reference Lake 114 two days prior to deployment. Males and females were housed together until the day of deployment, at which time the sexes were separated. Only males were used in the deployment study. Males were deployed in cages in Lakes 114 and 260. The cages were suspended several feet below the water surface and held in place by anchors and buoys. Fish were provided no food during the period of deployment. The cages allowed for the free movement of water and suspended materials. Minnows were retrieved from cages on days 1, 3, 7 and 13 of deployment. Agarose gel-based RT-PCR was performed on these samples. There was an insufficient number of fish to continue the study in Lake 114 through day 14, so all seven remaining fish were removed on day 7. The experiment was also terminated early in Lake 260 since only four fish were remaining on day 13. Male fathead minnows exhibited an increase in vitellogenin mRNA levels after only 1 day of deployment in Lake 260 (Fig. 4). Vitellogenin mRNA levels remained high throughout the study to day 13. Response to EE2 by males was variable, with some fish showing high levels of expression and others showing very little expression. The standard deviations for these samples are quite high. Males in Lake 114 showed no significant expression on days 3 and 7 (Fig. 4). However, on day 1 there was a single fish in Lake 114 that had elevated levels of vitellogenin mRNA. The other four fish showed no Vg expression.

In order to determine the kinetics of vitellogenin expression during the initial period of exposure in 2001, male FHM were collected from Lake 260 after 7 weeks, 9 weeks and 3 months of dosing. Male fathead minnow were collected at the same time from reference Lake 114. Agarose gel-based RT-PCR was performed on samples and vitellogenin expression quantified relative to 18S ribosomal RNA (rRNA)

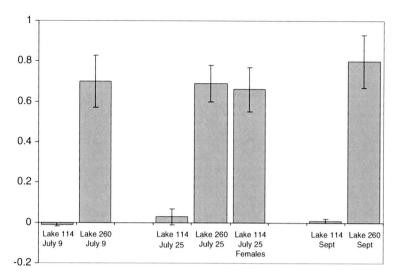

Fig. 5 2001 Summer and fall results of indigenous male and female fathead minnows

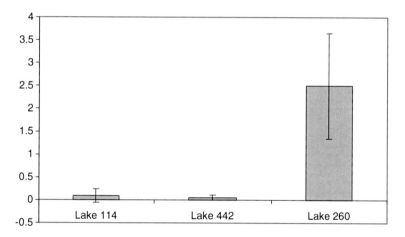

Fig. 6 Vitellogenin gene expression in male pearl dace collected in May 2003

expression. Vitellogenin was induced in males collected from Lake 260 at all time points (Fig. 5). Males collected from Lake 114 had little to no vitellogenin mRNA. Vitellogenin expression in males was comparable to that of females collected from Lake 114 on July 25. The level of expression of vitellogenin in male fathead min-nows collected from Lake 260 was statistically different from that of males from Lake 114.

Similar results were found in male Pearl Dace collected from reference lakes 114 and 440 and dosed Lake 260 in 2003 (Fig. 6). One interesting result found in both female Pearl Dace and fathead minnows was an increased level of Vg gene

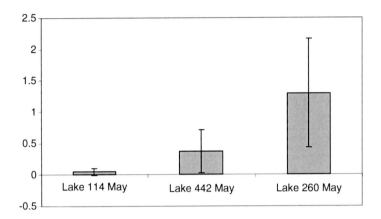

Fig. 7 Vitellogenin gene expression in female pearl dace collected in 2003

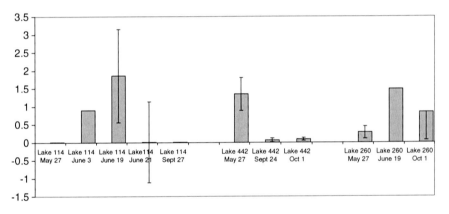

Fig. 8 Vitellogenin gene expression in female fathead minnows collected in 2002

expression compared to that observed in reference lakes (Figs. 7 and 8). Female FHM from Lake 260 had elevated Vg levels beyond the spawning season, September and October 2002 (Fig. 8).

Kidd et al. [21] published results of whole fish homogenates analyzed for vitellogenin protein for fish collected during the same time as those analyzed for vitellogenin gene expression. Whole fish vitellogenin protein analyses showed results similar to found using the gene expression assay. Male and female fathead minnows collected from the EE2 dosed lake had elevated levels of vitellogenin protein when compared to reference Lakes, 114 and 442—most strikingly during the second and third year of whole lake dosing, years 2002 and 2003.

Kidd et al. [21] also presented histological results of fish collected during similar times as those analyzed for vitellogenin gene expression and whole fish homogenate vitellogenin protein. Testicular tissues of all of the male fathead minnow collected the first spring after EE2 additions began displaying delayed spermatogenesis, widespread fibrosis and malformations of the tubules. Testicular germinal tissue from all EE2 exposed males consisted primarily of spermatogonia instead of the spermatocytes that would be the norm during the given time of year. Gonad size in the spring of 2002 averaged $0.40 \pm 0.21\%$ ($n = 10$) of the body weight for fathead minnow from Lake 260. This was approximately one third of the mean value for reference Lake 114 fish ($1.39 \pm 0.38\%$; $n = 15$) and only one fifth of that from reference Lake 442 fish ($2.27 \pm 0.41\%$; $n = 10$) collected at the same time of year [21].

Kidd et al. [21] published results that showed the fathead minnow population in Lake 260 collapsed in the fall of 2002, after the second season of EE2 additions, because of a loss of young-of-the-year. This reproductive failure was also observed in the third season of amendments and continued for an additional 2 years after the EE2 additions had ceased, although a few small individuals were caught each year, indicating some reproduction was occurring. The loss of smaller size classes of fathead minnow was not observed in reference Lake 442, a system with similar species composition, water volume, and trophic status.

The whole lake dosing experiment demonstrates the relationship of estrogenic exposure from the molecular level to population effects. It illustrates that gene expression can be exploited as an earlier indicator of exposure and an ecologically relevant tool that can be used to reconstruct exposures to endocrine disrupting compounds.

In 2005 EPA examined the results of exposure to 17α-ethynylestradiol in the Experimental Streams Facility (ESF) located in Milford, Ohio. ESF has channels that are fed continuously with water from the East Fork of the Little Miami River, southwestern Ohio. Streams were dosed with three concentrations of EE2; 2.5, 12.5, and 62.5 ng L^{-1}. Male FHM were placed in minnow traps in the tail tanks (located at the terminus of each stream), of each dosed stream, and a non dosed stream, for a period of 4 days. In addition, male fathead minnows were placed in minnow traps in a ditch outside the facility that receives water from all ESF channels to assess whether EE2 was being discharged to the East Fork of the Little Miami River by way of the receiving stream. This ditch discharges into the effluent of a nearby WWTP, so fish were also placed in the effluent of the WWTP downstream following mixture with ditch water, and in the East Fork of the Little Miami River below the WWTP discharge. Figure 9 contains vitellogenin gene expression profiles of liver-specific mRNA collected from this study. Vitellogenin gene transcription indicates a dose-dependent response to experimental concentrations of EE2, intermediate response from exposure to ditch water and no expression in the WWTP or receiving stream.

Our final example demonstrates that exposure-induced vitellogenin gene expression has potential to assess estrogenicity in receiving streams below potential sources of estrogenic compounds. Figure 10 contains the results of a study conducted in the Eagle Creek Watershed, near Indianapolis Indiana. Male fathead minnows were caged for 7 days below two effluents, sites 2 and 4, and an animal feedlot, Site 1 during spring and fall of 2008. The two dotted lines illustrate the EE2 equivalents of 2.5 and 5.0 ng L^{-1} as estimated from laboratory studies. The results are not indicative of the presence of estrogenic compounds. Chemical analyses were

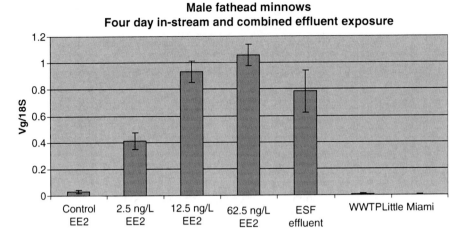

Fig. 9 Results of a 4-day EE2 exposure to male fathead minnows in EPA's Experimental Stream Facility and outside deployments to monitor potential estrogenic discharge to Little Miami River

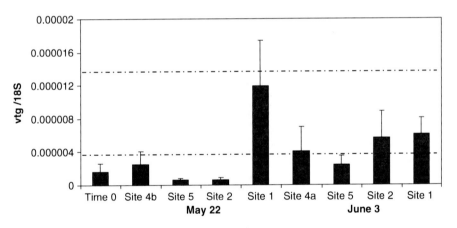

Fig. 10 Results of 7-day deployments of male fathead minnows within the Eagle Creek Watershed in spring and summer. *Top dashed line* indicates 5 ng EE2 L^{-1} equivalent exposure lower dashed line indicates 2.5 ng EE2 L^{-1} equivalent exposure

performed on water samples collected during these exposures; however neither EE2 nor estradiol (E2) was detected. The ramification of this study is the possibility for biologic detection of estrogenicity in the absence of detectable chemical analyses. This could indicate mixtures of estrogenic compounds that might be less than limits of chemical detection but not below the necessary level to induce vitellogenin gene expression.

The studies outlined in this section are the support and foundation for the concept of using gene expression as an approach to reconstructing exposures to estrogenic compounds. Current research is underway using a toxicity identification evaluation (TIE) approach to identify estrogenic compounds using gene arrays. Specific gene fingerprints of these compounds may be developed that can be used on effluents and surface waters found to be estrogenic to identify the specific chemical or chemicals. In turn this information can be used to help develop a control strategy to reduce concentrations of estrogenic compounds in the environment.

Acknowledging the Dynamics and Complexities

Exposure is the result of a stressor intersecting with a receptor. To understand the phenomenon of exposure we must consider the characteristics and behavior of the stressor and the characteristics and behavior of the receptor. For example, once the stressor is released into the environment physical conditions and variables move it across and through the landscape, during which time the stressor may be transformed by myriad processes, including chemical, photo-, and microbiological influences. Conversely the receptor might also move across the landscape and acquire morphological or physiological adaptations at various life stages that confer protection from stressors by way of induced polyphenisms or, in some cases in the course of early development, succumb to greater vulnerability and mortality. Environmentally induced polyphenisms in early development [26, 27], can yield a consequence of population-specific exposure threshold. These phenomena, in conjunction with unknown and possible highly variable genetic backgrounds of geographically distinct communities, further confounds molecular measurements of ecological exposure to parallel stressors—especially in populations where polyphenisms modulate susceptibility to exposure. Reconstructing ecological exposure is to attempt an understanding of the history, genetics and inestimable processes that influenced the fateful juncture of a deleterious stressor and populations of biologic receptors. Subsequent to exposure, the risk of an adverse outcome to wildlife communities and populations is a function of magnitude, frequency, duration, and cumulative aspects of exposure—all of which are amenable to retrospective inference using existing models.

For exploitation of any ecologically based biomarker, there is an absolute need to consider geophysical parameters and nutrient conditions when applying indicators to risk analyses of environmental toxicants. Attempting to establish a relationship between biomarker and external concentration of contaminant, for retrospective analysis or immediate survey, investigators have an obligation to consider and characterize the oftentimes disregarded temporal, spatial and geochemical and other physical conditions of the proximal study area [28]. Physical conditions that have overwhelming implication on bioavailability of xenobiotic stressors, in field and laboratory studies [29] include, but are not limited to, total nitrogen and phosphorus, pH, dissolved oxygen, alkalinity, dissolved minerals, microbiologic communities, temperature, turbidity and regional meteorology.

In a recent study using surrogate ecosystems, we investigated the effect on relative trophic levels on EE2 induced expression of the gene for vitellogenin in fathead minnows [30]. Male fathead minnows, were exposed to a single dose of 17α-ethynylestradiol at a nominal concentration of 20 ng EE2 L^{-1} in fiberglass mesocosm tanks containing either a carrier vehicle (DMSO) or water control or secondary nutrient treatment reflecting oligotrophic (0.012 mg L^{-1} total phosphorous (TP)), mesotrophic (0.025 mg L^{-1} TP), or eutrophic systems (0.045 mg L^{-1} TP). A total of 21 tanks were used in a random treatment/control design of three replicates per treatment/control. In preconditioned mesocosms that were designed to replicate 17α-ethynylestradiol (EE2) exposure in respective trophic systems, results suggested that the level of vitellogenin gene expression was inversely related to the nutrient load, with the highest expression observed in mesocosms lacking in plant nutrients and having an abundance of dissolved oxygen.

Given the dynamic nature of primary productivity in aquatic ecosystems, investigators who embark upon exposure reconstruction and retrospective quantification of external dose must rely heavily on fate and transport models and climatological records, in addition to taking into account the physical–chemical parameters of the system (watershed). Such might be the case in a scenario of recent application of one of the most widely used herbicides in the United States, followed closely by a precipitation event. Atrazine (2-chloro-4-ethylamino-6-isopropylamino-s-triazine) is arguably one of the most pervasive stressors detected in groundwater and aquatic ecosystems throughout lower 48 states [31]. The regulated concentration of this herbicide in US drinking water is 3 ppb (3 μg L^{-1}); however, in stream levels on the order of 224 ppb have been reported in areas dominated primarily by agricultural activities.

Although previous studies have suggested that atrazine is not estrogenic by virtue of failure to induce vitellogenin transcription in mature male goldfish (*Carassius auratus*) [32], the controversial chemical nevertheless has been determined to be an endocrine disrupting compound (EDC). Recent investigations have indicated that atrazine binds directly to steroidogenic factor-1 (SF-1), an "orphan" receptor, facilitating enhanced SF-1 binding to the aromatase promoter [33], resulting in over expression of the aromatase gene. Data supports the long held speculation that atrazine functions as an endocrine disruptor in wildlife, with possibilities of reproductive impairment and skewed gender ratios, and is capable of initiating reproductive neoplasia in experimental animals and multiple human cell lines. By way of latest data, reconstructing what is presumed to be atrazine exposure might transpire similarly to the previously described lotic environment scenarios, again using GIS, fate models, chemistry and aromatase (CYP19) expression as the initial screening biomarker.

Down the Primrose Pathway

Use of nucleic acid technologies, including thermal cycle amplification (PCR) and microarray hybridization, to inform ecological exposure and hazard assessment has added significantly to the observations rendered by toxicological testing over the past 3 decades; however, the collective benefits might not shift the balance from pitfalls and indiscreet interpretation of biologic observations. The promise to inform ecological risk assessment with a molecular crystal ball that would allow us to faithfully divine clear-cut linkages between molecular triggers initiating toxic pathways, and biologically relevant adverse endpoints, has largely been caught up in myriad complex interactions that occur as organisms simply attempt to make it through another day in an environmental mélange. This combined with the anthropocentric supposition that gene expression is necessarily the lone inevitable consequence of antecedent environmental events resulting in all biologic and phenotypic outcomes. Exposure outcomes represent the complex interplay between genetics, the action of many genes, behavior and the environment.

Earlier the suggestion was made that when attempting to describe biologic processes at any level of organization, investigators are advised to consider combined hydrogeologic and geophysical parameters—replicating them in laboratory/mesocosm studies to the extent feasible. In combination, and absence of xenotoxicant agents, physical and geological factors are, indeed, a mixture of individual stressors—continually altering molecular responses based on the ever-changing nature of ecological systems. Adding to this potential exposure medium are one or more synthetic xenoestrogens; then, exposure-specific transcript patterns in any number of aquatic species arising from microarray analysis become theoretically untenable. The challenge that ecotoxicogenomics has yet to meet is discrimination among the biologic networks and gene sets that account for preponderance of physiological stress, directing a homeostatic condition to conclude in allostatic overload. This is no small informatics feat given the number of biologic influences. The above difficulties do not include all too common technical discrepancies such as reproducibility, use of multiple array platforms, and transformation of raw data.

In the sphere of ecological genomics, much has been noted regarding compensatory and adaptive responses to xenobiotic exposure. Recent finding in *Daphnia magna* suggest [34] that local environmental conditions can lead to genetic adaptation of natural populations. Demonstrable selection pressure occurring in a local habitat suggests that physical conditions with added stress of xenotoxicants might significantly contribute to genetic attrition in natural populations of *Daphnia*. This phenomenon further exemplifies investigators' need to be judicious when parsing data from differential gene expression profiles.

The concept of hormesis has in recent years resurfaced as a general model for physiological response to exposure. There is vociferous argument that most, if not all, experimental exposures in animal models occur as a function of hormesis—the endpoints described by either resulting in either a J-shaped or an inverted-U dose–response curve [35]. Results from experiments using synthetic estrogenic compounds dispute the classic notion of hormesis as an adaptive response [36]. The investigators argue that in the case of manmade xenoestrogens exposure, observed apical endpoints result from an adverse stimulatory response, radically dissimilar from the notion that toxicity pathways, elicited by exposures over a range of concentrations, make corrections for low level internal doses. Outcomes of adverse stimulatory responses are detrimental to populations and communities, redirecting energy essential for homeostatic processes which results in diminished fitness. If such adverse response is the case in human health as well as ecological exposures; this concept will undoubtedly confront current means of performing risk assessments. Mechanisms accounting for such dose related biologic activity have been described for numerous pharmacologic agents at the level of intercellular receptors in mammalian models. Research finding to date have yet to suggest that there is a specific hormetic mechanism; however, such conclusions offer strong inference for an all-purpose strategy the aims of which are conservation of resources across biological systems. Considering increasing numbers of natural habitats in decline throughout past decades, resource conservation would have far reaching implications for xenobiotic exposures in aquatic and terrestrial ecosystems. Hormesis indicates that potential lack of correspondence among gene networks at low and high doses might lead to altered strategies for biomarker development, in addition to a radical change regarding the way in which the business of risk assessment is conducted.

The path from ecological genomics to resource management and regulatory policy is hindered for the time being by a seemingly insurmountable gap of uncertainty. Mindful ecological monitoring and retrospective ecological analyses will serve to reduce uncertainty in exposure science. Linking available knowledge bases including those containing characterizations on physiology, cellular and molecular mechanisms, histopathology and behavior, in conjunction with spatial information, will assist in narrowing the number of candidate toxicants, in addition to providing possible scientific insights regarding timing and duration of exposure. As a point of departure, investigators engaged in ecological monitoring could begin to assemble an inventory of preliminary molecular biomarkers by exposing aquatic test model organisms, in mesocosms, to high production volume chemicals (agrochemicals, etc.) or chemicals of emerging concern (CECs; pharmaceuticals) to discriminate exposure-specific gene expression markers. Exposure media that reflects known degrees of primary productivity (total nitrogen and phosphorus; oligo-, meso- and eutrophic systems), might make possible identification of a restricted suite of primary and correlative mode of action gene-based biomarkers. Using microarray platforms to discern patterns of transcription or for isolation of individual gene products from which synthetic PCR oligonucleotides could be generated; this collection of indicators could then serve as the initial screening line up to taper suspected agents of exposure in a reconstructive scenario. The use of whole genome microarrays for initial site-specific exposure surveys might, by virtue of environmental complexity, portray more qualitative profiles; however, distillation of results will in all probability lead to platforms designed for more targeted transcriptomic read outs.

Beyond Genes and Proteins

One of the main objectives of exposure reconstruction is estimation of the temporal aspects of exposure. In all likelihood, only a small set of molecular processes have the capability to reveal the duration component of ecological exposure—the gene and protein for vitellogenin, a structural precursor to yolk protein in oviparous animals, being one of the current most notable examples. As seen in Figs. 2 and 3, induced transcription of the vitellogenin (*vtg*) gene detects the immediacy of exposure to environmental estrogens or estrogen mimics. Circulating vitellogenin protein, following downstream translation of the processed message, is detected approximately 14 days post transcription. Relative numbers of vitellogenin transcripts increase proportionally with concentration of estrogenic compound and with time. Magnitude of response can be indicative of either. Absolute quantification of induced vitellogenin transcripts permits investigators to back calculate from biomarker to presumptive concentration in EE2 equivalents. If there is a presumption that a xenoestrogenic exposure event has occurred in an aquatic system and the liver-specific gene indicates no active transcription, then investigators would test for the circulating protein using an available ELISA technique. This then allows for a possible first approximation of external concentration in addition to time at which exposure to the estrogenic compound occurred. Again, this temporal estimate must be made in context of other available data, such as identified point-sources, recent meteorological events and regional land use information such as seasonal application of agrichemicals. In cases such as this, investigators and modelers must assume the roles of *ecosleuths*™.

Epigenetics; Methylation and microRNA

Epigenetics represents most recently observed cellular phenomena destined to transform the landscape of exposure science, particularly as related to ecotoxicology and biomarker discovery in aquatic sentinels. The term "epigenetic" refers to heritable changes in gene expression that occur in the absence of structural modifications in sequence of DNA. These heritable instructions in transcriptional machinery can become fixed in the genome mitotically, meiotically, or during both genetic events [37, 38]. Epigenetic processes, which are reflected in levels of gene expression, comprise the following known mechanisms; nucleotide-specific DNA methylation, modifications in chromatin structure mediated by both acetylation and methylation of histone proteins, and the expression of noncoding, small (micro) RNAs (miRNAs) in posttranscriptional regulatory control programs.

Given the suite of diverse epigenetic mechanisms, the one most likely to yield experimental clues to immediate or preceding xenobiotic exposure, is methylation of DNA at the cytosine-5 site within CpG dinucleotides (CpG islands)—particularly in 5′ upstream promoter regions of transcriptional units, and *cis*-regulatory elements that often function distally with respect to the site of transcription initiation. Methylated CpG dinucleotides establish distinguishable epigenetic features that are commonly associated with transcriptionally silent, condensed chromatin. Such biochemical modifications generate genomically spatial and functional controls that harmonize with *trans*-regulatory mechanisms. A straightforward description of methylated promoters would posit that methyl groups projecting from abundant cytosine nucleotides act to sterically impede the binding activity of soluble transcription factors and cofactors; therefore, it follows that a state of hypomethylation results in accessible conformation and increased levels of gene transcription. The analogous structural modification on chromatin is histone hyperacetylation—the epigenetic switch usually associated with an upsurge in transcriptional activity.

Epigenetic features have long been known to play an important role in developmental plasticity [39, 40]. Molecular mechanisms responsible for bringing about epigenetic modifications, that are manifest in phenotypes, are becoming increasingly well characterized. Methylation of nucleic acids and the DNA packaging histone proteins is established as a primary orchestrator in early embryogenesis in metazoa and is the principal mechanism of X-chromosome inactivation (Barr bodies) in addition to the phenomenon of imprinting during early development [41]. Ongoing epigenetic investigations in the area of human health point to incontrovertible data indicating that environmental exposures, particularly during early development, can provoke long-term epigenetic changes. These fixed structural alterations which can be transgenerationally inherited, appear to be primary etiology in various disease states that arise in later life stages [42]. Recent investigation into the stable and lasting consequences of genomic DNA methylation support the long held strong inference that presence or absence of cytosine-5 methyl groups has substantial influence regarding aspects of physiology and behavior. Detrimental postpartum events imposed upon mice suggests that early life stress (ELS) was associated with sustained DNA hypomethylation [43] of a critical DNA regulatory region that was shown to withstand age-dependent conversion in methylation states. This state of hypomethylation resulted in hypersecretion of the glucocorticoid corticosterone, resulting in changes of ability to effectively manage stress and with memory function. One might presume that analogous ELS, resulting from unintended contact with environmental stressors and xenobiotics, could occur in wildlife populations— leading to long-term effects with consequences ranging from individuals to communities.

Methylation states of DNA, in context of environmental exposure, will provide the most accessible biologic window into the inadvertent loss, or gain, of gene function, not only during early development but at any life stage—in any aquatic organism. Altered states of DNA methylation triggered by environmental exposure, analogous to the identical trends in disease progression, will lead to unscheduled transcription and premature initiation, or deactivation, of certain genes. Since diet, xenobiotics and behavior can produce changes in the assorted epigenetic organization [44], it follows that environmental causation might systematically change epigenetic profiles and influence future responses to environment stressors. Some endocrine disrupters have been shown to exert genome-wide effects on the state of DNA methylation [45]. In one study that doubtless has ramifications for all oviparous animals and ecological analyses, hormone treatment of immature White Leghorn roosters resulted in a demethylation of upstream estradiol-receptor binding site of the gene for vitellogenin [46]. This change of methylation state, observed on only one of the two strands of DNA, is referred to as hemimethylation and was directly correlated with immediate onset of the vitellogenin gene primary transcript.

The epigenetic phenomenon of DNA methylation, because of stability over time, will provide direct linkage for interpretation of ecotoxicogenomic data. Although the majority of studies in area of epigenetic methylation focus on single gene events [47], postexposure global and anonymous methylation patterns might yield insight not only into the initiating stressor but also, through transgenerational analysis, into the temporal range wherein the observed exposure-driven methylation states occurred. Investigating the interplay between the environment and epigenome [48] has the potential to yield MOA-specific exposure biomarkers in aquatic and terrestrial ecosystems.

The second mode of epigenesis that has implications for ecological monitoring, and possibly exposure reconstruction, is the recently described gene regulatory process mediated by a class of small, noncoding RNA molecules [49] that function in genomes of eukaryotes. Most small RNA candidates identified to date, in human, mouse and rat, exhibit conservation in other vertebrates, including dog, cow, chicken, opossum, and zebrafish [50], and conceivably throughout the animal kingdom. Two primary categories of these small RNAs have thus far been described: short interfering RNAs (siRNAs) and microRNAs (miRNAs) [51]. MicroRNAs comprise a genus of 20–30 nucleotide moieties that bind to sequence-specific, complementary regions of processed mRNA transcripts in a double-stranded conformation, appropriating the message, thereby decreasing or eliminating production of the corresponding protein product. This posttranscriptional mode of regulating gene expression gene is mediated not only by formation of miRNA-target hybrids, but also target mRNA degradation [52].

Predictions have been made that higher Eukaryotes express thousands of miRNAs, and although only a fraction of those have been identified, there is corroborating data to suggest that this class of small RNAs play an important role in numerous developmental processes and critical response pathways. As the numbers of characterized small RNAs in diverse genomes continue to expand, modeled conjecture suggests that miRNAs can regulate a substantial fraction of the genome. In 2005, computational predictions held that 10% of all protein-coding transcripts were subject to regulatory control by miRNAs; however, recent data indicates that this fraction will likely expand considerably. Single miRNAs are hypothesized to regulate multiple gene products, and there are suggestions that, in genomes of higher eukaryotes, the functional importance of miRNAs in regulatory programs of gene expression could surpass that of soluble *trans*-acting factors.

One ecological investigation endeavored to link stressor induced effects observed in a wild population of teleosts, to epigenetic causation. Investigators observed that *Fundulus heteroclitus* (killifish) inhabiting a creosote-contaminated system in the Elizabeth River, Virginia, exhibit what is termed, "refractory CYP1A phenotype." Contrary to conventional wisdom, the population lacked the expected induction of cytochrome P4501A (CYP1A) mRNA [53], an essential enzymatic component of phase I xenobiotic and drug metabolism induced by aromatic hydrocarbons. Additionally, the population lacked immunodetectable catalytic P450 protein.

Although the refractory CYP1A phenotype indicted heritability, strictly genetic bases did not seem to be linked to causation. The hypothesis that cytosine methylation at CpG sites in the promoter region of CYP1A underlies the refractory CYP1A phenotype was tested using the technique of bisulfite sequencing. Liver-specific genomic DNA was isolated from wild-caught adult killifish and from pools of laboratory reared F_1 embryos. Analyses of DNA isolated from indigenous fish taken from both the contaminated and reference site indicated that there was no detectable cytosine-5 methylation at any of the 34 CpG sites examined, including three regions that are considered integral to the putative xenobiotic response element (XRE). The investigators also noted that the "refractory CYP1A phenotype" gradually diminished in the course of development in laboratory reared F_1 generation fish.

Although in the above study, promoter methylation has been excluded as a causal factor for the described phenotype, an epigenetic program might well be at play. The deficiency of functional catalytic protein and a time-dependent gain of function in F_1 generation and in subsequent life stages are consistent with and offer a compelling argument that small RNAs might serve as the regulatory mechanism for the refractory CYP1A phenotype.

The study of microRNAs, and roles in posttranscriptional regulation, is still considered to be in preliminary stages, although the small RNA network has already been recognized in the area of human health as targets for biomarker and therapeutic development. If it is the case that miRNA–mRNA hybrids are determined to maintain stability over time—that is, greater duration than xenobiotic-induced translatable messages—or if, under given environmental pressures, microRNAs are constitutively expressed with developmental or tissue specificity, then these molecules will offer another inroad into ecological retrospective exposure analysis.

Otolith Geochemistry

Eco investigators taking yet a different approach to exposure monitoring make a forceful argument for using the novel approach of otolith geochemistry [54] as a tool for making strong inference for environmental conditions and exposure history in individual and populations of teleosts. Otoliths are structures of the inner ear of fish located just behind the eyes, also referred to as "ear bone" or "ear stone." Calcium carbonate (generally aragonite), the primary constituent of otoliths, is derived from water and components present therein which bind to the mineral structure otoliths with continual deposition. Because of acellular biomineralization, otolith structures are not subject to resorption during periods of starvation or stress and eliminate confounding variables such as size, age, and gender. As the otolith expands in mass, new calcium carbonate crystals form and, as with most crystal structures, lattice vacancies are a consequence of crystal formation. Analyses of the trace elemental composition or isotopic signatures of trace elements within a fish otolith provide insight into the water bodies, and associated conditions, in which individuals have previously been inhabitants. The otolith method, analyzing for organochlorine pesticides and PCBs, has recently been exploited to distinguish spatiotemporal variability and origins between North Atlantic and Mid Atlantic populations of Bluefin tuna [55]. Results of this study essentially describe a successful effort to reconstruct the toxicant profile to which populations had been exposed.

Additionally, fish scales—which develop from dermal mesenchyme—might offer a path to examine a xenobiotic gradient, revealing an ordered history of toxicant exposure. This method has been applied to detect presence of mercury in largemouth bass [56], comparing relationship between total Hg concentration in scale samples and muscle tissue in the same organism.

Coda: Reconstructing Ecological Exposures

In most cases, selection of the right biomonitoring assays as well as back-prediction of exposure patterns, environmental concentrations, or sources of contaminants will be heavily dependent on data from other sources. As methods for ecological monitoring proliferate, it will become increasingly inefficient to run every possible biomonitoring assay in every case. In some cases, lists of "usual suspects" will be targeted for monitoring, while at other times site-specific information will suggest particular assays to be exploited. Usually, the measurements from these assays will be consistent with a range of exposure routes, timings, and concentrations. Additional data will then be needed in order to narrow these ranges sufficiently to provide actionable information for environmental decision-making.

The development and continued update of "usual suspects" inventory can support exposure reconstruction in cases where suspected contaminant stressors are not immediately evident. Previous successful reconstructions as well as existing chemical monitoring data can provide convenient and reasonable starting points for developing lists of candidate toxicants. In addition, data on import, production, distribution, usage, and disposal allows estimation of possible introduction rates of different contaminants into the environment through a variety of pathways. Measured and predicted physiochemical properties of potential contaminants can be used to predict transport and eventual fate, including important exposure processes such as biomagnification, thereby transforming estimated environmental introduction rates into potential external concentrations and rates of biologic exposure. Preexisting data on potency, differential susceptibility, and modes of action can be combined with potential exposure rates to prioritize contaminants for placement on candidate contaminant lists, based on the likelihood of harmful outcome resulting from exposure. This information can also be used to subselect biomonitoring approaches based on site-specific conditions, such as observations on modes of toxicity, range of species affected, local transport processes, and proximity to potential sources of contamination.

Once assembled, biomonitoring results are typically found to be consistent with a range of exposure scenarios, and additional information often plays a critical role in narrowing this range to the point that useful science policy decisions can be made. Any given set of biomonitoring results should be consistent with a range of duration–concentration combinations, the size of which is dependent on the shape of the dose–response curve, the frequency of sampling, and the kinetics of the biomonitoring signal following exposure. Individual duration–concentration combinations may in turn be consistent with a variety of potential contaminant sources and routes of environmental fate and transport. Information on potential introduction rates from near and remote sources, as well as fate and transport properties will often provide the information pivotal to correctly identify the source and route, as well as provide important corroboration of the proposed contaminant identity and the exposure duration–concentration profile. Monitoring for chemicals or manufactured products may depend less on the risks posed by the end-state product than on the risks induced by extraction, processing, and transportation of raw materials or by the wastes generated during manufacturing processes. In such situations, biomonitoring assessments should integrate risks from the entire product life cycle.

Identifying and minimizing exposures to contaminants plays a critical role in environmental protection. Current biomonitoring studies provide minimal information to identify the what, when, where, and how much of exposures. The concept of exposure reconstruction can provide a framework to guide the development of new biomarkers and the design of future biomonitoring studies that will shift the research focus from detection of exposures to elucidating the mechanisms of exposures from sources to internal measurements. Human and ecosystem exposures can be better understood through the strategic development of biomarkers of exposure that can inform the exposure reconstruction process. Significant research in the area of exposure reconstruction is necessary to advance the protection of humans and ecosystems, and research in this area presents numerous opportunities and challenges. A recent publication [57] highlights a number of these issues including the variation in performance and computational complexity of inversion techniques, the multiplicity of potential real-world exposure scenarios, and the impact of biochemical properties and sampling characteristics related to biomarkers.

Gene–environment interactions are extremely complex and irrefutably nonlinear. No existing ecological risk models are informed with the capability of predicting exposure dose relationships and outcomes that arise from the intersection of inestimable environmental conditions and complex biologic responses; however, as the number of well-characterized genomes becomes greater, our understanding of byzantine processes will enhance not only predictive risk assessment but also the ability to describe retrospective exposures.

Exposure reconstruction demands that we essentially shift modes of thinking from what was previously *deductive* reasoning to the "strong inference" *inductive* interpretation, the flow of which is depicted below [58].

- Observation; determination of chemical and biologic patterns
  - Speculative multiple hypotheses based on incremental, retrospective data from multiple sources
      Strong inference; extrapolative external concentration; hypothesis elimination and *causal reconstruction*

The above approach will permit the formulation of *conditional inductive trees* that provide the foundation for rebuilding an exposure phenomenon. If the cumulative information that arises from a reconstruction scenario is sufficient, then a "bench-scale" experimental reconstruction can be designed with replicated ecological parameters, using mesocosms, artificial streams, or other surrogate ecosystems. This will facilitate further development of genomic indicators for continued monitoring and site surveys.

References

1. Aylward LL, Brunet RC, Starr TB, Carrier G, Delzell E, Cheng H, Beall C (2005) Exposure reconstruction for the TCDD-exposed NIOSH cohort using a concentration- and age-dependent model of elimination. Risk Anal 25:945–956
2. Bauchinger M (1998) Retrospective dose reconstruction of human radiation exposure by FISH/chromosome painting. Mutat Res 404:89–96
3. Roy A, Weisel CP, Gallo MA, Georgopoulos PG (1996) Studies of multiroute exposure/dose reconstruction using physiologically based pharmacokinetic models. Toxicol Ind Health 12:153–163
4. Williams PR, Paustenbach DJ (2003) Reconstruction of benzene exposure for the Pliofilm cohort (1936-1976) using Monte Carlo techniques. J Toxicol Environ Health A 66:677–781
5. CDC (2003) Second national report on human exposure to environmental chemicals. US Department of Health and Human Services, Center for Disease Control and Prevention, Atlanta, GA
6. CDC (2005) Third national report on human exposure to environmental chemicals. US Department of Health and Human Services, Center for Disease Control and Prevention, Atlanta, GA
7. USEPA (2005) National lake fish tissue study. US Environmental Protection Agency, Washington, DC
8. NRC (2006) Human biomonitoring for environmental chemicals. National Research Council of the National Academies, The National Academies Press, Washington, DC
9. Tan YM, Liao KH, Conolly RB, Blount BC, Mason AM, Clewell HJ (2006) Use of a physiologically based pharmacokinetic model to identify exposures consistent with human biomonitoring data for chloroform. J Toxicol Environ Health A 69:1727–1756
10. Lattier DL, Gordon DA, Burks DJ, Toth GP (2001) Vitellogenin gene transcription: a relative quantitative exposure indicator of environmental estrogens. Environ Toxicol Chem 20:1979–1985
11. Korte JJ, Kahl MD, Jensen KM, Pasha MS, Parks LG, LeBlanc GA, Ankley GT (2000) Fathead minnow vitellogenin: Complementary DNA sequence and messenger RNA and protein expression after 17 beta-estradiol treatment. Environ Toxicol Chem 19:972–981
12. Schmid T, Gonzalez-Valero J, Rufli H, Dietrich DR (2002) Determination of vitellogenin kinetics in male fathead minnows (*Pimephales promelas*). Toxicol Lett 131:65–74
13. Paules R (2003) Phenotypic anchoring: linking cause and effect. Environ Health Perspect 111:A338–A339
14. Moggs JG, Tinwell H, Spurway T, Chang HS, Pate I, Lim FL, Moore DJ, Soames A, Stuckey R, Currie R, Zhu T, Kimber I, Ashby J, Orphanides G (2004) Phenotypic anchoring of gene expression changes during estrogen-induced uterine growth. Environ Health Perspect 112:1589–1606
15. Lattier DL, Reddy TV, Gordon DA, Lazorchak TM, Smith ME, Williams DE, Wiechman B, Flick RW, Miracle AL, Toth GP (2002) 17 α-ethynylestradiol-induced vitellogenin gene transcription quantified in livers of adult males, larvae, and gills of fathead minnows (*Pimephales promelas*). Environ Toxicol Chem 21:2385–2393
16. Biales AD, Bencic DC, Lazorchak JL, Lattier DL (2007) A quantitative real-time polymerase chain reaction method for the analysis of vitellogenin transcripts in model and nonmodel fish species. Environ Toxicol Chem 26:2679–2686

17. Arukwe A, Goksoyr A (2003) Eggshell and egg yolk proteins in fish: hepatic proteins for the next generation: oogenetic, population, and evolutionary implications of endocrine disruption. Comp Hepatol 2:4
18. Sumpter JP, Jobling S (1995) Vitellogenesis as a biomarker for estrogenic contamination of the aquatic environment. Environ Health Perspect 103(suppl 7):173–178
19. Ankley GT, Bencic DC, Breen MS, Collette TW, Conolly RB, Denslow ND, Edwards SW, Ekman DR, Garcia-Reyero N, Jensen KM, Lazorchak JM, Martinovic D, Miller DH, Perkins EJ, Orlando EF, Villeneuve DL, Wang RL, Watanabe KH (2009) Endocrine disrupting chemicals in fish: developing exposure indicators and predictive models of effects based on mechanism of action. Aquat Toxicol 92:168–178
20. Herman RL, Kincaid HL (1988) Pathological effects of orally administered estradiol to rainbow trout. Aquaculture 72:165–172
21. Kidd KA, Blanchfield PJ, Mills KH, Palace VP, Evans RE, Lazorchak JM, Flick RW (2007) Collapse of a fish population after exposure to a synthetic estrogen. Proc Natl Acad Sci USA 104:8897–8901
22. Schwaiger J, Spieser OH, Bauer C, Ferling H, Mallow U, Kalbfus W, Negele RD (2000) Chronic toxicity of nonylphenol and ethinylestradiol: haematological and histopathological effects in juvenile Common carp (*Cyprinus carpio*). Aquat Toxicol 51:69–78
23. Hutchinson TH, Barrett S, Buzby M, Constable D, Hartmann A, Hayes E, Huggett D, Laenge R, Lillicrap AD, Straub JO, Thompson RS (2003) A strategy to reduce the numbers of fish used in acute ecotoxicity testing of pharmaceuticals. Environ Toxicol Chem 22:3031–3036
24. Panter GH, Hutchinson TH, Lange R, Lye CM, Sumpter JP, Zerulla M, Tyler CR (2002) Utility of a juvenile fathead minnow screening assay for detecting (anti-)estrogenic substances. Environ Toxicol Chem 21:319–326
25. Scholz S, Mayer I (2008) Molecular biomarkers of endocrine disruption in small model fish. Mol Cell Endocrinol 293:57–70
26. Gilbert SF (2002) The genome in its ecological context: philosophical perspectives on interspecies epigenesis. Ann N Y Acad Sci 981:202–218
27. Nijhout HF (2003) Development and evolution of adaptive polyphenisms. Evol Dev 5:9–18
28. Petersen DG, Sundbäck K, Larson F, Dahllöf I (2009) Pyrene toxicity is affected by the nutrient status of a marine sediment community: implications for risk assessment. Aquat Toxicol 95(1):37–43
29. Persoone G, Van de Vel A, Van Steertegem M, De Nayer B (1989) Predictive value of laboratory tests with aquatic invertebrates: influence of experimental conditions. Aquat Toxicol 14:149–167
30. Gordon DA, Toth GP, Graham DW, Lazorchak JM, Reddy TV, Knapp CW, deNoyelles J, Frank CS, Lattier DL (2006) Effects of eutrophication on vitellogenin gene expression in male fathead minnows (*Pimephales promelas*) exposed to 17α-ethynylestradiol in field mesocosms. Environ Pollut 142:559–566
31. Barbash JE, Thelin GP, Kolpin DW, Gilliom RJ (2001) Major herbicides in ground water: results from the National Water-Quality Assessment. J Environ Qual 30:831–845
32. Spanò L, Tyler CR, Rv A, Devos P, Mandiki SNM, Silvestre F, Thomé J-P, Kestemont P (2004) Effects of atrazine on sex steroid dynamics, plasma vitellogenin concentration and gonad development in adult goldfish (*Carassius auratus*). Aquat Toxicol 66:369–379
33. Fan W, Yanase T, Morinaga H, Gondo S, Okabe T, Nomura M, Komatsu T, Morohashi K, Hayes TB, Takayanagi R, Nawata H (2007) Atrazine-induced aromatase expression is SF-1 dependent: implications for endocrine disruption in wildlife and reproductive cancers in humans. Environ Health Perspect 115:720–727
34. Coors A, Vanoverbeke J, De Bie T, De Meester L (2009) Land use, genetic diversity and toxicant tolerance in natural populations of *Daphnia magna*. Aquat Toxicol 95(1):71–79
35. Calabrese EJ, Baldwin LA (2003) Toxicology rethinks its central belief. Nature 421:691–692
36. Weltje L, vom Saal FS, Oehlmann J (2005) Reproductive stimulation by low doses of xenoestrogens contrasts with the view of hormesis as an adaptive response. Hum Exp Toxicol 24:431–437

37. Metivier R, Gallais R, Tiffoche C, Le Peron C, Jurkowska RZ, Carmouche RP, Ibberson D, Barath P, Demay F, Reid G, Benes V, Jeltsch A, Gannon F, Salbert G (2008) Cyclical DNA methylation of a transcriptionally active promoter. Nature 452:45–50
38. Goldberg AD, Allis CD, Bernstein E (2007) Epigenetics: a landscape takes shape. Cell 128:635–638
39. Dolinoy DC (2007) Epigenetic gene regulation: early environmental exposures. Pharmacogenomics 8:5–10
40. Dolinoy DC (2008) The agouti mouse model: an epigenetic biosensor for nutritional and environmental alterations on the fetal epigenome. Nutr Rev 66(suppl 1):S7–S11
41. Vire E, Brenner C, Deplus R, Blanchon L, Fraga M, Didelot C, Morey L, Van Eynde A, Bernard D, Vanderwinden JM, Bollen M, Esteller M, Di Croce L, de Launoit Y, Fuks F (2006) The Polycomb group protein EZH2 directly controls DNA methylation. Nature 439:871–874
42. Tost J (2009) DNA methylation: an introduction to the biology and the disease-associated changes of a promising biomarker. Methods Mol Biol 507:3–20
43. Murgatroyd C, Patchev AV, Wu Y, Micale V, Bockmuhl Y, Fischer D, Holsboer F, Wotjak CT, Almeida OF, Spengler D (2009) Dynamic DNA methylation programs persistent adverse effects of early-life stress. Nat Neurosci 12(12):1559–1566
44. Cooney CA (2007) Epigenetics–DNA-based mirror of our environment? Dis Markers 23:121–137
45. Tabb MM, Blumberg B (2006) New modes of action for endocrine-disrupting chemicals. Mol Endocrinol 20:475–482
46. Saluz HP, Jiricny J, Jost JP (1986) Genomic sequencing reveals a positive correlation between the kinetics of strand-specific DNA demethylation of the overlapping estradiol/glucocorticoid-receptor binding sites and the rate of avian vitellogenin mRNA synthesis. Proc Natl Acad Sci U S A 83:7167–7171
47. Reamon-Buettner SM, Mutschler V, Borlak J (2008) The next innovation cycle in toxicogenomics: environmental epigenetics. Mutat Res 659:158–165
48. Kovalchuk O (2008) Epigenetic research sheds new light on the nature of interactions between organisms and their environment. Environ Mol Mutagen 49:1–3
49. Carthew RW, Sontheimer EJ (2009) Origins and mechanisms of miRNAs and siRNAs. Cell 136:642–655
50. Berezikov E, Guryev V, van de Belt J, Wienholds E, Plasterk RH, Cuppen E (2005) Phylogenetic shadowing and computational identification of human microRNA genes. Cell 120:21–24
51. Valeri N, Vannini I, Fanini F, Calore F, Adair B, Fabbri M (2009) Epigenetics, miRNAs, and human cancer: a new chapter in human gene regulation. Mamm Genome 20(9–10):573–580
52. Hudder A, Novak RF (2008) miRNAs: effectors of environmental influences on gene expression and disease. Toxicol Sci 103:228–240
53. Timme-Laragy AR, Meyer JN, Waterland RA, Di Giulio RT (2005) Analysis of CpG methylation in the killifish CYP1A promoter. Comp Biochem Physiol C Toxicol Pharmacol 141:406–411
54. Campana SE, Thorrold SR (2001) Otoliths, increments, and elements: keys to a comprehensive understanding of fish populations? Can J Fish Aquat Sci 58:30–38
55. Dickhut RM, Deshpande AD, Cincinelli A, Cochran MA, Corsolini S, Brill RW, Secor DH, Graves JE (2009) Atlantic bluefin tuna (*Thunnus thynnus*) population dynamics delineated by organochlorine tracers. Environ Sci Technol 43:8522–8527
56. Lake JL, Ryba SA, Serbst JR, Libby AD (2006) Mercury in fish scales as an assessment method for predicting muscle tissue mercury concentrations in largemouth bass. Arch Environ Contam Toxicol 50:539–544
57. Georgopoulos PG, Sasso AF, Isukapalli SS, Lioy PJ, Vallero DA, Okino M, Reiter L (2009) Reconstructing population exposures to environmental chemicals from biomarkers: challenges and opportunities. J Expo Sci Environ Epidemiol 19:149–171
58. Chamberlin TC (1965) The method of multiple working hypotheses: with this method the dangers of parental affection for a favorite theory can be circumvented. Science 148:754–759

Considerations and Criteria for the Incorporation of Mechanistic Sublethal Endpoints into Environmental Risk Assessment for Biologically Active Compounds

Richard A. Brain and Bryan W. Brooks

Introduction

Awareness about the presence and unintended consequence of biologically active organic contaminants in the environment was largely borne out of seminal observational works concerning pesticides in the early 1960s [19], eventually culminating in the establishment of protective legislation, government regulatory bodies and a rigorous, continually improving risk assessment paradigm [79, 80]. As a result of associated and necessary scientific advancement, a mounting inventory of novel sublethal endpoints has materialized, which has also precariously and inadvertently highlighted the critical lack of comprehensive and cohesive regulatory position and process with respect to consideration of these metrics in environmental risk assessment (ERA). Although recent attention concerning this issue has been championed largely by a relatively new term; "biomarkers," the concept and importance of sublethal endpoints is certainly not new [18]. Moreover, the fundamental concept of biological context and causality concerning sublethal effects has been emphasized for decades, seemingly in concert with the environmental awareness movement itself [7] with reviews on the subject dating back to the early 1970s [76]. Yet with nearly 50 years of knowledge and experience, the maturing field of sublethal effects appears to be impeded by a corresponding ecological risk paradigm that is still comparatively less developed [89]. Although the conceptual barriers for consideration are well known [29, 36, 56, 76, 79], the actual process and criteria

R.A. Brain(✉)
Ecological Risk Assessment, Syngenta Crop Protection LLC, 410 Swing Road,
Greensboro, NC 27409, USA
e-mail: richard.brain@syngenta.com

B.W. Brooks
Department of Environmental Science, Center for Reservoir and Aquatic Systems Research,
The Institute of Ecological, Earth and Environmental Sciences, and Institute of Biomedical
Studies, Baylor University, One Bear Place #97266, Waco, TX 76798, USA

B.W. Brooks and D.B. Huggett (eds.), *Human Pharmaceuticals in the Environment:*   139
Current and Future Perspectives, Emerging Topics in Ecotoxicology 4,
DOI 10.1007/978-1-4614-3473-3_7, © Springer Science+Business Media, LLC 2012

for consideration, incorporation, and integration of mechanistic sublethal effects remain poorly defined, and in many cases completely lacking, leading to considerable uncertainty and subjectivity.

The term "sublethal" endpoint(s) encompasses a vast array of effects representing a spectrum of biological complexity ranging from biochemical to physiological, and as stated by Sprague [76] nearly 40 years ago: "understanding physiological action of a toxicant is the key to predicting important sub-lethal effects." Moreover, Sprague [76] also asserted that biochemistry, considered as a basic level (biological strata), should be related to higher levels of organization whenever possible in order to address the question of whether performance of an individual, or ultimately success of groups of individuals, is in turn affected. Echoed by Walker [90] some 20 years later, if the molecular mechanism of toxicity is known then the degree to which the chemical interacts with the target site can potentially be evaluated and possibly related to the nature and degree of toxic effect. However, within this seemingly intuitive and logical exercise in extrapolation and correlation lies the fundamental proviso that must be satisfied in order for sublethal effects to be formally and effectively considered for inclusion in ERA; causally and plausibly relating effects across biological strata. This is a critical consideration, and as identified by Bradbury et al. [12], while studies at lower biological strata are used to determine mechanism of action (MOA), interpretation of the relevant toxicological events for risk-management decisions is typically associated with adverse responses observed at higher biological strata. Among the pantheon of synthetically produced chemicals, no classes have more intensively characterized and exploited biochemical pathways than those affected specifically by pesticides and pharmaceuticals.

Pesticides and pharmaceuticals are both classified as "biologically active"; however, substantial disparity among ERA paradigms exists, owing primarily to fundamental differences in nature, development, application and use, which ultimately dictates how these compounds enter the environment; conceptual and perceptual differences also exist. Notwithstanding these obvious differences, which do need to be considered, certain fundamental issues are more systematic in nature and do not segregate uniquely or exclusively based on class. Consequently, due to their collective, yet respective, biologically active natures, conceptual issues and challenges common to both pharmaceutical and pesticide ERA should be considered in concert. And of particular joint interest is the recurring issue concerning utilization and incorporation of sublethal effects, particularly MOA-specific data. Although the foundation for any stressor induced effects cascade is first manifested at the biochemical level, knowing and understanding the pathway, causal linkages, and ultimate consequence across the spectrum of biological complexity requires intensive experimental characterization. Consequently, few classes of compounds demonstrate the requisite data intensive biological profile to support such analyses; pesticides and pharmaceuticals are unique exceptions. Thus, here we will explore considerations and criteria for the incorporation of MOA-specific sublethal effects into ERA, focusing particularly on biologically active compounds with well established pharmacological or toxicological MOAs.

Overview of Sublethal Effects in Risk Assessment

Sublethal effects language and utilization has begun to permeate the risk assessment lexicon for biologically active compounds. As a component of the US pesticide registration process, select sublethal effects are currently considered for the purpose of risk assessment, though as outlined in Table 1, these effects are largely gross morphological, physiological, pathological, or histological. However, the US Environmental Protection Agency (EPA) does exercise the option to consider additional sublethal effects on a case-by-case basis, with the caveat that careful consideration of the nature of the sublethal effect measure is provided and that a plausible clear causal relationship has been established with the assessment endpoint, namely survival and reproduction [79]. This language is commensurate with the Bradford Hill criteria for establishing causality [44] and consistent with current opinion regarding

Table 1 List of current sublethal effect measures considered by the US Environmental Protection Agency (EPA) for use in ecological risk assessment (ERA) of pesticides for the purposes of registration as outlined in the ERA overview document

Organism	Test-type	Sublethal measurement endpoints
Invertebrate	Life-cycle	Production of young by first generation
		Length of first generation
Fish	Early life-stage	Embryo hatch rate
		Time to hatch
		Time to swim-up
		Growth (length and weight)
		Pathological or histological effects
		Observations of other clinical signs
	Life-cycle	Embryo hatch rate
		Time to hatch
		Growth (length)
		Exposed adult egg production
		Second generation hatch rate
		Second generation growth
Birds	Reproduction	Maternal weight
		Eggs laid/hen
		Eggs cracked
		Eggshell thickness
		Viable embryos
		Hatchling number 14-day survivors
		Gross necropsy (organ lesions, fat and muscle deterioration)
		Observations of other clinical signs
Mammals	Two-generation reproduction	Total panel of reproduction parameters including: histopathology, parental and offspring growth, weight, mating, lactation, gonadal development milestones, sexual organ performance, and offspring production

Adapted from the USEPA [79]

biomarkers [36]. Not surprisingly, the Agency anticipates further advancement in this area as part of the continually improving state-of-the-science [79]. However, beyond those outlined in Table 1, there has been little to no consideration, detail or guidance concerning the nature or characteristics of potential sublethal endpoints, particularly concerning those related to MOA. That is not to say of course that MOA-specific endpoints are not considered at all, as is evidenced upon review of the toxicological data requirements for acetylcholinesterase (AChE) inhibiting compounds under 40 CFR Part 158 [80]; developmental neurotoxicity (DNT) studies are conditionally required based on weight of evidence. Based on a review of 20 DNT studies, 13 of which evaluated AChE inhibition, this sublethal endpoint was found to be the most sensitive metric [80]. The MOA and physiological consequences of organophosphate-mediated inhibition of AChE are well understood [53] satisfying the causal criteria. Moreover, as a requirement of FIFRA Part 158 Section 3(c)(2)(B), endangered species assessments are currently being required for all new pesticide registrations and registration reviews [80], where the impacted "action area" can potentially be defined based on sublethal endpoints [83] under consultation with the US Fish and Wildlife Service and National Marine Fisheries [79].

Borne out of considerations stemming from both the Food Quality Protection Act (FQPA) and amendments to the Safe Drinking Water Act (SDWA) passed in 1996, the Endocrine Disruptor Screening Program (EDSP) was developed based on provisions calling for the screening and testing of chemicals and pesticides for possible endocrine disrupting effects. Not surprisingly, considering the highly specific receptor-mediated nature of endocrine active compounds, current protocols under the EDSP incorporate MOA-specific sublethal endpoints as an integral component of whole organism testing [82]. As outlined in OPPTS EDSP Test Guideline 890.1350 [82] measurement of plasma vitellogenin content, an indicator of estrogenic agonists when expressed in males, is required in conjunction with histological/physiological endpoints and mortality.

According to European guidance concerning birds and mammals [31] under "Relevance of endpoints in long-term toxicity tests" and regarding prediction of effects at the population level, only endpoints which are related to survival rate, reproduction rate and development (collectively termed "key factors of population dynamics") are considered ecotoxicologically relevant. Stated vaguely in the same section, although some sublethal endpoints assessed in mammalian tests are not ecologically relevant, it is suggested that before disregarding biochemical effects, lab to field extrapolative uncertainty should be considered [31]. However, no insight regarding specific criteria for inclusion are outlined, though it is stated that transient or reversible sublethal effects are less relevant than those that are continuous or irreversible after exposure termination [31]. The validity of terminology within this contention, however, falls under question in circumstance where irreversible effects sustained subsequent to exposure termination at higher biological strata are incurred as a consequence of a reversible sublethal endpoint such as enzyme inhibition (e.g., paralysis as a result of AChE inhibition). Under such a circumstance, the inclusion of a reversible sublethal effect could be convincingly argued.

Under the EU aquatic ecotoxicology guidance [32], if short-term exposure leads to sublethal effects, which are not covered by acute toxicity testing, further evaluations might be needed in such special cases. However, again, no specific criteria concerning the nature, orientation, or appropriateness of sublethal mechanistic effects (MOA) is provided. Only under the EU terrestrial ecotoxicology guidance [33] is any mention or consideration of MOA made. For nontarget arthropod testing it is recommended that for substances suspected to have a "special" MOA (e.g., insect growth regulators; IGRs) tests "should include sub-lethal endpoints and may need modifications" [33], though again no specific details or guidance concerning the criteria mentioned above are offered rendering decision nebulous and incorporation circumstantial.

For pharmaceuticals sublethal considerations are even less well defined. For example, in the USA, if a trigger value (≥ 1 µg/L) is exceeded and an EA is required [84], based on a number of assumptions, albeit somewhat vague, guidance does explicitly state that alternative, scientifically justified approaches can also be used [84], though it is not stated whether this includes MOA-specific sublethal considerations. Based on the inverse RQ methodology employed (PNEC/PEC), if the appropriate assessment factor (AF; 1,000, 100, and 10 for Tiers 1, 2, and 3, respectively) is not exceeded then no further testing is required, "unless sub-lethal effects are observed at the maximum expected environmental concentration (MEEC)" [84]. Furthermore, it is recommended that sublethal effects (observed effects) at the MEEC indicate that chronic toxicity testing (Tier 3) should be performed [84]. Under Tier 3 US EPA and OECD ("or other peer-reviewed literature") methods and organisms are cited for reference [84], though unfortunately no further guidance is offered, particularly concerning MOA. With respect to European guidance concerning human medicinal products no specific mention is made concerning sublethal effects outside of what is contained in cited test protocols [28]; the same is true for the internationally harmonized guidance on veterinary medicinal products [87, 88].

In addition to these efforts, US EPA has recently examined use of MOA-specific sublethal endpoints under the statutory requirements of the US Clean Water Act. Specifically, US EPA developed a "white paper" examining the role of biomarkers related to MOA when developing National Ambient Water Quality Criteria (NAWQC; [81]). Historically, NAWQC derivation relied on standardized ecotoxicity responses such as survival, short-term growth or invertebrate reproduction [77]. Minimum data requirements for acute and chronic NAWQC include toxicity data from eight different organisms representing various trophic levels of an aquatic food web [77]. The impetus for developing this document included considering the use of MOA related endpoints for contaminants of emerging concern, which include pharmaceuticals and endocrine disrupting chemicals (EDCs). EPA specifically stated that "the 'Good Science' clause of the *Guidelines* (e.g., [77]) provides the flexibility to adopt procedures that will produce a technically rigorous and protective criterion" [81]. They further concluded,

> Chronic test data and other data should be examined to determine whether, for the specific chemical or MOA, endpoints beyond those traditionally used for criteria derivation may

have intrinsic "biological importance" and therefore could be used as a basis for defining threshold of effect (e.g., sex ratio). Specifically, in the context of EDCs:

- Other "endocrine-sensitive endpoints" (e.g., VTG, testis-ova) should be examined to determine whether they can be relied upon as definitive indicators of other biologically important endpoints (e.g., reproduction), with the idea that they may be incorporated into calculation of the criterion. Important sources of this information would include full lifecycle tests in which these other endpoints were measured alongside traditional chronic endpoints, and may include tests with other chemicals with the same MOA (e.g., E2 for EE2).
- If endpoints, such as VTG or testis-ova, are used as direct or indirect indicators of effect, it is critically important that the baseline condition (e.g., variation during normal development) be understood sufficiently to define when changes are biologically meaningful.
- Selection of appropriate endpoints (and their associated effect thresholds) may, in some instances, transcend "biological importance" (the focus of the Guidelines) to reflect societal concerns (e.g., physical appearance of wild-caught fish) [81].

Following a favorable review by the US EPA Science Advisory Board, the US EPA plans on developing a technical support document on deriving aquatic life criteria for CECs.

Establishing Causality and Confidence

Forty years ago Sprague [76] indicated that "it is relatively easy to document small changes within an animal, but there is often a question whether the changes are deleterious, or merely within the normal range of adaptation of the animal." Given that the targets of protection in ecological risk assessment are characteristically populations, communities, and ecosystems, but only rarely individuals, sublethal mechanistic responses must be consistently, systematically, and plausibly linked to responses at these higher strata if such metrics are to be used as reliable and consequential effects indicators [36]. Bridging this span of uncertainty in the effects spectrum requires convincingly demonstrating causality, the principals for which were first established by Bradford Hill and Sir Richard Doll [25, 44]. Commensurate with these principals, the US EPA maintains that clear, reasonable, and plausible links between sublethal effects and survival or reproductive capacity of organisms in the field must be firmly established in order to influence the confidence of the overall risk assessment conclusions [79]. However, addressing this caveat biologically requires substantial improvements in our understanding of how mechanistic processes at each level are functionally integrated in terms of whole-organism performance [36]. This is the so called burden of proof, and the more intensively characterized the response is mechanistically, the greater the potential weight of evidence for establishing causal linkages.

For any given toxicological response, effects cascades are initially manifested at the biochemical level regardless of how specific (e.g., receptor mediated) or generic (e.g., narcosis) the MOA. If biological complexity is conceptualized as an inverted

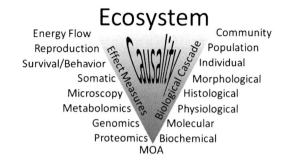

Fig. 1 Inverted biological cascade pyramid and associated effect measures demonstrating the concept of causality in weight of evidence approaches employed with ecological risk assessments of biologically active molecules (e.g., pharmaceuticals, pesticides)

pyramid (Fig. 1) with the most fundamentally basic organic building blocks (e.g., biomolecules) at the apex, progressing upwards from this foundation, tissues are derived, followed by organs, organ systems, organisms, and finally collections of similar and diverse organisms [12]. Consequently, effects at the apex or hypothetical biological foundation have the potential to cascade or reverberate across higher levels of biological organization. However, the degree to which higher strata are affected depends entirely upon the significance (biological consequence) of effects at lower strata and ultimately how tightly these tiered or stratified effects are related biologically; this in effect defines the "biological cascade" (Fig. 1). The term "sublethal" can be defined at various points along the inverted biological cascade pyramid (i.e., below "individual"; Fig. 1), and the greater the separation of strata along this axis the more difficult it is convincingly, consistently and reliably to establish causality. Not surprisingly, the greater the number of strata measured in a given assay, the greater the potential to establish causality, thus dictating the strength in weight-of-evidence, and consequently, the confidence in the sublethal endpoint(s) of interest. Moreover, testing effects at different strata over a range of exposures facilitates establishment and comparison of concentration–response trends, which accordingly also dictates the strength of causality based on the characteristics of the aforementioned trends (shape, range, comparative sensitivity, etc.). In this context, we identify essentially five critical or core elements that need to be established in order to foster confidence in sublethal effects for ERA consideration, (1) biological plausibility (causality), (2) sensitivity, (3) biological consequence, (4) effect concurrence, and (5) diagnostic capacity (Fig. 2). First, the more closely response trends evaluated at multiple strata simultaneously track one another (respond or scale proportionally with concentration as either congruent or inverse functions), the greater the causal strength of their collective association (for comparison see Fig. 3a, c). Second, the greater the resolution afforded by a sublethal response in comparison to higher biological strata (with the caveat that differential sensitivity becomes bound maximally by relevance), the greater the utility in predictive power (for comparison see Fig. 3a, b). Third, the greater the biological consequence conveyed to higher strata (biologically meaningful effects) resultant from the sublethal response, the more consequential the endpoint. Fourth, the greater the synchronization in response manifestation between effects at different biological strata (synchronized temporally and/or concentration-dependently), the stronger the relationship. Finally, the

Fig. 2 Five general criteria or core elements required to confer confidence in selecting a sublethal endpoint of interest (e.g., mechanism of action (MOA)) in relation to effects at higher biological strata during ecological risk assessments of biologically active molecules (e.g., pharmaceuticals, pesticides)

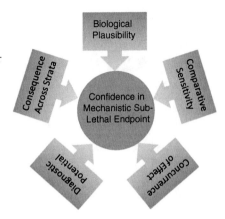

more unique in nature the sublethal biological effects "signature" the greater the potential diagnostic capacity, and consequently, the greater the utility within the risk assessment paradigm. If a sublethal concentration–response trend does not proportionally reflect, and concurrently manifest with those derived at higher strata and with convincing biological consequence, or if the sublethal effect measure does not afford greater sensitivity (predictive ability), it is of little value to RA beyond potential toxicological diagnostics, which, however, can be extremely important. The sensitivity criterion is somewhat subjective and requires judgment concerning the upper bound where relevance (biological consequence) becomes questionable, particularly when considering transient or fully reversible effects. For example, reasonable lower and upper bounds of sensitivity between MOA specific- and higher-strata effects (survival and reproduction as a reference point) may be considered as ≥ 1.5 and $\leq 10\times$ for enzyme inhibition, based on the subset of examples detailed subsequently in the present evaluation. For downstream metabolite reduction or upstream metabolite accumulation as a surrogate of enzyme inhibition the range may be considered broader, for example ≥ 1.5 and $\leq 50\times$, though the broader the range the greater the burden of strength in the causal relationship. Clearly such proposed criteria are useful for developing testable hypotheses and thus require further study for validation for various MOAs and organisms. Moreover, depending on the nature of the MOA-specific effect the dynamic range of sensitivity vs. relevance may vary, and requires expert judgment.

Nowhere else has this paradigm of causal scrutiny been highlighted more than in the arena of biomarkers. The term "biomarker" is somewhat ambiguous and can be applied broadly and generally but encompasses biochemical, physiological, or ecological structures or processes (including MOA) that have been correlated or causally linked to biological effects measured at one or more levels of biological organization [56]. No biomarker can by itself offer a complete solution and a battery of biomarkers evaluated across the spectrum of biological resolution will likely be necessary in order to convincingly evaluate chemical hazards [29]. Thus, how precisely an effect can be identified and/or characterized depends upon a multiparametric approach which includes biomarkers of general stress and more specific

Fig. 3 Hypothetical concentration–response curves for effect measures representing biological strata extremes (e.g., enzyme inhibition vs. mortality) demonstrating differential sensitivity with proportional response tracking (**a**), similar sensitivity with proportional response tracking (**b**), and differential sensitivity with unproportional response tracking (**c**)

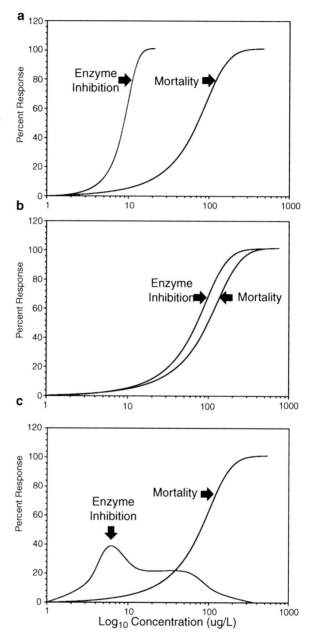

biomarkers such as MOA [34]. However, measuring a suite of biomarkers will only be useful if they are integrated into a mechanistic model with obvious links to fitness [36] fulfilling the conditions of causality and conferring requisite weight-of-evidence relating effects along the inverted biological cascade pyramid (Fig. 1).

Mechanistic Causality in Environmental Toxicology

As eloquently stated by Bartholomew [7] and highlighted by Sprague [76] "...there are a number of levels of biological integration and...each level finds its explanation of mechanisms in the levels below, and its significance in the levels above." Due to the hierarchy in biological complexity, investigations concerning MOA require higher resolution sophisticated experimental procedures and equipment, and are typically not measured concurrently with effects at higher levels of biological organization, but rather autonomously from tissue extracts, cell cultures or purified proteins. As a consequence. it is often difficult to convincingly demonstrate that biochemical ripples result in physiological, population or community waves. Much like an object which breaks the water's surface, the origin of the ripples can be traced, but the impact will not be fully understood and appreciated unless the resultant waves are measured on the shoreline, simultaneously. Implicit in this analogy is the ability to not only identify the source of the ripples, but to also chronicle and characterize the process or processes turning them into waves. Although numerous studies have evaluated MOA at the biochemical level or measured impacts at higher biological strata, comparatively few studies have actually evaluated both in tandem, rendering causality more often than not speculative or extrapolative in nature due to fundamental gaps in the weight-of-evidence case. Often, the paucity of mechanistic data (MOA) owes to inherent difficulties in elucidating, characterizing, and even measuring effects at the biochemical level. Case in point, a few years after the publication of *Silent Spring* [19] the relationship between eggshell thinning and exposure to chlorinated organics (most notoriously dichlorodiphenyltrichloroethane; DDT) was first established in two landmark studies [43, 66]. However, the actual causal agent (diphenyldichloroethylene; DDE) was not identified until nearly a decade later [54], and to this day the exact causative biochemical mechanism underlying the eggshell thinning phenomena is not completely understood, nearly 50 years after the first anecdotal reports of population decline. Thus, even when visually distinct impacts at the population level are recognized and the causative agent identified, the underlying process or mechanism(s) triggering the biological cascade can prove elusive even with the highest profile and most intensively researched examples. Notwithstanding, examples concurrently or pseudoconcurrently (identical experimental conditions) measuring and relating effects at the extremes of biological strata do exist in the realm of ecotoxicology. Reviews concerning "biomarkers" have been compiled for both plants [13, 29, 34] and animals [1, 36, 56–58, 90]; however, the following examples will focus exclusively on MOA-specific sublethal endpoints in relation to higher biological strata.

Examples in Vertebrates (Fish) and Invertebrates

Perhaps the most prominent and compelling example causally demonstrating weight of evidence in the biological effects cascade, indicative of the inverted complexity pyramid (biochemical peak and population base), concerns vitellogenin,

17α-ethinylestradiol (EE2), and fish. Mechanistically, vitellogenin is a lipoprotein precursor of egg yolk proteins occurring naturally in the liver of oviparous vertebrate female fishes, activated through estrogen receptors by 17β-estradiol [59]. Not expressed under normal physiological conditions in males, induction of vitellogenin (mRNA and protein) can signal exposure to estrogen-receptor agonists, which may result in renal pathology, death, and hypothesized at the time, reproductive consequences [35, 92]. Although the proposed biological cascade made logical sense mechanistically, a convincing example causally linking vitellogenin induction to survival, development, reproduction (fecundity), and ultimately population-level impacts remained elusive until a group of scientists decided to dose an entire experimental lake with EE2 in Northern Canada [50]. During 3 years of seasonal exposure to 5–6 ng/L EE2, male fathead minnows (*Pimephales promelas*) demonstrated sustained vitellogenin protein levels three orders of magnitude greater than male reference samples, with mRNA levels over a magnitude higher in exposed males than in reference females [50]. Moreover, numerous associated histological abnormalities were documented in exposed male testicular tissues including delayed spermatogenesis, fibrosis, tubule malformations, arrested testicular development, and increased incidence of ova-testes (intersex) with the presence of primary-stage oocytes [50]. As a consequence, subsequent to the second seasonal application, the fathead minnow population in the EE2 exposed lake "collapsed," with reproductive failure continuing for an additional 2 years after EE2 additions had ceased [50]. Although it can be argued in this case that vitellogenin induction is not an explicit MOA-related biochemical product toxicologically, the response is unequivocally manifested from an estrogen receptor-mediated cascade resultant from EE2 exposure, leading to histological malformations (intersex); ultimately reflected across the biological continuum at the population-level. Given that only one concentration of EE2 was tested, an artifact of experimental design, it is not possible to establish definitive concentration–response trends for comparison of relative sensitivity among biological strata within the inverted pyramid scheme (Fig. 1). However, the associated causal linkages are robust, even though not all five major criteria for inclusion (Fig. 2) were fulfilled.

Estrogen antagonists have also been showed to influence vitellogenin content with causal linkages to reduced fecundity and model-forecast population declines in *P. promelas* [59]. In female fish exposed to estrogen antagonists the vitellogenin response is conceptually inverted compared to the response in males exposed to estrogens, typified by a reduction in content, which can potentially forecast reproductive success [59]. Miller et al. [59] found that exposing fathead minnows to several estrogen antagonists (17 β-trenbolone, 17 α-trenbolone, prochloraz, fenarimol, and fadrozole) resulted in concentration-dependent inhibition of vitellogenin plasma content with strongly associated concentration-dependent reductions in fecundity. Population modeling (based on fecundity) revealed ominous projections commensurate with those identified by Kidd et al. [50], where a 25% reduction in female vitellogenin content would potentially exhibit a nearly 35% population decrease after just 2 years of exposure [59]. Although population declines were model forecast, and histology was not assessed thereby weakening the causal strength along the axis of the inverted biological effects pyramid (Fig. 1), the concomitant, and strongly correlated, concentration-dependent reductions in vitellogenin and fecundity provides strong causal evidence between these effects strata. Although the sublethal response (vitellogenin content) proved to be of similar sensitivity to higher strata effects (fecundity) in this case, the response tracked nearly identically and afforded valuable diagnostic information thus largely fulfilling the five major criteria for inclusion. In contrast to the study by Kidd et al. [50], vitellogenin content in this case can be considered directly related to MOA. Estrogen antagonist induced reduction of vitellogenin in exposed female fish resulting in reduced lipoprotein content, and consequently reduced egg production and viability, is mechanistically consequential, whereas in males exposed to estrogen agonists, increased vitellogenin production is not directly MOA related or necessarily consequential; oocyte formation is ultimately the mechanistic manifestation of exposure in male fish.

Among biologically active compounds, perhaps no MOA has been more intensively studied than AChE inhibition, and the corresponding body of literature concerning MOA-specific effects for organophosphates (OPs) alone is immense. However, relatively few studies actually measure AChE inhibition relative to survival concurrently in the same test (see [38] for a comprehensive review), even with mechanistic evolutions in vivo (in fish) dating back over 50 years [91]. In vertebrates and invertebrates AChEs (and in some cases butyrylcholinesterases; BChEs) are critically responsible for deactivating the neurotransmitter acetylcholine (ACh) via a hydrolysis reaction into choline and acetic acid [41]. In vertebrates, ACh performs numerous functions as a neurotransmitter; excitatory action in the somatic nervous system involved in voluntary muscle control, preganglionic and postganglionic functions in the parasympathetic nervous system, and preganglionic functions in the sympathetic nervous system [41]. The function of ACh in invertebrates is comparatively less well characterized, though its primary function is as a neurotransmitter for afferent nerve fibers [38]. Excess build-up of ACh in the synaptic cleft can results in overstimulation (excitation) of the post-synaptic neuron eventually leading to paralysis and potentially death [41]. However, considerable tissue-specific (e.g., brain vs. muscle) variability in sensitivity can exist [61, 78].

Perhaps the earliest study relating AChE inhibition with survival was performed with Sheepshead minnows (*Cyprinodon variegatus*) exposed to Guthion, phorate, parathion, phosphamidon, Cygon, malathion, EPN, Dursban, dichlorvos, diazinon, Dibrom, and methyl parathion, where inhibition to below 20% functionality was associated with median (40–60%) mortality [22]. Similarly exposure of several estuarine fish species including spot, *Leiostomus xanthurus*; Atlantic croaker, *Micropogon undulatus*; sheepshead minnows; and pinfish, *Lagodon rhomboids* to malathion, naled, Guthion, and parathion at median lethal concentrations (LC_{40-60}) resulted in mean AChE inhibition between 70 and 96% [23]. Coppage and Matthews [23] also found that pinfish (*L. rhomboids*) exposed to malathion exhibited constantly measured AChE inhibition at 72–79% in replicate exposed groups with 40–60% lethality at 3.5, 24, 48, and 72 h at mean exposure concentrations of 575, 142, 92, and 58 μg/L, respectively. Moreover, mean AChE activity was found to decrease in a concentration-dependent manner with increasing exposure at multiple time-points [24]. Although these studies do not convincingly fulfill the five major criteria for inclusion, they represent pioneering studies relating MOA-specific effects (AChE inhibition) with consequences at higher biological strata.

One caveat worth mentioning in the context of the current discussion with AChE inhibition is that although the relationship between effects at varying biological strata may be causally unequivocal, mechanistically speaking, based on decades of research, disparity in sensitivity is not always uniform. For example, Fulton [37] as described in [38] found a nearly 40-fold difference between concentrations resulting in acute lethality (96-h $LC_{50} = 32.16$ µg/L) and brain AChE inhibition (24-h $EC_{50} = 0.81$ µg/L) in the estuarine mummichog *Fundulus heteroclitus* exposed to azinphosmethyl. Thus, even when sensitivity exceeds the subjective divide or threshold such that biological consequence cannot be convincingly demonstrated, the mechanistic cascade may in fact be sound. In such cases a collective multispecies weight of evidence approach may be necessary, acknowledging tissue and species specific variability.

Considering the degree of historically intensive mechanistic study, there are surprisingly few current examples relating AChE inhibition beyond behavior to survival; however, Rao [65] has recently demonstrated a strong relationship in *Oreochromis mossambicus* (Tilapia) exposed to chlorpyrifos. In fish exposed to respective 24, 48, 72, and 96-h LC_{50}s of 44, 36, 31, and 26, AChE median inhibitory times (IT_{50}s) increased sequentially with concentration (8, 18, 27, and 35 h, respectively) indicating temporal dependence of inhibition on concentration. Moreover , inversely related uniform increases in AChE activity across four recovery durations (3, 7, 14, and 21 days, respectively) were also found, where time to recovery was strongly and negatively correlated ($r^2 = -0.99$) to increasing chlorpyrifos concentration [65]. A separate AChE inhibition experiment evaluated over varying durations (4, 8, 12, 16, 20, and 24 h, respectively) of a single exposure concentration (40 µg/L) also indicated strong temporal dependence ($r^2 = 0.99$), where percent inhibition increased sequentially with exposure duration [65]. Although this example does not establish proportional biological concurrence in effect manifestation and biological consequence across strata directly, the weight of evidence indirectly satisfies these criteria. It is not possible, however, to compare relative sensitivity due to the differing metrics utilized (time vs. concentration), though the diagnostic capacity is excellent.

Utilizing an alternative strategy Muniswamy et al. [61] measured ACh accumulation in relation to AChE inhibition in the freshwater fish *Labeo rohita* exposed to fenvalerate. Exposure of *L. rohita* to experimentally determined lethal ($LC_{50} = 6$ µg/L) and sublethal (one eighth of the $LC_{50} = 0.75$ µg/L) exposures were found to result in time- and concentration-dependent inhibition in the activity of AChE and consequent accumulation of ACh in multiple tissues (brain, gill, liver, and muscle). Although only two concentrations were evaluated over differing durations (1, 2, 3, and 4 days for lethal, and 1, 5, 10, 15, and 20 days for sublethal exposures, respectively) in separate experiments, the difference in effect response was proportionally consistent for both AChE inhibition and resultant Ach accumulation. Thus, much like the previous example [65], several criteria for MOA-specific sublethal effects consideration/inclusion are addressed indirectly, while others are satisfied directly.

As consistently illustrated with AChE inhibition, numerous studies have evaluated the relationship between MOA and behavioral endpoints [48, 49, 71], though the incorporation of these metrics into ERA is somewhat contentious [68]. Justifiable arguments concerning the relevance, diagnostic capacity, and comparative sensitivity of behavioral endpoints are evident [68], however, relating such effects to those accepted as consequential (survival and reproduction) remain a fundamental requirement. With this caveat in mind Kavitha and Rao [48, 49] demonstrated time-dependent reduced locomotor behavior (distance moved) in mosquito fish (*Gambusia affinis*) exposed to median lethal concentrations of monocrotophos ($LC_{50} = 20.49$ mg/L) and chlorpyrifos ($LC_{50} = 297$ µg/L) for 96 h in separate experiments. During subsequent recovery evaluations similar time-dependent increases in AChE activity and swimming speed were concurrently detailed for both compounds [48, 49]. In addition, for both chlorpyrifos and monocrotophos exposures, antioxidant enzymes (catalase, superoxide dismutase, and glutathione reductase) were all found to recover from initial inhibition in a time-dependent manner concurrently with AChE; lipid peroxidation also decreased concurrently with AChE recovery in a time-dependent manner [48, 49]. The concurrent and proportional response of multiple related sublethal effects after exposure to median lethal concentrations of AChE inhibitors again provides indirect evidence concerning causal criteria and validates the utility of behavioral endpoints in the requisite context of biological consequence.

An exemplary study demonstrating multistrata effect-correspondence in aquatic invertebrates was conducted by Duquesne [26], where the stated goal was to specifically address the issue of translating effects through different levels of biological organization. Duquesne [26] exposed *Daphnia magna* to paraoxon-methyl, the metabolite of parathion, for 24 h and monitored AChE activity, survival, body size, reproductive performance, and population growth rate, representing multiple strata in the inverted biological effects cascade pyramid scheme (Fig. 1). Exposure of *D. magna* to paraoxon-methyl was found to pseudoconcurrently (separate experiments but identical conditions) inhibit AChE activity and decrease survival in a concentration-dependent manner [26]. At exposures of 1.0 µg/L paraoxon-methyl or above AChE activity was reduced by $\geq 70\%$, whereas survival decreased significantly at or above exposures of 2.2 µg/L; respective EC_{50} and LC_{50} values after 24 h of exposure were 0.7 and 2.3 µg/L, comparatively, indicating that the sublethal response (AChE inhibition) was three times more sensitive than survival. Exposure to 1.5 µg/L resulted in significantly reduced body size 6 days after 24-h exposure and reproduction (number of offspring per surviving individual) was impaired 8–9 days subsequent with an associated 14-day EC_{50} of 2.0 µg/L [26]. Hence, as indicated by Duquesne [26] the suborganismal effects (e.g., transient inhibition of AChE) were concomitantly accompanied by effects at the organismal (survival, reduction in reproductive performance, decrease in body size) and population (reduced population growth rate) levels. In this case AChE inhibition was measured in a separate experiment than survival and reproduction, and although the experiments were done with the same test organism under the same conditions, ideally it is preferable to evaluate effects at multiple strata simultaneously in the same test. Nevertheless, by evaluating effects at multiple biological strata in a concentration-dependent and physiologically plausible manner and demonstrating differential sensitivity the causality case is compelling and four of the major criteria for sublethal effects (MOA) inclusion are robustly fulfilled.

A Similar example of multistrata effect-correspondence has also been demonstrated in terrestrial invertebrates [62]. In brown planthoppers (*Nilaparvata lugens*) exposed to azadirachtin (0.25, 0.5, and 1.0 ppm), mortality in adult females was found to increase in a concentration-dependent manner with an LC_{50} of 0.47 ppm. In a separate exposure series (0.1, 0.25, and 0.5 ppm) under the same conditions AChE activity, female body weight, and fecundity all demonstrated concentration-dependent reductions [62], although AChE activity of adult females was significantly inhibited at only 0.25 and 0.5 ppm, whereas fecundity and female-weight were found to be significantly reduced at all exposure concentrations. In addition, evaluation of ovary histology indicated disruption to follicle epithelial cells at 0.25 ppm and destruction at 0.5; morphological abnormalities were also noted at 0.5 and 1 ppm azadirachtin. In this case, the results indicate that the biochemical endpoint (AChE inhibition) demonstrated sensitive similar to that of survival, histology and gross morphology, though less sensitive than somatic or reproductive endpoints. Thus, the mechanistic endpoint does not convincingly fulfill all five criteria for inclusion here; however, the causal relationship across biological strata is sound and the diagnostic capacity is valuable.

Examples in Plants

Commensurate with AChE inhibition in vertebrates and invertebrates, inhibition of photosynthesis is arguably the most intensively studied MOA in plants. Research concerning the MOA of photosystem II (PSII) inhibiting herbicides such as *s*-triazines date back to the 1950s [60], where it was first posited that these compounds disrupted the Hill reaction (photoreduction of an electron acceptor by electrons and protons originating from water and resulting in the evolution of oxygen). Subsequent research specifically suggested inhibition of noncyclic photophosphorylation [75]; however, the ultimate target site was not confirmed until 1980s [47] as blockage of electron flow between the primary acceptor (Q) and the secondary acceptor (B), now formally known as the Q_B binding site of the D1 protein and plastoquinone, respectively. Considering the previously mentioned protracted timeframe concerning MOA discovery, there are few examples actually detailing concurrent measurement of PSII and morphological (growth) inhibition, until recently. Utilizing a relatively new technique (chlorophyll fluorescence) to evaluate PSII inhibition Magnusson et al. [55] found very similar concentration–response curve shapes and slopes among growth rate, biomass and PSII efficiency (effective quantum yield) for estuarine diatoms (*Navicula* sp.) and green algae (*Nephroselmis pyriformis*) exposed to diuron, hexazinone, and atrazine. Photosystem II efficiency is a parameter which essential measures the proportion of chlorophyll absorbed light associated with PSII that is used in photochemistry and provides a measure of the linear electron transport Maxwell and Johnson [94].

Among the three metrics evaluated, biomass was the most sensitive for *Navicula* sp. whereas PSII efficiency was the most sensitive for *N. pyriformis*; growth rate was the least sensitive in all cases [55]. After 72 h of exposure the relationships between growth rate, biomass and PSII efficiency were linear and correlated with $r^2 \geq 0.90$ for each species and herbicide concentration [55]. Moreover the correlation regression slopes were near unity for both species indicating good agreement between endpoints over concentrations spanning three orders of magnitude and for two very different organisms [55]. As stated by Magnusson et al. [55] these results directly link inhibition of PSII mechanistically with declines measured in endpoints at higher strata (growth rate and biomass), which are used routinely by industry and regulators. These results also convincingly satisfy the five consideration criteria, particularly concurrence, consequence and biological plausibility of effect. Diagnostic resolution is afforded by measuring a mechanistic surrogate for plastoquinone competition (electron transfer), and although sensitivity of the mechanistic endpoint was comparable to biomass, PSII efficiency was up to 1.8-fold, and on average 1.5-fold more sensitive than the standard regulatory endpoint (growth rate) based on comparison of EC_{50}'s [55].

Perhaps considered comparatively less intuitive in nature than the case with herbicides, fairly robust examples systematically linking metabolite accumulation and depletion to growth inhibition have recently been demonstrated in plants exposed to pharmaceuticals. Much like the previous example [55], Brain et al. [14] characterized very similar concentration–response curve shapes between inhibition of sterol biosynthesis (stigmasterol and β-sitosterol) and biomass production in *Lemna gibba* exposed to statin blood lipid regulators (atorvastatin and lovastatin). In higher plants the target enzyme 3-hydroxy-3-methylglutaryl coenzyme-A reductase (HMGR; analogous to the human receptor) regulates cytosolic isoprenoid biosynthesis in the mevalonic acid (MVA) pathway [6], ultimately responsible for the synthesis of sterols, which are critical components of plant membranes also regulating morphogenesis and development [39, 73]. As a consequence of statin-induced, concentration-dependent reductions in sterol production in exposed plants *in vivo*, biomass (growth) was similarly and concurrently inhibited; however, the mechanistic endpoint (sterol reductions as a result of HMGR inhibition) was two (atorvastatin exposed) to nearly ten times (lovastatin exposed) as sensitive based on comparison of respective EC_{50}'s [14]. This relationship established between biologically stratified endpoints [14] is supported by a thoroughly characterized biochemical pathway [74] with known implications resulting from disruption [39, 73] thus satisfying the remaining confidence and incorporation criteria of plausibility and diagnostics.

In a conceptually similar study Brain et al. [16] showed an inverse relationship between metabolite production and growth inhibition in *L. gibba* exposed to the sulfonamide antibiotic sulfamethoxazole. In bacteria sulfonamides specifically target the enzyme dihydropteroate synthase (DHPS) in the folate biosynthetic pathway, which was recently established as identical to that of plants [8]. In plants, as in all living organisms, folates (Vitamin B9) are responsible for a host of functions [42, 67], particularly as essential molecules mediating the transfer of one-carbon units

(C1 metabolism) in metabolic pathways that are of paramount importance to cellular viability. As an indirect diagnostic surrogate measure of DHPS inhibition, Brain et al. [16] quantitatively assessed an upstream metabolite (*p*-aminobenzoic acid; *p*ABA), and found accumulation levels that were 18 and 39 times more sensitive to sulfamethoxazole exposure in vivo than biomass and frond number morphological endpoints, respectively, based on EC_{50}'s [16]. Moreover, the characterized exponential rise in *p*ABA accumulation was manifested concurrently and proportionally with growth inhibition thus satisfying biological consequence and rounding out satisfaction of the five confidence and incorporation criteria.

Approach and Strategy: Orienting the ERA

According to the decision tree outlined in Fig. 4, MOA must first be characterized biochemically, and second identified (or likely present) in the nontarget organism(s) of interest, prior to further consideration in the ERA process (see [4, 15]). In the context of predictive modeling and intelligent testing [21], pragmatically orienting, focusing, and prioritizing testing efforts requires an in-depth understanding of MOA [12]. However, implicitly underlying this contention is the qualification of adequately discerning MOA, ranging in nature from receptor-mediated to narcosis (baseline toxicity). As a generality, the more specific the MOA the narrower the biological range of potentially affected nontarget organisms; conversely, the more general the MOA, the broader the biological range of potential nontarget organisms. Consequently, a more fundamentally important question is whether or not the nontarget organism of interest expresses an appropriate and susceptible receptor (see [46]). Perhaps the most definitive methodology to address this question lies within the technology known as "omics" (e.g., genomics and proteomics) [52]. Detailing the intensive process of MOA discovery during chemical development is beyond the scope of the current discussion; hence we shall focus explicitly on methodological strategies for MOA identification in potentially susceptible nontarget organisms here. Integral to this process is the concept of evolutionary conservation (receptor homology; genetic similarity), within and among different taxa [20, 21, 27, 40, 51, 52]. As outlined by Kwekel et al. [52], "the availability of complete genome sequences for multiple species provides unprecedented opportunities for comprehensive comparative analysis in support of mechanistic and predictive toxicology..." However, simply comparing gene sequences is not enough; rather, investigations of orthologous gene relationships based on sequence similarity (reciprocal BLAST (Basic Local Alignment Search Tool) best hit), synteny (conserved order of genes), phylogenetic tree matching (organism-level relatedness), and functional complementation (conservation of molecular function) across species are a recommended approach [52]. This process is most pragmatically pursued using focused, gene-specific, and hypothesis-driven investigations, and several queriable genetic and proteomic resources are currently available for this purpose and reviewed by Kwekel et al. [52].

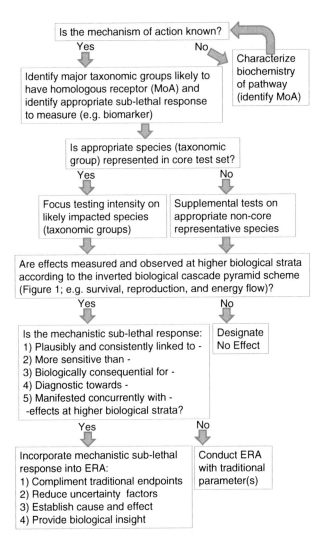

Fig. 4 Decision flowchart concerning the incorporation of MOA-specific considerations and sublethal effects into the collective ecological risk assessment process for biologically active compounds (e.g., pharmaceuticals, pesticides)

An exemplary illustration of receptor homology is provided by Gunnarsson et al. [40], where best BLASTP hit (protein-based BLAST search) was utilized to evaluate possible sequence alignments for 1,318 unique human target orthologs (135 drug targets for a total of 1,152 drugs) in 16 species with representation from vertebrates, invertebrates, arthropods, plants, yeast, and bacteria. Among 5 functional categories (enzyme, receptor, ion channel, transporter, and "other") enzyme groups were found to have the largest intersections, sharing 53 drug targets, suggesting these targets are highly conserved and ecotoxicologically relevant [40].

Among classes, vertebrates were found to have the highest similarity in target homology with human drug target orthologs. In the context of species currently used for aquatic environmental risk assessments, fish and frogs were predicted to have by far the greatest number with the highest degree of similarity [40], which has also been corroborated by Christen et al. [20]. For example, Zebrafish had orthologs to 86% of the drug targets, whereas only 61% were conserved in *Daphnia* and 35% in green algae [40]. Thus, orthology prediction can be used as an effective tool to identify potential receptors in nontarget organisms, and additionally refine and orient effects testing efforts to the most relevant species. As with any technology, however, caution must be exercised as low sequence homology is not an absolute predictor of receptor-mediated effect or lack of effect. As an example HMG-CoA reductase was only found to be 30% homologous between humans and plants [40], yet research suggests conservation of function as observed for strong inhibition by human targeted drugs [14]. Thus, after identifying whether a conserved drug target is present in nontarget organisms, it is critical to define whether the function is also conserved [4]. Notwithstanding, -omics approaches hold promise and will only become more refined and specific as advances are made. In fact, the Adverse Outcome Pathways (AOP) concept represents a robust approach for ecological risk assessments of pharmaceuticals, pesticides and other contaminants [5, 45]. For example, Villeneuve and Garcia-Reyero [89] further explore the utility of -omics and AOP in several predictive ecotoxicology applications.

Prior to pursuing "omics" technology, however, addressing the more obvious and intuitive consideration of taxonomy (or life history) provides an initial, albeit coarse, point of reference. For example, nontarget effects of herbicides would generally be most logically pursued among plants, whereas the effects of estrogenic compounds would more appropriately be pursued among nontarget animals (e.g., vertebrates). Fortunately, between these two extremes lie intermediate methodologies to further aid and address these questions, for example, quantitative structure activity relationships (QSARs) and acute to chronic ratios (ACRs).

If the MOA is unknown QSARs afford a means to at least suggest the nature of effect by classifying a compound into four major potential classes: nonpolar narcotic compounds (I), polar narcotic compounds (II), reactive compounds (III), and compounds with a specific mode of action (IV; e.g., pharmaceuticals, pesticides) [86]. This general classification strategy has been applied to numerous databases representing a diverse array of compounds [10–12, 30, 70, 85, 86]. Using toxic ratios (the ratio of predicted and measured toxicity) Vaal et al. [85] found that reactive compounds and compounds with a specific mode of action were a factor of 10–100,000 more toxic than predicted, which was found to be seemingly independent of corresponding log Kow values, a critical determinant in QSAR modeling. Thus for a given compound, higher toxic ratios are potentially indicative or at least suggestive of a specifically acting MOA. Moreover, toxic ratios applied to a base subset of organisms (e.g., fish, invertebrates, and plants) can potentially provide insight into which major taxonomic groups are likely to be impacted, or those most likely to contain an appropriate receptor. Similar in concept to QSARs, chemical "read-across" approaches, which facilitate inferences about potential toxicity based on structural similarity to chemicals with known toxicity profiles, are also gaining favor, particularly for prioritization [21] and potentially regulatory efforts in the USA [45] and in the EU under REACH [93].

In a conceptually similar fashion, ACRs (e.g., ratio of the LC_{50} and NOEC or LOEC) have demonstrated utility in distinguishing between specific-acting and generally acting MOAs [69]. In an analysis of nonpolar narcosis (baseline toxicity), polar narcosis, a-specific reactivity, specific reactivity (receptor mediated), and heavy metals, Roex et al. [69] demonstrated that specifically acting chemicals typically have larger average ACRs, though variability was also large. Ahlers et al. [2] found less definitive trends; although narcosis (polar and particularly, nonpolar narcosis) was considered a useful predictor for low ACRs, a nonnarcotic MOA was not considered a reliable indicator of high ACR. However, partitioning the dataset according to specific "structural alerts" (SAs; defining chemical groups such as phenols, amines, esters etc.), compounds containing at least one SA had a substantially increased probability for a high ACR [2]. Thus it was suggested that a scheme combining both MOA and SA knowledge could potentially better discriminate between low and high ACRs [2].

When an ACR value is unknown, default values are often employed for regulatory purposes. For example, Raimondo et al. [63] identified a 90th centile ACR value of 79.5 for aquatic contaminants. In fact, Rand [64] suggested that the larger the size of an ACR, the greater the likelihood of a chemical acting through a specific MOA. When ACRs were considered for pharmaceuticals, Sanderson and Thomsen [72] suggested that an ACR of 100 may be adequate for estimating the chronic responses of *Daphnia* sp. and algae because nonspecific, narcosis MOAs may be appropriate. In these organisms, nonspecific, narcosis MOA are more likely to be observed based on relatively lower conservation of drug targets than aquatic vertebrates [4, 17, 40, 46], though, as noted above, the effects of antibiotic effects to plants and algae represent noticeable exceptions [15, 16]. However, it is critical to note that much higher ACR values (e.g., >1,000,000) have been reported for some pharmaceuticals when sublethal chronic fish responses (e.g., not 7 day juvenile *P. promelas* growth) are plausibly linked to pharmacological MOAs [3]. Berninger and Brooks [9] noted that a default ACR value of 100 could represent just a 20th centile value for chronic effects of pharmaceutical on fish, when MOA related endpoints are used to calculate an ACR. Here again, it may be possible to leverage mammalian therapeutic information using biological "read-across" approaches, which could identify classes of pharmaceuticals presenting the greatest potential hazards to fish. For example, maximizing the pharmacological margin of safety (MOS) is an important consideration during the development of pharmaceuticals. However, as demonstrated by Berninger and Brooks [9], compounds with larger MOS values are often more potent such that higher MOS values may be predictive of larger ACRs in fish.

Once a chemicals MOA has been characterized and candidate nontarget species with known or suspected (target) susceptibility have been identified, effects testing can be taxonomically focused and the "relationship nature" between MOA and effects at higher biological strata established (Fig. 4). Establishing or defining "relationship nature" requires satisfaction of the five criteria of MOA incorporation outlined previously (Fig. 2) and providing a robust causal weight-of-evidence case according to the biological effects cascade pyramid scheme (Fig. 1). If the sublethal mechanistic response is purely transient and without relevant consequence for higher biological strata (e.g., fitness and survival) consideration for regulation, or establishment of life-criteria cannot be justified. Ultimately, this debate must address the "so what" question; if an affect measured at lower strata is temporary and/or reversible without direct consequence at the community, population, or even organismal level why is it important? Unfortunately, the threshold at which sublethal responses become consequential or relevant is unavoidably subjective, but can be defined as the concentration beyond which irreparable impact to higher biological strata are predicted to occur via biological chain-reaction or cascade. As a conservative estimate suggested here, a bracketing range of MOA sensitivity between 1.5 and 10× is subjectively considered as being predictively useful for, yet still consequentially relevant to, effects at higher biological strata (based on enzyme inhibition). However, justification for basing ERA thresholds on sublethal mechanistic effects instead of lethality is fundamentally and causally limited, and thus direct and absolute substitution for traditional threshold is not explicitly recommended. Mechanistic sublethal effects, or for the sake of argument any sublethal effect in general, cannot singularly be used to replace traditionally measured effects at higher-strata (survival and reproduction) for the purposes of ERA given the requisite need to establish cause-and-effect relationships (relate) to these metrics. Thus, sublethal responses, by virtue of requiring validation in effects of regulatory consequence (survival and fecundity), are not capable of circumventing traditional test metrics, rather their predictive and diagnostic capability provides a powerful and invaluable foundation for understanding process, cause-and-effect, and addressing uncertainty surrounding true threshold tolerance. Toxicity values typically employed in ERA range considerably in terms of tolerable or acceptable impact criteria from highly conservative (e.g., NOAELs and LOAELs or LC_5s and LC_{10}s) to less conservative (e.g., LC_{25}s and LC_{50}s) depending on the nature and goals of the assessment. Thus the case could be made that simply using a more conservative value based on traditional metrics and traditional uncertainty assessment could be an equally effective approach, depending on the magnitude of the uncertainty factors used for extrapolation from acute to chronic effects. However, this contention systematically ignores the value in mechanistically based cause-and-effect relationships across multiple biological strata. In a practical sense then, incorporation of well defined sublethal effects values could more appropriately be considered for the purposes of reducing uncertainty (addressing and alleviating arbitrary application factors) rather than strictly replacing the traditional endpoint out of principal to further proliferate unnecessary and unrealistic conservative precaution. The fundamental caveat inherent in the previous statement of course is explicitly contingent on convincingly demonstrating the five hypothetical criteria for inclusion outlined here. If ACR values are developed for various MOAs, then ERAs of biologically active molecules will benefit from the application of science-based, rather than simple default, uncertainty factors.

Conclusions

Under the current ERA paradigm for biologically active compounds accepted and required sublethal effects data are largely composed of fecundity-based, gross-necropsy, and pathology measures. Conversely, incorporation of sublethal effects data from lower biological strata remains a highly contentious issue due to lack of established guidance concerning formal criteria for acceptance. In the present assessment a methodological framework is proposed which consists of five formal criteria, intended as acceptability considerations concerning the causal weight of evidence supporting the incorporation of MOA-specific data into ERA process; plausibility and consistent linkage, comparative sensitivity, biological consequence, diagnostic capacity, and temporal, concentration–response concurrence. In fact, the approach presented here is consistent with the AOP concept just recently developed by Ankley et al. [5]. Although our criteria manifested specifically in consideration of MOA, the methodology can be broadly applied to effects assessed at lower biological strata in general. Depending on the adequacy or degree to which the suggested criteria are satisfied experimentally dictates the strength of the causal weight-of-evidence case, ultimately providing justification for incorporation. The underlying fundamental premise underlying this methodology derives from the concept of biological effects cascading. It is argued that invariably every effect realized and measured at higher biological strata is first manifested at the biochemical level and essentially resonated and magnified up through higher biological tiers conceptually analogous to an inverted pyramid. If effects measured at lower biological tiers cannot be relevantly and consequentially linked to those measured at higher strata, there is effectively no justification for incorporation or further pursuit of the effect(s) in question. Once causal strength has been established, the nature of sublethal effects incorporation is suggested to be predictive, pre-emptive and diagnostic. Rather than empirically and systematically replacing traditional endpoints for the purposes of conservatism, utilization of sublethal effects data is recommended to reduce uncertainty, address and alleviate arbitrary application factors, and emphasize cause-and-effect relationships across multiple biological strata. The process suggested here can be preemptively refined based on intelligent testing methodologies where taxonomy, QSARs, ACRs, and more specifically genomics and proteomics technologies can be effectively utilized to orient, focus, and refine testing efforts by predicting candidate nontarget organisms or groups of organisms most likely to be susceptible to a given stressor of interest. Ultimately, the fundamental goal of sublethal effects generation and incorporation should center on uncertainty reduction not propagation.

References

1. Adams SM, Giesy JP, Tremblay LA, Eason CT (2001) The use of biomarkers in ecological risk assessment: recommendations from the Christchurch conference on biomarkers in ecotoxicology. Biomarkers 6:1–6

2. Ahlers J, Riedhammer C, Vogliano M, Ebert R-U, Kuhne R, Schuurmann G (2006) Acute and chronic ratios in aquatic toxicity-variation across trophic levels and relationship with chemical structure. Environ Toxicol Chem 25:2937–2945

3. Ankley GT, Black MC, Garric J, Hutchinson TH, Iguchi T (2005) A framework for assessing the hazard of pharmaceutical materials to aquatic species. In: Williams R (ed) Science for assessing the impacts of human pharmaceutical materials on aquatic ecosystems. SETAC Press, Pensacola, pp 183–238

4. Ankley GT, Brooks BW, Hugget DB, Sumpter JS (2007) Repeating history: pharmaceuticals in the environment. Environ Sci Technol 41:8211–8217

5. Ankley GT, Bennett RS, Erickson RJ, Hoff DJ, Hornung MW, Johnson RD, Mount DR, Nichols JW, Russom CL, Schmieder PK, Serrano JA, Tietge JE, Villeneuve DL (2010) Adverse outcome pathways: a conceptual framework to support ecotoxicology research and risk assessment. Environ Toxicol Chem 29:730–741

6. Bach TJ, Lichtenthaler HK (1982) Inhibition of mevalonate biosynthesis and of plant growth by the fungal metabolite mevinolin. In: Wintermanns JFGM, Kuiper PJC (eds) Biochemistry and metabolism of plant lipids. Elsevier Biochemical Press, Amsterdam, pp 515–521

7. Bartholomew G (1964) The roles of physiology and behavior in the maintenance of homeostasis in the desert environment. In: Hughes GM (ed) Symposia of the society for experimental biology, vol 18, Homeostasis and feedback mechanisms (p 460). Academic, New York, pp 7–29

8. Basset GJC, Quinlivan EP, Gregory JF, Hanson AD (2005) Folate synthesis and metabolism in plants and prospects for biofortification. Crop Sci 45:449–453

9. Berninger JP, Brooks BW (2010) Leveraging mammalian pharmaceutical toxicology and pharmacology data to predict chronic fish responses to pharmaceuticals. Toxicol Lett 193:69–78

10. Bradbury S (1994) Predicting modes of toxic action from chemical structure: an overview. SAR QSAR Environ Res 2:89–104

11. Bradbury S (1995) Quantitative structure-activity relationships and ecological risk assessment: an overview of predictive aquatic toxicology research. Toxicol Lett 79:229–237

12. Bradbury S, Feijtel T, van Leeuwen C (2004) Meeting the scientific needs of ecological risk assessment in a regulatory context. Environ Sci Technol 38:463A–470A

13. Brain RA, Cedergreen N (2009) Biomarkers in plants: selection and utility. Rev Environ Contam Toxicol 198:49–110

14. Brain RA, Reitsma TS, Lissemore LI, Bestari B-J, Sibley PK, Solomon KR (2006) Herbicidal effects of statin pharmaceuticals in *Lemna gibba*. Environ Sci Technol 40:5116–5123

15. Brain RA, Hanson ML, Solomon KR, Brooks BW (2008) Aquatic plants exposed to pharmaceuticals: effects and risks. Rev Environ Contam Toxicol 192:67–115

16. Brain RA, Ramirez AJ, Fulton BA, Chambliss CK, Brooks BW (2008) Herbicidal effects of sulfamethoxazole in *Lemna gibba*G3: using pABA as a biomarker of effect. Environ Sci Technol 42:8965–8970

17. Brooks BW, Ankley GT, Hobson JF, Lazorchak JM, Meyerhoff RD, Solomon KR (2008) Assessing the aquatic hazards of veterinary medicines. In: Crane M, Barrett K, Boxall A (eds) Effects of veterinary medicines in the environment. CRC Press, Boca Raton, pp 97–128

18. Cairns J (1966) Don't be half-safe-the current revolution in bio-assay techniques. In: Proceedings of the 21st industrial waste conference, Purdue University, Engineering Extension series, Richmond, pp 559–567

19. Carson R (1962) Silent spring. Houghton Mifflin, Boston

20. Christen V, Hickmann S, Rechenberg B, Fent K (2010) Highly active human pharmaceuticals in aquatic systems: a concept for their identification based on their mode of action. Aquat Toxicol 96:167–181

21. Combes R, Barratt M, Balls M (2003) An overall strategy for the testing of chemicals for human hazard and risk assessment under the EU REACH system. Altern Lab Anim 31:7–19

22. Coppage D (1972) Organophosphate pesticides: specific level of brain AChE inhibition related to death in sheepshead minnows. Trans Am Fish Soc 101:534–536

23. Coppage D, Matthews E (1974) Short-term effects of organophosphate pesticides on cholinesterases of estuarine fishes and pink shrimp. Bull Environ Contam Toxicol 11:483–488

24. Coppage D, Matthews E, Cook G, Knight J (1975) Brain acetylcholinesterase inhibition in fish as a diagnosis of environmental poisoning by malathion, O, O-dimethyl s-(1,2-dicarbethoxyethyl) phosphorodithioate. Pestic Biochem Physiol 5:536–542

25. Doll R, Hill A (1950) Smoking and carcinoma of the lung: preliminary report. Br Med J 4682:739–748

26. Duquesne S (2006) Effects of an organophosphate on *Daphnia magna* at suborganismal and organismal levels: implications for population dynamics. Ecotoxicol Environ Saf 65:145–150

27. Eads B, Andrews J, Colbourne J (2008) Ecological genomics in *Daphnia*: stress responses and environmental sex determination. Nat Heredity 100:184–190

28. EMEA (2005) Guideline on the environmental risk assessment of medicinal products for human use. The European Agency for the Evaluation of Medicinal Products: Committee for Medicinal Products for Human Use, London

29. Ernst WHO, Peterson PJ (1994) The role of biomarkers in environmental assessment (4). Terrestrial plants. Ecotoxicology 3:180–192

30. Escher B, Hermens J (2002) Modes of action in ecotoxicology: their role in body burdens, species sensitivity, QSARs, and mixture effects. Environ Sci Technol 36:4201–4217

31. EUC (2002) Working document, Guidance document on risk assessment for birds and mammals under Council Directive 91/414/EEC. SANCO/4145/2000—final. European Commission Health & Consumer Protection Directorate-General, Directorate E—Food safety: plant health, animal health and welfare, international questions E1—Plant health, Brussels

32. EUC (2002) Working document, Guidance document on aquatic ecotoxicology in the context of the Directive 91/414/EEC. Sanco/3268/2001 rev.4 (final). European Commission Health & Consumer Protection Directorate-General, Directorate E—Food safety: plant health, animal health and welfare, international questions E1—Plant health, Brussels

33. EUC (2002) Working document, Guidance document on terrestrial ecotoxicology under Council Directive 91/414/EEC. SANCO/10329/2002 rev 2 final. European Commission Health & Consumer Protection Directorate-General, Directorate E—Food safety: plant health, animal health and welfare, international questions E1—Plant health, Brussels

34. Ferrat L, Pergent-Martini C, Roméo M (2003) Assessment of the use of biomarkers in aquatic plants for the evaluation of environmental quality: application to seagrasses. Aquat Toxicol 65:187–204

35. Folmar L, Gardner G, Schreibman M, Magliulo-Cepriano L, Mills L, Zaroogian G, Gutjahr-Gobell R, Haebler R, Horowitz D, Denslow N (2001) Vitellogenin-induced pathology in male summer flounder (*Paralichthys dentatus*). Aquat Toxicol 51:431–441

36. Forbes VE, Palmqvist A, Bach L (2006) The use and misuse of biomarkers in ecotoxicology. Environ Toxicol Chem 25:272–280

37. Fulton M (1989) The effects of certain intrinsic and extrinsic variables on the lethal and sublethal toxicity of selected organophosphorus insecticides in the mummichog, *Fundulus heteroclitus*, under laboratory and field conditions. Thesis, University of South Carolina, Columbia

38. Fulton M, Key P (2001) Acetylcholinesterase inhibition in estuarine fish and invertebrates as an indicator of organophosphorous insecticide exposure and effects. Environ Toxicol Chem 20:37–45

39. Grandmougin-Ferjani A, Schuller-Muller I, Hartmann MA (1997) Sterol modulation of the plasma membrane H+-ATPase activity from corn roots reconstituted into soyabean lipids. Plant Physiol 113:163–174

40. Gunnarsson L, Jauhiainen A, Kristiansson E, Nerman O, Larsson J (2008) Evolutionary conservation of human drug targets in organisms used for environmental risk assessments. Environ Sci Technol 42:5807–5813

41. Habig C, DiGiulio R (1991) Biochemical characteristics of cholinesterases in aquatic organisms. In: Mineau P (ed) Cholinesterase inhibiting insecticides: their impact on wildlife and the environment, vol 2, Chemicals in agriculture. Elsevier, New York, pp 19–34

42. Hanson AD, Roje S (2001) One-carbon metabolism in higher plants. Annu Rev Plant Physiol 52:119–137

43. Hickey J, Anderson D (1968) Chlorinated hydrocarbons and eggshell changes in raptorial and fish-eating birds. Science 162:271–273

44. Hill A (1965) The environment and disease: association or causation? Proc R Soc Med 58:295–300

45. Hoff D, Lehmann W, Pease A, Raimondo S, Russom C, Steeger T (2010) Predicting the toxicities of chemicals to aquatic animal species (White Paper). U.S. Environmental Protection Agency, Washington, DC, 127 p

46. Hugget DB, Cook JC, Ericson JF, Williams RT (2003) A theoretical model for utilizing mammalian pharmacology and safety data to prioritize potential impacts of human pharmaceuticals to fish. Hum Ecol Risk Assess 9:1789–1799

47. Jursinic P, Stemler A (1983) Changes in [14C]atrazine binding associated with the oxidation-reduction state of the secondary quinone acceptor of photosystem II. Plant Physiol 73:703–708

48. Kavitha P, Rao V (2007) Oxidative stress and locomotor behaviour response as biomarkers for assessing recovery status of mosquito Wsh, *Gambusia* aYnis after lethal eVect of an organophosphate pesticide, monocrotophos. Pestic Biochem Physiol 87:182–188

49. Kavitha P, Rao V (2008) Toxic effects of chlorpyrifos on antioxidant enzymes and target enzyme acetylcholinesterase interaction in mosquito fish, *Gambusia affinis*. Environ Toxicol Pharmacol 26:192–198

50. Kidd K, Blanchfield P, Mills K, Palace V, Evans R, Lazorchak J, Flick R (2007) Collapse of a fish population after exposure to a synthetic estrogen. Proc Natl Acad Sci U S A 104:8897–8901

51. Köhler H-R, Kloas W, Schirling M, Lutz I, Reye A, Langen J-S, Triebskorn R, Nagel R, Schönfelder G (2007) Sex steroid receptor evolution and signalling in aquatic invertebrates. Ecotoxicology 16:131–143

52. Kwekel J, Burgoon L, Zacharewski T (2008) Comparative toxicogenomics in mechanistic and predictive toxicology. In: Brown J (ed) Comparative genomics basic and applied research. CRC Press, Taylor and Francis Group, Boca Raton, p 400

53. Levine R (1991) Recognized and possible effects of pesticides in humans. In: Hayes W, Laws E (eds) Handbook of pesticide toxicology, vol 1. Academic, New York, pp 275–360

54. Lincer J (1975) DDE-induced eggshell-thinning in the American Kestrel: a comparison of the field situation and laboratory results. J Appl Ecol 12:781–793

55. Magnusson M, Heimann K, Negri A (2008) Comparative effects of herbicides on photosynthesis and growth of tropical estuarine microalgae. Mar Pollut Bull 56:1545–1552

56. McCarty LS, Munkittrick KR (1996) Environmental biomarkers in aquatic toxicology: fiction, fantasy, or functional? Hum Ecol Risk Assess 2:268–274

57. McCarthy JF, Shugart LR (1990) Biomarkers of environmental contamination. Lewis Publishers, Boca Raton

58. McCarty LS, Power M, Munkittrick KR (2002) Bioindicator vs biomarkers in ecological risk assessment. Hum Ecol Risk Assess 8:159–164

59. Miller D, Jensen K, Villeneuve D, Kahl M, Makynen E, Durhan E, Ankley G (2007) Linkage of biochemical response to population-level effects: a case study with vitellogenin in the fathead minnow (*Pimephales promelas*). Environ Toxicol Chem 26:521–527

60. Moreland D, Gentner W, Hilton J, Hill K (1959) Studies on the mechanism of herbicidal action of 2-chloro-4,6-bis(ethylamino)-S-triazine. Plant Physiol 34:432–435

61. Muniswamy D, Patil V, Chebbi S, Marigoudar S, Chittaragi J, Halappa R (2009) Gas-liquid chromatography for fenvalerate residue analysis: in vivo alterations in the acetylcholinesterase activity and acetylcholine in different tissues of the fish, *Labeo rohita* (Hamilton). Toxicol Mech Methods 19:410–415

62. Nathan S, Choi M, Seo H, Paik C, Kalaivani K, Kim J (2008) Effect of azadirachtin on acetyl-cholinesterase (AChE) activity and histology of the brown planthopper *Nilaparvata lugens* (Stal). Ecotoxicol Environ Saf 70:244–250

63. Raimondo S, Montague BJ, Barron MG (2007) Determinants of the variability in acute to chronic toxicity ratios for aquatic invertebrates and fish. Environ Toxicol Chem 26:2019–2023

64. Rand GM (1995) Fundamentals of aquatic toxicology: effects, environmental fate, and risk assessment. Taylor & Francis, Washington, DC

65. Rao V (2008) Brain acetylcholinesterase activity as a potential biomarker for the rapid assessment of chlorpyrifos toxicity in a euryhaline fish, *Oreochromis mossambicus*. Environ Bioindic 3:11–22

66. Ratcliffe D (1967) Decrease in eggshell weight in certain birds of prey. Nature 215:208–210

67. Rébeillé FD, Macherel D, Mouillon JM, Garin J, Douce R (1997) Folate biosynthesis in higher plants: purification and molecular cloning of a bifunctional 6-hydroxymethyl-7,8-dihydropterin pyrophosphokinase/7,8-dihydropteroate synthase localized in mitochondria. EMBO Rep 16:947–957

68. Robinson P (2009) Behavioural toxicity of organic chemical contaminants in fish: application to ecological risk assessments (ERAs). Can J Fish Aquat Sci 66:1179–1188

69. Roex E, van Gestel C, Van Wezel AP, Van Straalen NM (2000) Ratios between acute aquatic toxicity and effects on population growth rates in relation to toxicant mode of action. Environ Toxicol Chem 19:685–693

70. Russom C, Bradbury S, Broderius S, Hammermeister D, Drummond R (1997) Predicting modes of toxicity action from chemical structure: acute toxicity in the fathead minnow (*Pimephales promelas*). Environ Toxicol Chem 16:948–967

71. Sandahl J, Baldwin D, Jenkins J, Scholz N (2005) Comparative thresholds for acetylcholinesterase inhibition and behavioral impairment in coho salmon exposed to chlorpyrofos. Environ Toxicol Chem 24:136–145

72. Sanderson H, Thomsen M (2009) Comparative analysis of pharmaceuticals versus industrial chemicals acute aquatic toxicity classification according to the United Nations classification system for chemicals: assessment of the (Q)SAR predictability of pharmaceuticals acute aquatic toxicity and their predominant acute toxic mode-of-action. Toxicol Lett 187:84–93

73. Schaller H (2004) New aspects of sterol biosynthesis in growth and development of higher plants. Plant Physiol Biochem 42:465–476

74. Schwender J, Seemann M, Lichtenthaler HK, Rohmer M (1996) Biosynthesis of isoprenoids (carotenoids, sterols, prenyl side-chains of chlorophylls and plastoquinone) via a novel pyruvate/glyceraldehyde 3-phosphate non-mevalonate pathway in the green algae *Scenedesmus obliquus*. Biochem J 316:73–80

75. Shimabukuro R, Swanson H (1969) Atrazine metabolism, selectivity, and mode of action. J Agric Food Chem 17:199–205

76. Sprague J (1971) Measurement of pollutant toxicity to fish-III sublethal effects and "safe" concentrations. Water Res 5:245–266

77. Stephan CE, Mount DI, Hansen DJ, Gentile JH, Chapman GA, Brungs WA (1985) Guidelines for deriving numerical national water quality criteria for the protection of aquatic organisms and their uses. US EPA ORD ERL. PB 85-227049, Duluth, pp 1–97

78. Strum A, Wogram J, Segner H, Liess M (2000) Different sensitivity to organophosphates of acetylcholinesterase and butyrylcholinesterase from three-spined stickleback (*Gasterosteus aculeatus*): application in biomonitoring. Environ Toxicol Chem 19:1607–1615

79. USEPA (2004) Overview of the ecological risk assessment process in the Office of Pesticide Programs, U.S. Environmental Protection Agency: endangered and threatened species effects determinations. United States Environmental Protection Agency, Office of Prevention, Pesticides and Toxic Substances Office of Pesticide Programs, Washington, DC

80. USEPA (2007) Federal Register. Part II. Environmental Protection Agency. 40 CFR Parts 9, 152, 156, 159 et al. Pesticides; data requirements for conventional chemicals, technical amendments, and data requirements for biochemical and microbial pesticides; final rules. United States Environmental Protection Agency, Washington, DC, 56 p

81. USEPA (2008) Aquatic life criteria for contaminants of emerging concern. Part I: general challenges and recommendations. U.S. Environmental Protection Agency, Washington DC. http://water.epa.gov/scitech/swguidance/waterquality/standards/upload/2008_06_03_criteria_sab-emergingconcerns.pdf. Accessed 04/2010
82. USEPA (2009) Endocrine disruptor screening program test guidelines OPPTS 890.1350: fish short-term reproduction assay. Office of Prevention, Pesticides and Toxic Substances (7101), Washington, DC, 93 p
83. USEPA (2009) Federal Register/Vol. 74, No. 84/Monday, May 4, 2009/Rules and regulations, p 20422. http://edocket.access.gpo.gov/2009/pdf/E9-10203.pdf. Accessed 04/2010
84. USFDA (1998) Guidance for industry; environmental assessment of human drug and biologics applications. U.S. Department of Health and Human Services, Food and Drug Administration, Center for Drug Evaluation and Research (CDER), Center for Biologics Evaluation and Research (CBER), Rockville, CMC 6, 39 p
85. Vaal M, Van Leeuwen C, Hoekstra J (2000) Variation in sensitivity of aquatic species to toxicants: practical consequences for effect assessment of chemical substances. Environ Manage 25:415–423
86. Verhaar H, van Leeuwen C, Bol J, Hermens J (1994) Application of QSARs in risk management of existing chemicals. SAR QSAR Environ Res 2:39–58
87. VICH (2000) Environmental impact assessment (EIAS) for veterinary medicinal products (VMPS)—phase I. VICH GL6 (ecotoxicity phase I). VICH International Cooperation on Harmonisation of Technical Requirements for Registration of Veterinary Medicinal Products, Brussels
88. VICH (2004) Environmental impact assessment for veterinary medicinal products: phase II guidance. VICH-GL38 (ecotoxicity phase II). VICH International Cooperation on Harmonisation of Technical Requirements for Registration of Veterinary Medicinal Products, Brussels
89. Villeneuve DL, Garcia-Reyero N (2011) Predictive ecotoxicology in the 21st century. Environ Toxicol Chem 30:1–9
90. Walker C (1995) Biochemical biomarkers in ecotoxicology—some recent developments. Sci Total Environ 171:189–195
91. Weiss C (1958) The determination of cholinesterase in the brain tissue of three species of fresh water fish and its inactivation *in vivo*. Ecology 39:194–199
92. Wheeler J, Gimeno S, Crane M, Lopez-Juez E, Morritt D (2005) Vitellogenin: a review of analytical methods to detect (anti) estrogenic activity in fish. Toxicol Mech Methods 15:293–306
93. Williams ES, Panko J, Paustenbach DJ (2009) The European Union's REACH regulation: a review of its history and requirements. Crit Rev Toxicol 39:553–575
94. Maxwell K, Johnson GN (2000) Chlorophyll fluorescence—a practical guide. J Exp Bot 51:659–668

Human Health Risk Assessment for Pharmaceuticals in the Environment: Existing Practice, Uncertainty, and Future Directions

E. Spencer Williams and Bryan W. Brooks

Abbreviations

5-FU	5-Fluorouracil
ADI	Acceptable daily intake
API	Active pharmaceutical ingredient
BCF	Bioconcentration factor
CBZ	Carbemazepine
COPC	Contaminant of potential concern
CPA	Cyclophosphamide
E2	Estradiol
EDC	Endocrine-disrupting compound
EE2	Ethinylestradiol
ERA	Ecological or environmental risk assessment
GAC	Granular activated carbon
GREAT-ER	Geography-referenced regional exposure assessment tool for European rivers
HHRA	Human health risk assessment
HQ	Hazard quotient
LOAEL	Lowest observed adverse effects level
LOEL	Lowest observed effects level
MEC	Measured or monitored environmental concentration
MOA	Mode of action
MOS	Margin of safety

E.S. Williams (✉) • B.W. Brooks
Department of Environmental Science, Institute of Biomedical Studies,
Center for Reservoir and Aquatic Systems Research, Baylor University,
One Bear Place, #97266, Waco, TX 76798-7266, USA
e-mail: Sp_Williams@Baylor.edu; Bryan_Brooks@Baylor.edu

B.W. Brooks and D.B. Huggett (eds.), *Human Pharmaceuticals in the Environment:*   167
Current and Future Perspectives, Emerging Topics in Ecotoxicology 4,
DOI 10.1007/978-1-4614-3473-3_8, © Springer Science+Business Media, LLC 2012

NOAEL	No observed adverse effects level
NOEL	No observed effects level
OTC	Over the counter
PEC	Predicted environmental concentration
PhATE	Pharmaceutical Assessment and Transport Evaluation
PIE	Pharmaceuticals in the environment
PNEC	Predicted no-effect concentration
POD	Point of departure
RfD	Reference dose
TTC	Threshold of toxicologic concern
UF	Uncertainty factor
WWTP	Wastewater treatment plant

Introduction

Globally, several thousand substances are produced for pharmaceutical and bio-medical applications in humans. The production tonnage of these compounds is astronomical, ranging to hundreds of tons annually. Based on data collected by the National Center for Health Statistics, individuals who visited their physician recorded an average of almost seven medications taken per person [1]. As expected, this number increases dramatically in older persons to almost 20 medications per person after age 65. As the global population ages, the use of pharmaceuticals to alleviate age-related conditions can reasonably be expected to increase. Further, the ongoing development of large markets such as China and India will further increase the magnitude of pharmaceutical consumption.

Scientists have been aware of the presence of active pharmaceutical ingredients (APIs) in the environment since the late 1970s [2]. Efforts to monitor the occurrence of these APIs more comprehensively began in earnest in the early 1990s, focused primarily on substances that appeared to modulate the activity of endocrine systems in humans and aquatic receptors (i.e., endocrine-disrupting compounds or EDCs). Copious effort has been devoted to understanding the potential risks to the environment associated with EDCs and other types of APIs, including analgesics, neuroactive substances, and cardiovascular drugs. As of the end of 2009, over 39,000 articles were found in a search of the ScienceDirect database using the keywords "pharmaceuticals," "risk," and "water" [3]. The focus of most of these studies has centered on risk to ecological receptors, and many environmental impacts of APIs have been identified [4–6]. In general, it has been believed that the environmental concentrations of APIs are too low to constitute a risk to human health in developed countries, and several studies have been conducted to assess this perspective. However, a recent poll among expert stakeholders reported that 62% of those interviewed believed that pharmaceuticals in the environment (PIE) represent a risk to human health [7].

Several investigations have been conducted to determine the concentrations of limited sets of substances in environmental compartments. Most of the studies that report measured concentrations of API in environmental compartments (PIE) focus only on a few analytes, though a handful of studies pursued more robust data sets. The most exhaustive of these studies, conducted by the USGS, analyzed water samples from 30 states for 48 prescription and nonprescription drugs, along with caffeine [8]. A comprehensive list of all detections of PIE would be difficult to compile, as the literature has expanded exponentially on this topic in recent years. APIs have been observed in a number of different environmental compartments and subcompartments, including surface, ground, and drinking water, as well as wastewater treatment plant (WWTP) influent and effluent. APIs have been also detected in soil and leachate from landfills.

Colborn et al. coined the term "endocrine disruption" in the early 1990s to describe the effects of some chemical substances found in the Great Lakes ecosystem. The observations indicated that diverse substances could impact reproductive development and health in ecological receptors [9]. It is perhaps not surprising that the wide variety of pharmaceutical substances designed to modulate the activity of human reproductive systems should also carry this potential and thus be described as an environmental endocrine-disrupting compound (EDC). Chief among these products are hormones designed to prevent pregnancy or to alleviate ongoing symptoms of menopause. Naturally, the presence of EDCs in the environment has been a source of concern with regard to public health [10]. More research on this area is definitely warranted, and one of the avenues for this research is by using risk assessment tools.

An increasing body of research is available on ecological impacts resulting from exposure to pharmaceutical substances. The effects associated with EDCs are best characterized and include reduced fertility and delayed embryonic development in fish [11, 12]. Another well-documented example of unanticipated toxicity in an ecological receptor is the deaths of large numbers of vultures in Central Asia linked to the presence of diclofenac in cattle carcasses on which the vultures were feeding [13]. For the most part however, APIs are present in the environment at concentrations that are orders of magnitude lower than that which would be expected to cause acute toxicity, even for those APIs designated as ecological hazards [5]. There have been notable exceptions, as eco-risks have been suggested for acetylsalicylic acid, paracetamol, ibuprofen, amoxicillin, oxytetracycline, and mefenamic acid, among others [6, 14, 15]. Currently, there are no monitoring or regulatory requirements pertaining to APIs in surface and drinking water, though the US EPA is moving in that direction [16].

Several studies have been conducted to assess the risk to human health arising from APIs in the environment [17–31]. These investigators have focused primarily on exposures through ingestion of drinking water and fish. The available studies have assessed risk from a limited set of PIE. This is to be expected with the huge number of PIE and the relatively small amount of data on the concentrations of the substances.

As a possible hazard, "pharmaceuticals" is a terribly broad term to employ in a risk assessment complex, analogous to using "chemicals" for an occupational risk assessment. According to Drugs@FDA, there are 5,986 approved substances for human use in the USA, representing a wide variety of different chemical structures and properties, even among those that have the same molecular targets (http://www. accessdata.fda.gov/scripts/cder/drugsatfda/). Some of these substances are also used for veterinary applications [32]. It is also worth noting that substances not approved by the FDA can enter the environment through disposal as a result of manufacture or testing, but the concentrations would be expected to be minimal in the developed world [33, 34]. There are many classes of pharmaceuticals, designed for many different purposes and with tremendous variation in physicochemical characteristics and structures.

The EPA does not currently regulate the concentrations of APIs in drinking water. However, several APIs have been listed in the newest Contaminant Candidate List (CCL3), including equilenin, equilin, estradiol (E2), ethinylestradiol (EE2), estrone, estriol, mestranol, norethindone, quinolin, and erythromycin (http://www.epa.gov/ogwdw000/ccl/ccl3.html). Nine of the ten APIs added to the list are hormones, believed to act as environmental EDCs. After further regulatory review, the EPA may determine that the presence of APIs in drinking water, beginning with these candidates, will be regulated under the Safe Water Act. The USGS has included 43 veterinary and human pharmaceuticals in their list of emerging contaminants for national reconnaissance studies in water bodies in the USA [35].

Detections of APIs in the Environment

Surveys of bodies of water in the USA, and sources of untreated drinking water, were conducted by the USGS in 1999–2001 [8, 36]. Kolpin et al. sampled 139 streams across the USA, and analyzed the samples for 31 human and veterinary antibiotics, 15 prescription drugs, 7 nonprescription drugs, and 18 steroids and hormones [8]. This study targeted bodies of water that were likely to be contaminated [37]. Many of these were detected in the ng/L range, and at relatively low frequency (1–30% of samples). Higher frequency of detection was observed for caffeine and nicotine metabolites, which were labeled as nonprescription drugs in this study. A follow-up study in 2001 by Focazio et al. analyzed pharmaceuticals in untreated surface and groundwater that are used for drinking water [36]. Unsurprisingly, the detection frequency for most analytes was lower than was seen in surface waters from the prior study. As with the prior study, the highest frequency of detection was observed for nonprescription drugs including caffeine (and its metabolite 1,7-dimethylxanthine) and cotinine.

It is perhaps to be expected that the concentrations of various APIs will vary across seasons. The most obvious example of this phenomenon is the expectation that antibiotics will be more likely to be released into the environment during cold and flu seasons in the spring and fall. However, these fluctuations may be difficult to track, as sampling between wet and dry seasons in some parts of the USA demonstrates [38].

Beyond over the counter (OTC) and prescription pharmaceuticals, illegal substances may be present in water that contributes to drinking water supplies [39]. In particular, methamphetamine was widely detected in a study of Nebraska waste- and surface waters [40]. Published concentrations of cocaine, heroin, morphine, amphetamine, methamphetamine, and LSD are summarized by Petrovic et al. [41] and Zuccato and Castiglioni [39]. The usage rates of these materials are not well understood, and thus there has been some interest in using effluent concentrations of illicit substances to characterize consumption and perhaps also to track consumers [42]. For example, Kasprzyk-Hordern et al. calculated the consumption of cocaine and amphetamine in South Wales by analyzing the concentrations of these substances in raw wastewater [43, 44].

A number of other API occurrence studies have been conducted around the world [174]. This research has focused on European countries such as Italy [42, 46], Germany [47, 48], France [49, 50], Switzerland, Greece, Sweden, Denmark, Finland, and the UK [15, 51–53]. Beyond Europe, studies are available for Australia [54–56], India [57, 58], Brazil [59], Korea [60], Japan [61–63], China [64, 65], Vietnam [66], and Taiwan [67].

Routes of Pharmaceutical Introduction into the Environment

The concentration of PIE may or may not be related to the total mass manufactured or the mass prescribed or purchased by consumers. This information may be obtained via several routes, including governmental agencies, manufacturers, and consulting groups like IMS Health [17, 21, 23–25, 30, 34]. Pharmaceutical substances are then marketed and sold through a number of avenues, primarily through over-the-counter (OTC) sales and prescriptions. Prescription drugs obviously are sold directly to pharmacies, who then sell them to consumers who have been prescribed these drugs by a physician. According to the National Association of Chain Drug Stores, 3.4 billion prescriptions were filled in the USA in 2006, amounting to $716 billion in sales (http://www.nacds.org/).

A conceptual model of the dominant route of APIs through their lifecycle is detailed in Fig. 1. There are several possible routes of APIs into the environment, including excretion from consumers (urinary, fecal, or dermal), disposal of medications, and manufacturing waste streams. An interesting analysis of the route of ibuprofen and metaprolol through usage and disposal in the environment was presented by Bound and Voulvoulis [68].

Fig. 1 The major lifecycle
pathways of human active
pharmaceutical ingredients
from manufacture to potential
points of human exposure

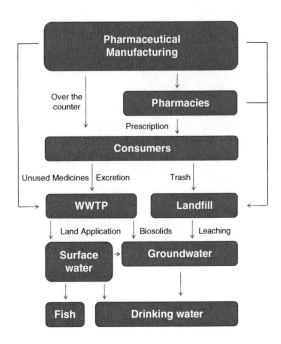

Excretion

It is generally believed that the primary route through which these substances enter
the environment is through excretion of native substance, metabolites, or conjugates
[4, 33, 69]. This perspective is supported by the observation that temporal variations
in the API mass emitted are accompanied by similar variations in nitrogen [69].
A major source of APIs in the environment is urine, to the extent that separate
wastewater collection for urine has been recommended by at least one author [70].
Once excreted, these substances and their metabolites enter waste streams that pass
through WWTPs and then into other aquatic compartments. The importance of
environmental contamination by disposal of unused medications or as a result of
manufacturing processes is unclear at this point, but these routes should not be
ignored as examined closely in Chapter 10 of this volume. Also, medications enter
the terrestrial environment and groundwater as a result of disposal of solid and
semisolid wastes from WWTPs.

Though excretion is believed to be the primary route of APIs to the environment,
understanding the nature of this route on a substance-by-substance basis is very
difficult. The pharmacological reality of these substances is that a range of metabolites and conjugates will be generated, and these by-products and the parent
compound will be excreted in varying magnitudes through urine and feces [42, 43,
68, 71]. Jjemba classified a large number of APIs based on the extent to which they
were excreted as parent compound [72]. The author noted a variety of compounds that
are excreted as greater than 70% parent, including amoxicillin, atenolol, cimetidine,

ciprofloxacin, codeine, furosemide, and valsartan [73]. Metoprolol, for example, is excreted primarily in urine, 77% as parent compound [74, 75]. However, even among the β-blockers, there is wide variation in the route of excretion (urine vs. feces) and whether parent or metabolites or conjugates are excreted [76]. Similar variation is noted among cancer chemotherapeutic drugs 5-fluorouracil (5-FU), cyclophosphamide (CPA), and doxorubicin [24]. The percentage of API excreted as the parent compound correlates strongly with its occurrence in the environment [77]. Further, APIs that are excreted as polar metabolites (such as glucuronides) can be cleaved in sewage treatment or in ecological compartments to the parent compound [78]. This variability is further compounded by differences in metabolism and excretion due to individual-specific factors such as gender, age, nutrition, endocrine function, and preexisting disease [79, 80].

Beyond urine and feces, Daughton suggest that excretion through skin may be a significant route for some medications [81]. The authors posit that these substances, especially those applied topically, may be washed from the skin and to wastewater. There is also the possibility of excretion of APIs in sweat [81]. Absorbent patches have been used to test for illicit drugs. Sweat has already been characterized as a mechanism to monitor for the use of illicit substances including cocaine, marijuana, amphetamines, and heroin (for a review, see ref. [82]). Daughton and Ruhoy suggest that for some APIs that are extensively metabolized, sweat may contribute three times as much to the total excreted load compared to urine [83].

Improper Disposal

It has been theorized that APIs which are primarily disposed of directly to wastewater (e.g., via sinks or toilets) could contribute a disproportionate amount of the parent compound to the environmental load, as these APIs would bypass the metabolic processes in the body [68, 84]. In fact, the loadings of some parent compounds by this practice relative to excretion could be very important if a therapeutic undergoes near complete metabolism and thus is excreted as metabolites. Some healthcare professionals endorse this method of disposal, but increasingly it is recognized that these disposals can constitute a significant contribution to the concentrations of API in aquatic environments [34, 80, 85]. Poison control centers often counsel callers to dispose of medications in sinks or toilets, to prevent children from coming into contact with the APIs. Bound and Voulvoulis reported the results of a survey of API consumers in the UK, which suggested that overall disposals to wastewater (i.e., using a sink or toilet) were relatively minor (0–16.7% of respondents, based on different drug types) and a significant portion (21.8%) were returned to the pharmacy [68]. Ibuprofen, acetaminophen, and diclofenac are among the unused drugs most frequently returned to pharmacies. However, a survey of Americans suggested that slightly more than 1% of medications were returned to the pharmacy, while 54% were disposed into trash, and 35% into sinks or toilets [86]. The low level of return is not surprising as in most US states, pharmacies are not allowed to accept returned medication from patients [68]. Beyond that, it has also been observed that medications returned to the pharmacy may be disposed of in similar ways used by consumers. In the case of contraceptive patches containing EE2, special disposal advisories were added to the packaging and safety leaflet to urge consumers to either return unused patches to the pharmacy and to seal used patches back in the original packaging before disposing in solid waste [87].

Canada has established the Medications Return Program (MRP), and similar programs have been enacted in Italy, France, and Australia [80]. Another role of this program is to identify the reasons why medications go unused and ultimately to tailor the prescriptions to both cut down on overprescribing and excessive disposal. In 2007, the White House Office of National Drug Control Policy (ONDCP) issued federal guidance on drug disposal by consumers, and the US Fish and Wildlife Service and the American Pharmacists' Association initiated a SMARxT Disposal program, each of which recommended disposing of unused medications in the trash, after removing any labeling and mixing the substances with unpalatable materials like kitty litter (presumably to deter those who would accidentally or intentionally come into contact with the APIs) [85]. However, the ODNCP guidance still recommends sink or toilet disposal for 13 APIs that are either highly toxic or likely to be abused. For illegal drugs, direct disposal to sinks or toilets may be a more significant pathway for parent compounds or by-products in the synthesis [33]. Specific information on pharmaceutical Take Back Programs is found in Chap. 10 of this book.

Manufacturing Waste Streams

There are hundreds of companies that manufacture pharmaceutical substances. In the USA, the largest and best known of these include AstraZeneca, Bristol-Myers Squibb, GlaxoSmithKline, Eli Lilly, Merck, Pfizer, Procter & Gamble, Roche, Schering-Plough, and Wyeth. Each of these companies manufactures an array of proprietary substances intended to treat a wide variety of conditions through prescriptions or over the counter. Formulations of these substances of course vary as widely as their uses and the associated process can be expected to result in some type of waste stream. It is also to be expected that these companies have multiple manufacturing facilities, spread across multiple continents, consistent with their need to provide pharmaceutical substances to a global community.

Until recently, it was generally understood that the mass of pharmaceutical substances discharged to surface waters streams is relatively in the ng/L to low μg/L range in the developed world [32, 34, 88]. However, observations in developing countries are less available. For example, a study conducted by Larsson et al. indicated that wastewater streams emanating from a WWTP serving 90 drug manufacturing facilities in India contained 21 pharmaceutical substances over 1 μg/L, some reaching as high as 31 mg/L [58, 89]. Further examination of the surface and groundwater connected to this WWTP revealed concentrations of several APIs, including cetirizine, ciprofloxacin, metoprolol, and trimethoprim [57]. The authors also demonstrated that many APIs were detectable in drinking water wells in the area in the ng to μg/L range. In China, Li et al. reported concentrations of oxytetracycline and related compounds at levels in excess of 1 μg/L in treated wastewater arising from pharmaceutical production facility and that these substances may be causing the proliferation of resistant microbial strains [90]. From this information, it can be inferred that API manufacturing facilities, when not managed properly, have the potential to contribute a significant proportion of total load to aquatic environments. In the USA, a recent study by Philips et al. [171] identified elevated levels of several APIs in surface waters receiving discharges from manufacturing facilities.

Waste streams from hospitals, which are most likely an amalgamation of patient excreta and disposed APIs, may also be a significant source of APIs in the environment. In a study in Sweden, as much as 12% of the total load of acetaminophen entering WWTPs was contributed by hospital effluent, while many other APIs fell in the 2–4% range [91]. A study in Australia suggested that hospital effluent may contribute as much as 25% of the total load of roxithromycin, 10% of trimethoprim, and approximately 5% of furosemide, ibuprofen, acetaminophen, ranitidine, and salicylic acid [56]. Hospitals may also be expected to be associated with significant output of anticancer medications, though these courses of treatment now frequently occur on an outpatient basis [73].

Seasonal Variability

The use patterns of individual pharmaceutical substances vary based on their intended target. This naturally leads to a seasonal variability of consumer usage of APIs, especially antibiotics and anti-inflammatory drugs, which would be expected to be used much more during winter months [78, 92]. Castiglioni et al. demonstrated that the wastewater loads of several APIs including ibuprofen, ciprofloxacin, ofloxacin, and sulfamethoxazole were lower during the summer [92]. Some pharmaceuticals are intended to moderate symptoms and thus are taken for extended periods [33]. Thus, it is not surprising that the loads of β-blockers, diuretics, and antiulcer medications, or the nonpharmaceutical substance caffeine, do not vary seasonally [92, 93]. In the absence of seasonal variations in usage, the concentrations at various times of the year could also be significantly affected by the flow conditions; concentrations of APIs will be higher in summer months when wastewater effluent represents a larger fraction of total flow [38, 50]. Degradation is also affected by seasonal factors such as irradiance, temperatures, and microbial activity, both in the natural environment and in WWTP systems [63, 94–96]. Of course, seasonal variations will therefore not be observed in all studies for all APIs [77, 97, 98].

APIs in Biosolids

Biosolids from WWTPs are frequently used to fertilize croplands. This use may allow both runoff of the APIs enriched in the sludge into surface waters or uptake into edible foodstuffs [99–101]. These biosolids can contain relatively high concentrations of

APIs [83, 102–105]. Due to the analytical challenges associated with quantification of APIs in biosolids, the significance of such materials with regard to hazard, exposure, and risk is poorly understood and thus remains a significant research need [34].

APIs in MSW/Landfills

As mentioned above, disposal of APIs leads the substances not only to aquatic environments; they also are disposed into landfills [68]. Aside from disposed medications, sewage sludge (which contains APIs) from WWTP may also be disposed of at a landfill [33]. Several studies have indicated that APIs that enter landfills can leach into surrounding groundwater [53, 106, 107]. In particular, Holm et al. observed in samples of landfill leachate several substances "originating from waste from the pharmaceutical industry" [53]. Clofibric acid, ibuprofen, and prophenazone were identified in leachate from a domestic landfill in Germany [108]. An understanding of APIs in leachate from landfills in the developing world is not known.

Veterinary Pharmaceuticals

Occasionally, APIs are used both for human and veterinary applications. These primarily include anti-infectives and hormones [32]. The routes for veterinary APIs (vAPIs) into the environment can include emissions from manufacturing and from disposal, as with human APIs. However, excreted urine and feces from livestock animals which contain vAPIs are directly discharged to land and thus have the potential to contaminate soil or surface waters (through runoff) [32]. It appears that hazard information for vAPI may be more readily available in some cases than for human APIs [109].

The presence of vAPIs in the environment has been observed [8]. Hamscher et al. noted the presence of tetracycline and chlortetracycline, APIs used in veterinary applications, in animal manure and in soil fertilized with manure [110]. Ivermectin, a substance commonly used to deworm livestock animals, has been found to persist in soil [87]. It has also been reported that tylosin, a veterinary antibiotic, has been detected in drinking water [46]. An important activity that appears to introduce vAPIs into the environment is that of aquaculture, the practice of raising aquatic animals. Often, these substances are given in food pellets. The majority of these vAPIs ultimately leave the aquaculture area and enter the surrounding aquatic environments. This pathway into the environment is well reviewed by Boxall et al. [111].

Fate and Behavior of APIs in the Environment

The route of APIs mostly passes through excretion by consumers, and thus it is to be expected that a large proportion of these APIs will move through a WWTP before entering other aquatic compartments. WWTPs are variably effective in removing APIs from wastewater; this may be a function of the technology type. Once the parent APIs, metabolites, or conjugates enter the environment, they are subject to normal processes of transport and degradation. For example, acetylsalicylic acid is degraded to salicylic acid following deacetylation and can also be converted to ortho-hydroxyhippuric acid and gentisic acid [78]. These metabolites have all been detected in wastewater influent [47]. However, it is worth noting that salicylic acid can also arise from nonpharmaceutical uses [78]. Degradation processes can best be divided into biotic (i.e., biotransformation) and abiotic (e.g., photodegradation, hydrolysis). Manufacturers of APIs have in some instances taken steps to retard biotic degradation processes so that their product will last longer [112].

Loffler et al. studied the fate and behavior of ten common APIs in water and sediment they gathered from a waterway in Germany [113]. The bottom line of these experiments was that, in the absence of the possibility of photodegradation, parent compounds in many cases seemed to persist in the environment, while their metabolites (as expected) remained in the system for shorter periods. Carbemazepine (CBZ) was found to be recalcitrant in the model system, with a 50% dissipation time (DT_{50}) of 328 days and moderate affinity for sediment. A metabolite, CBZ-diol, was also suspected of persisting in aquatic environments. Clofibric acid was also stable in the experiment ($DT_{50} = 119$ day), but had low sediment affinity. Diazepam also persisted strongly in the environmental model, though its human metabolite oxazepam degraded somewhat more quickly. Ivermectin sorbed strongly into sediments and the potential for accumulation in that compartment appeared high. Ibuprofen and its metabolite 2-hydroxibuprofen however were converted almost completely to CO_2 by the end of the experiment. Acetaminophen was also degraded relatively rapidly. Further experimentation suggested that the rapid degradation of these two APIs was due to biotic processes.

Photodegradation appears to be an important process for APIs in aquatic environments, both through direct and indirect pathways [112, 114, 115]. In some locations, wastewater treatment processes include a UV irradiation step. Under these techniques, it appears that some types of APIs will be completely degraded, while others may be somewhat resistant [116, 117]. The experiments of Lin and Reinhard highlight the importance of other factors in degradation, as photolysis experiments conducted in river water produced much faster degradation rates than experiments conducted in purified water [115]. Under environmentally relevant conditions, the natural photosensitizers may hasten the photodegradation of APIs, possibly as a result of the formation of reactive oxygen species which react with the compound [51]. This type of indirect mechanism is also important to the degradation of cimetidine, clofibric acid, and ibuprofen [114, 118]. However, direct photochemical degradation is also in play, for ranitidine, naproxen, CBZ, and diclofenac [114]. As noted by Brooks et al. [173], clearly this is an area requiring more attention. Consideration of the environmental fate of APIs are more thoroughly examined in Chapter 4 of this volume.

Removal of APIs in WWTP and Drinking Water Treatment

WWTP treatment is somewhat effective at removing or at least lowering the concentrations of a broad spectrum of APIs, though undoubtedly a number of APIs pass through standard treatment without significant reductions in concentration [103, 119, 120]. The rate of removal varies widely between APIs, though the relationship between removal rate and structure or physicochemical properties is unclear [69]. Removal rate may also vary seasonally [92]. The technology and techniques used also dictate the removal efficiency during treatment, to a significant extent on a substance-by-substance basis [92, 102, 121]. There appears to be a level of variability between experiments, as well, as removal rates from 0 to 69% have been published for diclofenac [102].

Water treatment processes appear to remove hydrophobic substances most efficiently [122]. This is not surprising, as hydrophobic APIs would be expected to bind to organic material that would predominately be removed by flocculation and sedimentation processes. Biosolids in the WWTP system can be a significant reservoir of APIs, however, which is relevant both for resuspension/dissolution of the substance and for soil contamination should the biosolids be deposited on land or in a landfill later [103]. Regardless, the processes of clarification, disinfection, and filtration through granular activated carbon (GAC) can reduce concentrations of APIs in source water by 75% or to a level below detection limits [122]. A study of the behavior of 13 APIs in a drinking water plant suggested that while many were eliminated by a combination of ferric salt coagulation, sand filtration, ozonation, GAC filtration, and UV disinfection, ciprofloxacin was able to pass through into finished drinking water [123]. Ozonation, a method first used for the removal of coliform bacteria and enteric viruses, appears to be a very highly effective process for the removal of APIs during processing [124–126].

It is worth noting that while APIs are consumed in every corner of the globe, wastewater treatment is not uniform in all countries. It would be interesting to study whether the lack of sophisticated wastewater treatment or a lower rate of API consumption plays a larger role in the concentrations of API in the environments of developing nations. Treatment technologies for APIs in drinking source waters and wastewater influents are examined in Chap. 9 of this book.

Human Exposure

A summary of potential major pathways for human exposure to environmental APIs is presented in Fig. 2. Some of the pathways described are not expected to be complete. For instance, it is unlikely that any APIs will volatilize sufficiently to produce an inhalation dose, though there are some indications of potential inhalation exposure to antibiotics and thus potentially other APIs sorbed to particulate matter in some circumstances [110]. Also, dermal exposures though possible are not likely to

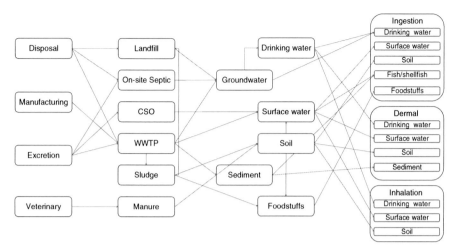

Fig. 2 Potential major exposure pathways for human active pharmaceutical ingredients to human receptors

reach a level of concern. Intuitively, ingestion of APIs in drinking water is expected to be the most relevant exposure pathway. Thus, sampling and analysis of potential sources of drinking water, as well as surface waters, seem to be the most informative. It is worth noting that sampling of these sources may not generate a reasonable measure of exposure, as sampling tends to focus on areas where contamination is suspected [22]. Indeed, risk assessments performed to date have focused primarily on this pathway, though exposure concentrations used in these exercises are more likely to rely on surface water data as a more conservative value. It seems likely that the magnitude of doses would be greatest in this pathway, but the importance of other potentially complete pathways must be a focus of future research efforts. Several studies have also assessed the potential for exposure through ingestion of fish caught in contaminated waters. Measured environmental and predicted environmental concentrations (MECs, PECs) used in all of the human health risk assessment (HHRA) exercises published to date are summarized in Table 1.

Measured Environmental Concentrations

Analytical measurements of APIs in drinking water are relatively scarce [73]. The available data do suggest that concentrations are lower in finished drinking water than in prior stages, suggesting that various processing technologies have varying effectiveness, but the vast majority of APIs are not detected in drinking water [122]. Webb et al. using data from monitoring studies in Germany estimated the exposure from 64 APIs in drinking water [31, 127, 128]. However, these monitoring studies

Table 1 Measured and predicted environmental concentrations of active pharmaceutical ingredients in risk assessment exercises published to date

Pharmaceutical substance	Concentration type	Concentrations	Unit	References
17α-Ethinylestradiol	$PEC_{DW\text{-}local}$	1.2	ng/L	[26]
	$PEC_{DW\text{-}regional}$	0.3	ng/L	[26]
	MEC_{max}	<0.5	ng/L	[32]
Acetominophen	$MEC_{SW\text{-}max}$	10,000	ng/L	[24]
	$PEC_{DW\text{-}max}$	220,000	ng/L	[24]
	$PEC_{SW\text{-}max}$	470,000	ng/L	[24]
Acetylsalicylate	$MEC_{DW\text{-}max}$	0.12	μg/L	[29]
	$MEC_{SW\text{-}max}$	0.065	μg/L	[29]
	$MEC_{SW\text{-}max}$	340	ng/L	[21]
	MEC_{max}	<10	ng/L	[32]
Albuterol	$MEC_{SW\text{-}max}$	35	ng/L	[24]
	$PEC_{DW\text{-}max}$	120	ng/L	[24]
	$PEC_{SW\text{-}max}$	250	ng/L	[24]
	MEC_{max}	<5	ng/L	[32]
Alkylating chemotherapeutic agents[a]	$PEC_{max\text{-}mean\ flow}$	10.72	ng/L	[28]
	$PEC_{max\text{-}high\ flow}$	19.99	ng/L	[28]
Anthracycline antibiotics[b]	$PEC_{max\text{-}mean\ flow}$	0.05	ng/L	[28]
	$PEC_{max\text{-}high\ flow}$	0.09	ng/L	[28]
Antimetabolite chemotherapeutics[c]	$PEC_{max\text{-}mean\ flow}$	2.02	ng/L	[28]
	$PEC_{max\text{-}high\ flow}$	3.76	ng/L	[28]
Atenolol	$MEC_{DW\text{-}max}$	0.02	μg/L	[30]
	MEC_{max}	<5	ng/L	[32]
Atomoxetine	PEC_{usage}	0.02	μg/L	[18]
	PhATE 99th PEC	0.12	μg/L	[18]
Atorvastatin	$MEC_{DW\text{-}max}$	<0.00025	μg/L	[30]
Benzylpenicillin	MEC_{max}	<50	ng/L	[32]
Betaxolol	MEC_{max}	<5	ng/L	[32]
Bezafibrate	MEC_{max}	27	ng/L	[32]
Bisoprolol	MEC_{max}	<5	ng/L	[32]
Carazolol	MEC_{max}	<5	ng/L	[32]
Carbamazepine	MEC_{max}	30	ng/L	[32]
	MEC_{max}	<50	ng/L	[32]
	$MEC_{90th\text{-}NA}$	150	ng/L	[23]
	$PEC_{90th\text{-}NA}$	333	ng/L	[23]
	$MEC_{DW\text{-}average}$	2.8	ng/L	[27]
	$MEC_{DW\text{-}max}$	5.7	ng/L	[27]
	$MEC_{DW\text{-}max}$	0.03	μg/L	[29]
	$MEC_{SW\text{-}max}$	0.227	μg/L	[29]
	$MEC_{DW\text{-}max}$	0.018	μg/L	[30]
Celiprolol	MEC_{max}	<5	ng/L	[32]
Chloramphenicol	MEC_{max}	<20	ng/L	[32]
Chlorotetracycline	MEC_{max}	<20	ng/L	[32]
Cimetidine	$MEC_{SW\text{-}max}$	580	ng/L	[24]
	$PEC_{DW\text{-}max}$	4,400	ng/L	[24]
	$PEC_{SW\text{-}max}$	9,300	ng/L	[24]

(continued)

Table 1 (continued)

Pharmaceutical substance	Concentration type	Concentrations	Unit	References
Ciprofloxacin	MEC_{SW-max}	30	ng/L	[24]
	PEC_{DW-max}	3,600	ng/L	[24]
	PEC_{SW-max}	7,600	ng/L	[24]
Clarithromycin	MEC_{max}	<20	ng/L	[32]
Clenbuterol	MEC_{max}	<10	ng/L	[32]
Clofibrate	MEC_{DW-max}	270	ng/L	[21]
	MEC_{max}	<20	ng/L	[32]
Clofibric acid[d]	MEC_{DW-max}	0.14	μg/L	[29]
	MEC_{SW-max}	0.091	μg/L	[29]
	MEC_{max}	70	ng/L	[32]
	MEC_{SW-max}	1,750	ng/L	[21]
Cloxacillin	MEC_{max}	<50	ng/L	[32]
Codeine	MEC_{SW-max}	1,000	ng/L	[24]
	PEC_{DW-max}	1,100	ng/L	[24]
	PEC_{SW-max}	2,400	ng/L	[24]
Cyclophosphamide	PEC_{DW}	2.5–8	ng/L	[26]
	$PEC_{local-average}$	5.6	ng/L	[19]
	MEC_{max}	4	ng/L	[19]
	MEC_{SW-max}	10.1	ng/L	[21]
	MEC_{max}	<10	ng/L	[32]
	MEC_{max}	<50	ng/L	[32]
Dehydrato-erythromycin	MEC_{max}	<20	ng/L	[32]
Dehydronifedipine[d]	MEC_{SW-max}	30	ng/L	[24]
	PEC_{DW-max}	1,100	ng/L	[24]
	PEC_{SW-max}	2,300	ng/L	[24]
Diazepam	MEC_{DW-max}	<0.00025	μg/L	[30]
	MEC_{max}	<20	ng/L	[32]
Diclofenac	MEC_{DW-max}	<0.00025	μg/L	[30]
	MEC_{max}	6	ng/L	[32]
Dicloxacillin	MEC_{max}	<50	ng/L	[32]
Digoxin	MEC_{SW-max}	130	ng/L	[24]
	PEC_{DW-max}	6.3	ng/L	[24]
	PEC_{SW-max}	13	ng/L	[24]
Digoxigenin[d]	MEC_{SW-max}	4	ng/L	[24]
	PEC_{DW-max}	6.3	ng/L	[24]
	PEC_{SW-max}	13	ng/L	[24]
Diltiazem	MEC_{SW-max}	49	ng/L	[24]
	PEC_{DW-max}	5,900	ng/L	[24]
	PEC_{SW-max}	12,000	ng/L	[24]
Doxycycline	MEC_{SW-max}	50	ng/L	[24]
	PEC_{DW-max}	990	ng/L	[24]
	PEC_{SW-max}	2,100	ng/L	[24]
	MEC_{max}	<20	ng/L	[32]
Duloxetine	PEC_{usage}	0.05	μg/L	[18]
	PhATE 99th PEC	0.13	μg/L	[18]
Enalapril	MEC_{DW-max}	<0.00025	μg/L	[30]

(continued)

Table 1 (continued)

Pharmaceutical substance	Concentration type	Concentrations	Unit	References
Enalaprilat[d]	$MEC_{SW\text{-}max}$	46	ng/L	[24]
	$PEC_{DW\text{-}max}$	30	ng/L	[24]
	$PEC_{SW\text{-}max}$	63	ng/L	[24]
Erythromycin-H_2O	$MEC_{SW\text{-}max}$	1,700	ng/L	[24]
	$PEC_{DW\text{-}max}$	3,500	ng/L	[24]
	$PEC_{SW\text{-}max}$	7,300	ng/L	[24]
Etofibrate	MEC_{max}	<20	ng/L	[32]
Fenofibrate	MEC_{max}	<20	ng/L	[32]
Fenofibric acid[d]	MEC_{max}	42	ng/L	[32]
Fenoprofen	MEC_{max}	<5	ng/L	[32]
Fenoterol	MEC_{max}	<5	ng/L	[32]
Fluoxetine	$MEC_{SW\text{-}max}$	46	ng/L	[24]
	$PEC_{DW\text{-}max}$	620	ng/L	[24]
	$PEC_{SW\text{-}max}$	1,300	ng/L	[24]
	$MEC_{DW\text{-}max}$	<0.00050	µg/L	[30]
Gemfibrozil	$MEC_{SW\text{-}max}$	1,550	ng/L	[24]
	$PEC_{DW\text{-}max}$	8,000	ng/L	[24]
	$PEC_{SW\text{-}max}$	17,000	ng/L	[24]
	$MEC_{DW\text{-}max}$	0.0021	µg/L	[30]
	MEC_{max}	<5	ng/L	[32]
Ibuprofen	$MEC_{SW\text{-}max}$	2,700	ng/L	[24]
	$PEC_{DW\text{-}max}$	63,000	ng/L	[24]
	$PEC_{SW\text{-}max}$	130,000	ng/L	[24]
	MEC_{max}	3	ng/L	[32]
Ifosfamide	MEC_{max}	<10	ng/L	[32]
	MEC_{max}	<50	ng/L	[32]
	$PEC_{local\text{-}average}$	10.9	ng/L	[19]
	MEC_{max}	206	ng/L	[19]
Indometacine	MEC_{max}	<5	ng/L	[32]
	$MEC_{SW\text{-}max}$	200	ng/L	[21]
Ketoprofen	MEC_{max}	<5	ng/L	[32]
Lincomycin	$MEC_{SW\text{-}max}$	730	ng/L	[24]
	$PEC_{DW\text{-}max}$	9.8	ng/L	[24]
	$PEC_{SW\text{-}max}$	21	ng/L	[24]
Meprobamate	$MEC_{DW\text{-}average}$	6.1	ng/L	[27]
	$MEC_{DW\text{-}max}$	13	ng/L	[27]
	$MEC_{DW\text{-}max}$	0.043	µg/L	[30]
Metaprolol	$MEC_{DW\text{-}max}$	2.1	µg/L	[29]
	$MEC_{SW\text{-}max}$	0.2	µg/L	[29]
Metformin	$MEC_{SW\text{-}max}$	150	ng/L	[24]
	$PEC_{DW\text{-}max}$	47,000	ng/L	[24]
	$PEC_{SW\text{-}max}$	98,000	ng/L	[24]
Methicillin	MEC_{max}	<50	ng/L	[32]
Metropolol	MEC_{max}	<5	ng/L	[32]
Nadolol	MEC_{max}	<5	ng/L	[32]
Nafcillin	MEC_{max}	<50	ng/L	[32]

(continued)

Table 1 (continued)

Pharmaceutical substance	Concentration type	Concentrations	Unit	References
Naproxen	$MEC_{DW\text{-}max}$	<0.00050	μg/L	[30]
Norfloxacin	$MEC_{SW\text{-}max}$	120	ng/L	[24]
	$PEC_{DW\text{-}max}$	74	ng/L	[24]
	$PEC_{SW\text{-}max}$	160	ng/L	[24]
Olanzapine	PEC_{usage}	0.01	μg/L	[18]
	PhATE 99th PEC	0.07	μg/L	[18]
Oxacillin	MEC_{max}	<50	ng/L	[32]
Oxytetracycline	$MEC_{SW\text{-}max}$	1,340	ng/L	[24]
	$PEC_{DW\text{-}max}$	0.92	ng/L	[24]
	$PEC_{SW\text{-}max}$	1.94	ng/L	[24]
	MEC_{max}	<20	ng/L	[32]
Pentoxifylline	MEC_{max}	<10	ng/L	[32]
	MEC_{max}	<20	ng/L	[32]
Phenazon	MEC_{max}	<50	ng/L	[32]
	$MEC_{DW\text{-}max}$	0.03	μg/L	[29]
	$MEC_{SW\text{-}max}$	0.11	μg/L	[29]
	MEC_{max}	50	ng/L	[32]
Phenoxymethylpenicillin	MEC_{max}	<50	ng/L	[32]
Phenytoin	$MEC_{DW\text{-}median}$	6.2	ng/L	[27]
	$MEC_{DW\text{-}max}$	19	ng/L	[27]
	$MEC_{DW\text{-}max}$	0.015	μg/L	[30]
Propranolol	MEC_{max}	<5	ng/L	[32]
Ranitidine	$MEC_{SW\text{-}max}$	39	ng/L	[24]
	$PEC_{DW\text{-}max}$	7,800	ng/L	[24]
	$PEC_{SW\text{-}max}$	16,000	ng/L	[24]
Risperidone	$MEC_{DW\text{-}max}$	0.00034	μg/L	[30]
Roxithromycin	MEC_{max}	<20	ng/L	[32]
Salicylic acid	MEC_{max}	<10	ng/L	[32]
Simvastatin	$MEC_{DW\text{-}max}$	<0.00025	μg/L	[30]
Sotalol	MEC_{max}	<5	ng/L	[32]
Sulfamethazine	MEC_{max}	<20	ng/L	[32]
Sulfamethoxazole	$MEC_{DW\text{-}max}$	0.03	μg/L	[29]
	$MEC_{SW\text{-}max}$	0.11	μg/L	[29]
	$MEC_{SW\text{-}max}$	1,900	ng/L	[24]
	$PEC_{DW\text{-}max}$	8,500	ng/L	[24]
	$PEC_{SW\text{-}max}$	18,000	ng/L	[24]
	$MEC_{DW\text{-}max}$	0.003	μg/L	[30]
	MEC_{max}	<20	ng/L	[32]
Sulfathiozole	$MEC_{SW\text{-}max}$	80	ng/L	[24]
	$PEC_{DW\text{-}max}$	13	ng/L	[24]
	$PEC_{SW\text{-}max}$	28	ng/L	[24]
Terbutalin	MEC_{max}	<10	ng/L	[32]
Tetracycline	$MEC_{SW\text{-}max}$	1,000	ng/L	[24]
	$PEC_{DW\text{-}max}$	3,100	ng/L	[24]
	$PEC_{SW\text{-}max}$	6,500	ng/L	[24]
	MEC_{max}	<20	ng/L	[32]

(continued)

Table 1 (continued)

Pharmaceutical substance	Concentration type	Concentrations	Unit	References
Timolol	MEC_{max}	<5	ng/L	[32]
Triclosan	$MEC_{DW\text{-}max}$	0.0012	µg/L	[30]
Trimethoprim	$MEC_{SW\text{-}max}$	710	ng/L	[24]
	$PEC_{DW\text{-}max}$	1,800	ng/L	[24]
	$PEC_{SW\text{-}max}$	3,700	ng/L	[24]
	$MEC_{DW\text{-}max}$	<0.00025	µg/L	[30]
	MEC_{max}	<20	ng/L	[32]
Warfarin	$MEC_{SW\text{-}max}$	0.5	ng/L	[24]
	$PEC_{DW\text{-}max}$	120	ng/L	[24]
	$PEC_{SW\text{-}max}$	250	ng/L	[24]

[a]Oxaliplatin, temozolomide, cisplatin, carboplatin, cyclophosphamide
[b]Epirubicin, doxorubicin
[c]Gemcitabine, fludarabine, capecitabine
[d]Metabolite

showed that only 10 of the 64 were detected in surface water at levels above the limit of quantitation and even within those that were detected the proportion of samples above the LOQ was 53% or below [31]. A summary of published concentrations for APIs in drinking water was reported by Jones et al. and indicated that of those detected, all fell in the ng/L range [129]. The detected compounds included bezafibrate, clofibric acid, CBZ, diazepam, diclofenac, and ibuprofen. A study of APIs and EDCs in source, finished, and delivered drinking water showed the presence of atenolol, CBZ, gemfibrozil, meprobamate, phenytoin, and several other APIs in drinking water [29, 130]. In Germany, several studies have indicated the presence of APIs in drinking water, including acetaminophen, acetylsalicylic acid, diclofenac, ibuprofen, CBZ, and naproxen [131]. Despite these data, it is clear that no systematic analysis of APIs in drinking water has been conducted on a national or global scale. The broad array of potential analytes is simply too broad. Thus, it is difficult to assess direct exposure through this pathway.

Cunningham et al. noted several published values for CBZ concentrations in drinking water, but ultimately opted to use surface water measurements that were significantly more numerous and regarded as a more conservative exposure value [22]. Schriks et al. used maximum concentration data from surface and groundwater, if primary data from the Rhine and Meuse rivers were not available [28]. For more information on the occurrence of APIs in the environment, an excellent review [174].

Predicted Exposure Concentrations

Many of the risk assessments performed to date include the use of modeling approaches for point-of-exposure concentrations. Environmental concentrations can be crudely estimated using a very simple equation, as described by Coetsier et al. [49]. This equation integrates total consumption, the excretion rate of the parent API, and the fraction that passes through WWTP, and divides this figure by the volume of wastewater generated per person, the number of persons, and the extent of dilution. This general approach appears to provide reasonable PECs for WWTP effluent, but the PECs for surface water are not in agreement with MECs [49] particularly in effluent-dominated surface waters [172].

More sophisticated computer models have been established to generate PECs for the USA and the EU. EUSES, a tool used in risk assessment for chemical substances, has been applied to provide a worst-case scenario of concentrations of APIs in the environment by Christensen [25]. However, two primary systems have been used to model the environmental concentrations of APIs in aquatic systems: the Pharmaceutical Assessment and Transport Evaluation (PhATE) model, and the Geography-Referenced Regional Exposure Assessment Tool for European Rivers (GREAT-ER). These models appear useful in addressing the possibility of data bias (resulting from sampling of areas suspected to be contaminated), filling the many gaps in monitoring data, and developing testable hypotheses for site-specific field studies.

PhATE was developed by the Pharmaceutical Research and Manufacturers of America (PhRMA) [132]. This model is designed to offer a PEC of pharmaceuticals discharged into surface waters through POTWs. PhATE was designed around 11 watersheds in the USA, covering approximately 19% of the nation's land area. Watersheds were selected in which drinking water supplies are derived from sources that can be impacted by WWTP discharges; hence, most metropolitan areas (Los Angeles, New York, Chicago, Miami, Denver) are not included in the model. PhATE is essentially a mass balance model which begins with the API use per capita, estimates the metabolism of the API, and then models the mass of API that enters the surface water compartment (as well as providing estimates of the amount of API lost from surface water through degradation and other processes). A preliminary validation exercise for this model was conducted using caffeine, triclosan, and linear alkylbenzene sulfonates (LAS), and the results suggested a reasonable relationship between PhATE-modeled PECs and measured environmental concentrations (MECs), i.e., within an order of magnitude [132]. However, PECs generated by the PhATE model deviated by multiple orders of magnitude when compared with MECs generated by Kolpin et al. [8]. These deviations were attributed to numerous factors, including variability in API removal in WWTP and possible difficulties in understanding accurate environmental concentrations due to analytical methodology. iSTREEM represents a similar model to PhATE that was originally developed for cleaning products, but has been applied to APIs (http://www.aciscience.org/iSTREEM.aspx)

In the EU, a similar model was developed in the late 1990s by a collaborative group under the auspices of the European Centre for Ecotoxicology and Toxicology of Chemicals (ECETOC) [133]. GREAT-ER was not necessarily designed to establish PECs for APIs, but rather was intended to be applied to a broader group of substances. However, it has been applied in many such exercises [76]. Input information for GREAT-ER can include excretion data as a source of input into WWTPs [76].

Several of the HHRAs published to date use some combination of MECs and PECs, either as validation or as measures of uncertainty. Before PhATE and

GREAT-ER were available, Christensen used publicly available data on consumption as a starting point for point-of-exposure concentrations generated by the EU computer program EUSES; the consumption data were obtained from the Danish Medicine Agency and pharmacy contacts [25]. Christensen used this program to model concentrations in drinking water, edible species, foodstuffs, and air. Kummerer and Al-Ahmad used a combination of the amount of CPA and ifosfamide used in Germany and measured concentrations of these two substances in WWTP influent and effluent to calculate regional and local PECs using a relatively simple equation [18]. The PECs were then calculated for surface water and used as a proxy for drinking water concentrations.

Johnson et al. also used consumption data, provided by the Health and Safety Executive (HSE) and Department of Health [24]. In the USA, Schwab et al. obtained information on the quantity of 26 APIs sold by consulting manufacturers and databases [23]. In contrast, Cunningham et al. used an estimate of the amount sold in the USA and six European countries coupled with the available data on environmental concentrations of their APIs of interest [21].

In the absence of monitored data, Cunningham et al. used the PhATE and GREAT-ER models to estimate PECs for 44 APIs marketed by GlaxoSmithKline [21]. The authors identified data for only nine of these APIs in the peer-reviewed literature, and these surface water data points were not informative for the purposes of risk assessment for drinking water and fish ingestion exposures. In contrast, Schulman et al. were able to find measured concentrations of their four APIs of interest (acetylsalicylic acid, clofibrate, CPA, and indomethacin) in multiple aquatic compartments including sewage effluent, surface water, and drinking water [20]. Schwab et al. used measured concentrations from Kolpin et al. and other peer-reviewed sources and supplemented their analysis of exposure concentrations using the PhATE model [8, 23]. Bercu et al. performed two exercises to estimate an exposure concentration to the three neuroactive compounds they studied; the first estimate was generated by dividing the total mass of the API that was sold in the USA by the annual volume of water discharged from all POTWs in the USA [17]. The second estimate was calculated using the PhATE model, again using the total mass of API sold as an input parameter. Johnson et al. in estimating risks for exposure to one chemotherapeutic agent, 5-FU, began by cataloging the consumption of the drug at all major cancer treatment centers in England and calculating the concentration of 5-FU in wastewater effluent [24]. These parameters were then incorporated in GREAT-ER to provide a reasonable estimate of surface water concentrations. Rowney et al. used a derivative model of GREAT-ER called LF2000-WQX and calculated per capita loading estimate that was modified by approximate removal efficiency by sewage treatment plants [27].

Exposure Through Food

Exposure to contaminant of potential concerns (COPCs) in fish tissue has become a target of recent research for a variety of chemicals and sites. Several studies indicate that APIs can accumulate in fish and other aquatic organisms [175] and thus exposure to these APIs through consumption by recreational or subsistence anglers must be considered. The examined compounds include fluoxetine, sertraline, ibuprofen, naptoxen, diclofenac, ketoprofen, gemfibrozil, diphenhydramine, diltiazem, CBZ, and paraoxetine [134–138]. Ramirez et al. conducted a reconnaissance study to determine levels of 24 APIs and metabolites in fish tissue in six effluent-dominated streams [138]. Only five of these analytes (norfluoxetine, sertraline, diphenhydramine, diltiazem, and CBZ) were detected in fillets and seven in liver (fluoxetine, gemfibrozil). Among the APIs not detected were acetaminophen, ibuprofen, propranolol, and warfarin. Interestingly, the authors noted that there was no clear association between lipid content and bioaccumulation of the detected APIs, though the pK_a of the detected substances appears to play a role in this observation.

Several of the published HHRAs have considered this exposure pathway [21, 23, 26]. As they are critical to understanding the potential for dosages through ingestion of fish tissue, bioconcentration factors (BCFs) have been estimated for several APIs [21, 23]. It appears critical, however, to account for site-specific pH influences on bioaccumulation of APIs in fish and other wildlife [176].

As with drinking water, the scarcity of data necessitates the use of conservative and/or modeling approaches to this pathway and invites further scientific effort in this area. The available data are also subject to the same limitation as that in surface and drinking water; fish have been sampled predominantly in effluent-dominated systems, where contamination would be expected to potentially represent worst-case scenarios in the developed world [134].

There is also a potential dietary pathway through crop foods. As mentioned above, APIs can enter the soil compartment through sewage sludge spreading or through manure from livestock. Experiments conducted by Boxall et al. indicate that vAPIs can arise in foodstuffs grown on lands which are contaminated with these substances [139]. Their findings indicate that florfenicol, levamisole, enrofloxacin, and trimethoprim were detected in lettuce or carrots grown on soil spiked with these substances.

Hazard and Dose–Response Assessment

In the context of a contaminated site, the process of hazard assessment is to identify chemicals that have the potential to cause adverse health effects in humans. This process is fairly straightforward when performed for industrial chemicals. However, pharmaceutical substances are designed to generate specific health effects at specific doses. Thus, it can be expected that all APIs designed for humans will have some type of effect at certain doses. Identification of a complete suite of APIs in an exposure scenario, as with COPCs at a contaminated site, may be impossible.

Selection of Substances for Risk Assessment

All of the HHRAs conducted to date began with a preselected set of APIs. One of the earliest HHRA exercises began with three representatives of API classes which were expected to cause effects at low exposure concentrations (i.e., E2, phenoxymethylpenicillin, CPA) [25]. Schwab et al. chose to study 26 APIs from 14 classes that were examined in national reconnaissance performed by Kolpin et al. [8, 23]. Schulman et al. selected four pharmaceutical compounds that "have been found more commonly or at the higher end of the spectrum of measured concentrations" in aqueous compartments [20]. Their selected compounds included acetylsalicylic acid, clofibrate, CPA, and indomethacin. The authors note CPA as the sole recognized carcinogen of the four, though there is limited animal evidence for carcinogenicity of clofibrate. Cunningham et al. chose to study CBZ and two metabolites, as CBZ has been frequently detected and has been observed in multiple aquatic compartments, including drinking water [22]. In Cunningham et al.'s prior study, they chose to study APIs manufactured by GlaxoSmithKline, as they worked for that manufacturer. These 44 APIs included amoxicillin trihydrate, digoxin, cimetidine, bupropion HCl, albuterol, metformin, and malphalan [21]. Bercu et al. studied three neuroactive drugs (atomoxetine, duloxetine, and olanzapine) for roughly the same reason [17]. Kummerer and Al-Ahmad focused on CPA and ifosfamide (IF), as these compounds were both known to be carcinogenic and to persist to a moderate degree in aquatic environments [18]. 5-FU was chosen by Johnson et al. for similar reasons [24]. Webb et al. assessed risk from 64 APIs for which monitoring data in Germany had been previously published [31]. Kumar and Xagoraraki assessed the potential risk from meprobamate, CBZ, and phenytoin, as they have previously been detected in water in the USA, and may have toxic effects in women and children [26]. In the USA, Snyder et al. developed a list of APIs to consider via a process designed to identify analytes which represented broader classes of compounds, while considering toxicity potential and likelihood of occurrence in raw and finished drinking water [29]. In a report to the Water Inspectorate of the United Kingdom, Crane et al. analyzed the potential risks arising from 396 APIs and 11 illegal drugs, though the methodology for selection of the substances was not explicitly stated [30].

vAPIs pose a slightly more complicated question with regard to human hazard. As mentioned before, some vAPIs are used in both humans and animals. However, for APIs designed only for veterinary applications, the human implications may be less clear. It is to be expected that these types of materials will have effects on humans (as they were developed for mammals), but understanding the dose–response relationships will be difficult. To date, no HHRA has been conducted for vAPIs; however, this represents an area of important research need.

It is also clear that some of the APIs that have been detected in the environment have genotoxic and/or carcinogenic properties. The most prominent of these are anticancer therapeutics such as CPA. APIs enter the environment at a low but continuous rate, though their levels may vary seasonally as described above. Thus, it is generally believed that acute effects are unlikely [140]. The possibility of long-term, chronic exposures is much more likely, and for most if not all APIs there is no toxicological data for this type of exposure duration.

The reality of the hazard assessment process is that it will not be effective in the absence of complete toxicity information. Such data are generally lacking for many chemicals, and for API this may be more complicated as the substances were engineered to be therapeutic for human or veterinary purposes. Certainly, many of these substances are expected to have side effects, which may be of interest in a toxicological investigation. When available, data on toxicological properties are used to determine whether adverse health outcomes can be expected from any dose of API.

Setting of Safety Values

The goal of dose–response assessment is to determine the quantitative relationship between doses received and health outcomes. The end result of the process is the development of one or more criteria values, which cover the gamut of likely or possible adverse health consequences, including carcinogenic and noncarcinogenic effects. Of course, it is highly likely that the primary consideration in development of criteria values is the toxicological mode of action (MOA) of the API. Many carcinogenic outcomes are expected to follow from a non threshold MOA, meaning that risk exists even at infinitesimally minute doses. Non carcinogenic health effects generally are not believed to follow a nonthreshold MOA.

The earliest example of HHRA used several different avenues to establish a safety value, against which intake values could be compared for risk characterization [25]. For 17α-EE2, a comparison with endogenous 17β-estrodiol (E2) was used qualitatively to compare to the intake value. With regard to phenoxymethylpenicillin, a value of 10 IU was used, the level below which no allergic responses would be expected.

Schulman et al. established safety values for acetylsalicylic acid, clofibrate, CPA, and indomethacin [20]. The point of departure (POD) for acetylsalicylic acid was a lowest observed effects level (LOEL) for anticoagulant therapy (30 mg/day), and safety factors included 3 to estimate a no observed effects level (NOEL) and 10 for interindividual variability. For clofibrate, the authors used the low end of the dose range (500 mg/day) for lowering of blood triglycerides as a LOEL and safety factors similar to those used for acetylsalicylic acid. In the case of indomethacin, a subtherapeutic dose of 37.5 mg/day was chosen as the POD, and as previously a safety factor of 30 was employed. Schulman et al. used these health-based limits to generate an ambient water quality criteria value using EPA methodology and compared published concentrations in surface and drinking water to that value. Therapeutic doses have been used as safety values in other risk assessment exercises as well [30, 31]; in one case, exposures within three orders of magnitude signaled a need for further study. This approach did not include the use of further uncertainty factors (UFs), though future efforts are clearly needed to develop and thus refine default UFs applied in HHRAs and ERAs.

Several HHRAs have set safety values using the concept of acceptable daily intake (ADI) (Table 2). Previously, the ADI has primarily been applied to food additives, pesticides, and veterinary drugs that are not genotoxic or carcinogenic [141, 142]. The ADI is very similar to the reference dose (RfD) in that it determines a POD and divides that value by UFs to impart a margin of safety (MOS). The POD for the ADI is often the no observed adverse effects level (NOAEL) or NOEL derived from the study in which toxicity was seen at the lowest dose [141, 142]. In some applications, the ADI is based on a NOEL as opposed to a NOAEL [31]. UFs can account for extrapolation from a lowest observed adverse effects level (LOAEL) to a NOAEL (when a NOAEL is not available), interindividual and interspecies differences, and for the possibility of an incomplete set of data. When the ADI has been set, it can be regarded as a "safe intake level (without an appreciable risk) for a healthy adult who is exposed to an average daily amount of the substance in question over a lifetime" [142]. ADIs have been determined for APIs in the environment in several HHRA exercises [17, 21, 22, 26, 28, 31, 139].

A procedure for generating ADIs for APIs was well articulated by Schwab et al. [23]. The authors chose the lowest therapeutic dose level as the POD for 26 APIs, including acetaminophen, codeine, fluoxetine, and tetracycline. Each of these PODs was then divided by up to five UFs, to account for extrapolation from a therapeutic dose to a NOAEL (UF_1), differences in exposure duration (UF_2), differences in sensitivity among species (UF_3), differences in susceptibility among individuals (UF_4), and quality of the data used to derive the POD (UF_5). For example, the POD for acetaminophen was 9.3 mg/kg/day, as the lowest effective daily therapeutic dose. This POD was divided by 3 to estimate a NOAEL, 3 to account for chronic exposures since the POD is based on an acute dose, and 3 to allow for differences in sensitivity among human individuals; thus the authors arrived at an ADI of 340 µg/kg/day for acetaminophen. These calculated ADIs were then used, along with estimates of BCFs, to calculate a set of predicted no-effect concentration (PNEC) values for ingestion of APIs through drinking water, fish tissue, and a combination of the two. A similar approach was taken by Kumar and Xagoraraki [26]. Kummerer and Al-Ahmad compared their exposure values to the lowest dose of CPA or ifosfamide given in anticancer therapy [18].

Cunningham et al. calculated ADIs following the general process employed by Schwab et al. [21–23]. The authors used LOELs, NOELs, or the lowest daily therapeutic dose as their POD and applied UFs to account for duration of exposure, interspecies variability, intraindividual susceptibility, and data quality. Ultimately, PNECs were assigned for ingestion of APIs through drinking water and fish consumption, as well as a combined metric. PNECs for three neuropharmaceuticals were also calculated using the ADI procedure set forth by Schwab [17]. An ADI-based value has also been offered for CPA, a chemotherapeutic agent which is known to be genotoxic [25]. Schulman et al., instead of employing the approach of Christensen, began with the cancer slope factor for CPA generated by California

Table 2 Adjusted daily intake and therapeutic daily intake values for active pharmaceutical ingredients in risk assessment exercises published to date

API	ADI/TDI	Value	Unit	References
17α-Ethinylestradiol	TDI	0.01	mg/day	[32]
Abacavir	ADI	57.1	μg/kg/day	[22]
Acetominophen	ADI	340	μg/kg/day	[24]
Acetylsalicylate	ADI	7	μg/kg/day	[29]
	HBL	1	mg/day	[21]
	TDI	30	mg/day	[32]
Acyclovir/valacyclovir	ADI	190	μg/kg/day	[22]
Albendazole	ADI	25.4	μg/kg/day	[22]
Albuterol	ADI	20.7	μg/kg/day	[22]
	ADI	2.8	μg/kg/day	[24]
	TDI	0.10	mg/day	[32]
Amoxycillintrihydrate	ADI	21.4	μg/kg/day	[22]
Amprenavir/fosamprenavir	ADI	133	μg/kg/day	[22]
Atenolol	ADI	2.7	μg/kg/day	[30]
	TDI	50	mg/day	[32]
Atomoxetine	ADI	1.4	μg/kg/day	[18]
Atorvastatin	ADI	0.54	μg/kg/day	[30]
Atovaquone	ADI	238	μg/kg/day	[22]
Beclomethasone	ADI	0.19	μg/kg/day	[22]
Benzylpenicillin	TDI	600	mg/day	[32]
Betamethasone	ADI	0.24	μg/kg/day	[22]
Betaxolol	TDI	10	mg/day	[32]
Bezafibrate	TDI	400	mg/day	[32]
Bisoprolol	TDI	2.5	mg/day	[32]
Bupropion	ADI	57.1	μg/kg/day	[22]
Carazolol	TDI	15	mg/day	[32]
Carbamazepine	TDI	400	mg/day	[32]
	ADI	15.9	μg/kg/day	[23]
	ADI, toxicological	7.5	μg/kg/day	[27]
	ADI, therapeutic (child)	78	μg/kg/day	[27]
	ADI, therapeutic (adult)	190	μg/kg/day	[27]
	ADI	0.34	μg/kg/day	[29, 30]
Carvedilol	ADI	3	μg/kg/day	[22]
Cefazolin	ADI	10	μg/kg/day	[22]
Ceftazidime	ADI	14.3	μg/kg/day	[22]
Cefuroxime	ADI	29	μg/kg/day	[22]
Celiprolol	TDI	200	mg/day	[32]
Chloramphenicol	TDI	3,000	mg/day	[32]
Chlorotetracycline	TDI	1,000	mg/day	[32]
Cimetidine	ADI	28.6	μg/kg/day	[22]
	ADI	29	μg/kg/day	[24]
Ciprofloxacin	ADI	1.6	μg/kg/day	[24]
Clarithromycin	TDI	500	mg/day	[32]
Clavulanic acid	ADI	90	μg/kg/day	[22]
Clenbuterol	TDI	0.02	mg/day	[32]

(continued)

Table 2 (continued)

API	ADI/TDI	Value	Unit	References
Clofibrate	HBL	16.7	mg/day	[21]
	TDI	500	mg/day	[32]
Cloxacillin	TDI	1,000	mg/day	[32]
Codeine	ADI	2	µg/kg/day	[24]
Cyclizine	ADI	71.4	µg/kg/day	[22]
Cyclophosphamide	RSD	0.014	µg/kg/day	[21]
	TDI	1	mg/day	[32]
Dehydrato-erythromycin	TDI	1,000	mg/day	[32]
Diazepam	ADI	0.16	µg/kg/day	[30]
Diazepam	TDI	6	mg/day	[32]
Diclofenac	ADI	1.6	µg/kg/day	[30]
	TDI	25	mg/day	[32]
Dicloxacillin	TDI	500	mg/day	[32]
Digoxin	ADI	0.071	µg/kg/day	[22]
Diltiazem	ADI	14	µg/kg/day	[24]
Doxycycline	ADI	30	µg/kg/day	[24]
Doxycycline	TDI	100	mg/day	[32]
Duloxetine	ADI	1.8	µg/kg/day	[18]
Dutasteride	ADI	0.0016	µg/kg/day	[22]
Enalapril	ADI	0.23	µg/kg/day	[30]
Erythromycin-H_2O	ADI	40	µg/kg/day	[24]
Fenofibrate	TDI	100	mg/day	[32]
	TDI	100	mg/day	[32]
Fenoprofen	TDI	900	mg/day	[32]
Fenoterol	TDI	0.50	mg/day	[32]
Fluoxetine	ADI	2.9	µg/kg/day	[24]
	ADI	1	µg/kg/day	[30]
Fluticasone	ADI	0.095	µg/kg/day	[22]
Gemfibrozil	ADI	0.56	µg/kg/day	[30]
	TDI	1,200	mg/day	[32]
Halofantrine	ADI	200	µg/kg/day	[22]
Hydrochlorothiazide	ADI	6	µg/kg/day	[22]
Ibuprofen	ADI	110	µg/kg/day	[24]
	TDI	1,200	mg/day	[32]
Ifosfamide	TDI	2,160	mg/day	[32]
Indomethacin	TDI	50	mg/day	[32]
	HBL	1.67	mg/day	[21]
Ketoprofen	TDI	100	mg/day	[32]
Lamivudine	ADI	15.9	µg/kg/day	[22]
Lamotrigine	ADI	11.9	µg/kg/day	[22]
Lincomycin	ADI	25	µg/kg/day	[24]
Melphalan	ADI	0.0021	µg/kg/day	[22]
Meprobamate	ADI, toxicological	58	µg/kg/day	[27]
	ADI, therapeutic (child)	130	µg/kg/day	[27]
	ADI, therapeutic (adult)	140	µg/kg/day	[27]
	ADI	6.1	µg/kg/day	[30]

(continued)

Table 2 (continued)

API	ADI/TDI	Value	Unit	References
Mercaptopurine	ADI	0.0021	µg/kg/day	[22]
Metaprolol	ADI	14	µg/kg/day	[29]
Metformin	ADI	79.4	µg/kg/day	[22]
Metformin	ADI	62	µg/kg/day	[24]
Methicillin	TDI	2,000	mg/day	[32]
Metropolol	TDI	25	mg/day	[32]
Nabumetone	ADI	476	µg/kg/day	[22]
Nadolol	TDI	40	mg/day	[32]
Nafcillin	TDI	1,000	mg/day	[32]
Naproxen	ADI	46	µg/kg/day	[30]
Naratriptan	ADI	1.2	µg/kg/day	[22]
Norfloxacin	ADI	190	µg/kg/day	[24]
Olanzapine	ADI	1.4	µg/kg/day	[18]
Ondansetron	ADI	2.38	µg/kg/day	[22]
Oxacillin	TDI	1,000	mg/day	[32]
Oxytetracycline	ADI	30	µg/kg/day	[24]
	TDI	1,000	mg/day	[32]
Pentoxifylline	TDI	1,200	mg/day	[32]
Phenazon	TDI	150	mg/day	[32]
	ADI	36	µg/kg/day	[29]
Phenoxymethylpenicillin	TDI	1,000	mg/day	[32]
Phenytoin	ADI, toxicological	10	µg/kg/day	[27]
	ADI, therapeutic (child)	42	µg/kg/day	[27]
	ADI, therapeutic (adult)	100	µg/kg/day	[27]
	ADI	0.083	µg/kg/day	[30]
Proguanil	ADI	95.2	µg/kg/day	[22]
Propranolol	TDI	30	mg/day	[32]
Ranitidine	ADI	10.7	µg/kg/day	[22]
	ADI	11	µg/kg/day	[24]
Risperidone	ADI	0.014	µg/kg/day	[30]
Ropinirole	ADI	0.12	µg/kg/day	[22]
Rosiglitazone	ADI	0.7	µg/kg/day	[22]
Roxithromycin	TDI	150	mg/day	[32]
Salicylic acid	TDI	3,000	mg/day	[32]
Salmeterol	ADI	0.05	µg/kg/day	[22]
Simvastatin	ADI	0.54	µg/kg/day	[30]
Sotalol	TDI	80	mg/day	[32]
Sulfamethazine	TDI	2,000	mg/day	[32]
Sulfamethoxazole	ADI	130	µg/kg/day	[24, 29]
Sulfamethoxazole	ADI	280	µg/kg/day	[30]
Sulfamethoxazole	TDI	800	mg/day	[32]
Sulfathiozole	ADI	50	µg/kg/day	[24]
Sumatriptan	ADI	23.8	µg/kg/day	[22]
Terbutalin	TDI	0.25	mg/day	[32]
Tetracycline	ADI	30	µg/kg/day	[24]
	TDI	1,000	mg/day	[32]

(continued)

Table 2 (continued)

API	ADI/TDI	Value	Unit	References
Timolol	TDI	20	mg/day	[32]
Topotecan	ADI	0.0021	μg/kg/day	[22]
Triamterene	ADI	71.5	μg/kg/day	[22]
Triclosan	ADI	12	μg/kg/day	[30]
Trimethoprim	ADI	9.4	μg/kg/day	[22]
	ADI	4.2	μg/kg/day	[24]
	ADI	100	μg/kg/day	[30]
	TDI	200	mg/day	[32]
Triprolidine	ADI	4.8	μg/kg/day	[22]
Vinorelbine	ADI	0.0021	μg/kg/day	[22]
Warfarin	ADI	0.16	μg/kg/day	[24]
Zanamivir	ADI	3.14	μg/kg/day	[22]
Zidovudine	ADI	14.3	μg/kg/day	[22]

EPA [20]. A dose level of 1 μg/day was established as a safety value, based on an excess cancer risk level of 1×10^{-5}. This figure was cited in risk characterization performed by Webb et al. [31].

As these ADIs accumulate in the literature, they have been and will continue to be used in further assessments of potential risks from APIs in the environment [28, 29]. In many cases, these ADIs and TDIs were extrapolated to ambient and drinking water concentrations that corresponded to safe levels (Table 3). These values included PNECs, drinking water equivalent levels (DWELs), ambient water quality guideline values (AWQCs), and proposed provisional guideline values (PGVs).

Thresholds of toxicologic concern (TTC) were also employed for 5-FU and other chemotherapeutics [21, 24]. In a rather qualitative approach, a potential range doses of 5-FU was compared to the general TTC of 1.5 μg/person/day recommended by Kroes et al. [143]. The TTC has also been also applied more quantitatively in HHRA practice, as it was used to set an ADI for four oncology drugs, for which too few data were available to assess the carcinogenic potency [21]. Schriks et al. and Rowney et al. also compared their predicted exposure values to the TTC, aside from the comparisons to other substance-specific safety values found in the peer-reviewed literature and elsewhere [27, 28].

Risk Characterization

In virtually all studies conducted to date, risk of adverse health effects from exposure to an API through drinking water or fish ingestion was judged to be negligible [17–30]. One exception was from Christensen's 1998 publication in which the risk

Table 3 Safety values for surface and drinking water for active pharmaceutical ingredients in risk assessment exercises published to date

COPCs	Environmental criterial value type	Values	Unit	References
Abacavir	$PNEC_{DW+F}$ (child)	8.20E+05	ng/L	[22]
Acetominophen	$PNEC_{DW+F}$	4.90E+06	ng/L	[24]
Acetylsalicylate	PGV	3.00E+01	μg/L	[29]
	$AWQC_{DW}$	4.80E+05	ng/L	[21]
Acyclovir/valacyclovir	$PNEC_{DW+F}$ (child)	2.70E+06	ng/L	[22]
Albendazole	$PNEC_{DW+F}$ (child)	3.60E+05	ng/L	[22]
Albuterol	$PNEC_{DW+F}$ (child)	3.00E+05	ng/L	[22]
	$PNEC_{DW+F}$	4.00E+04	ng/L	[24]
Amoxycillintrihydrate	$PNEC_{DW+F}$ (child)	3.10E+05	ng/L	[22]
Amprenavir/fosamprenavir	$PNEC_{DW+F}$ (child)	1.90E+06	ng/L	[22]
Atenolol	DWEL	8.10E+01	μg/L	[30]
Atomoxetine	PNEC (adults)	3.43E+01	μg/L	[18]
	PNEC (children)	2.57E+01	μg/L	[18]
Atorvastatin	DWEL	1.60E+01	μg/L	[30]
Atovaquone	$PNEC_{DW+F}$ (child)	1.10E+06	ng/L	[22]
Beclomethasone	$PNEC_{DW+F}$ (child)	2.00E+03	ng/L	[22]
Betamethasone	$PNEC_{DW+F}$ (child)	3.40E+03	ng/L	[22]
Bupropion	$PNEC_{DW+F}$ (child)	8.20E+05	ng/L	[22]
Carbemazepine	$PNEC_{DW+F}$ (child)	2.26E+05	ng/L	[23]
	$DWEL_{DW\text{-}toxicity}$	3.50E+05	μg/L	[27]
	PGV	1.00E+00	μg/L	[29]
	DWEL	1.00E+01	μg/L	[30]
Carvedilol	$PNEC_{DW+F}$ (child)	4.30E+04	ng/L	[22]
Cefazolin	$PNEC_{DW+F}$ (child)	1.40E+05	ng/L	[22]
Ceftazidime	$PNEC_{DW+F}$ (child)	2.00E+05	ng/L	[22]
Cefuroxime	$PNEC_{DW+F}$ (child)	4.20E+05	ng/L	[22]
Cimetidine	$PNEC_{DW+F}$ (child)	4.10E+05	ng/L	[22]
	$PNEC_{DW+F}$	4.10E+05	ng/L	[24]
Ciprofloxacin	$PNEC_{DW+F}$	2.30E+04	ng/L	[24]
Clavulanic acid	$PNEC_{DW+F}$ (child)	1.30E+06	ng/L	[22]
Clofibrate	$AWQC_{DW}$	2.20E+05	ng/L	[21]
Clofibric acid[a]	PGV	3.00E+01	μg/L	[29]
Codeine	$PNEC_{DW+F}$	2.90E+04	ng/L	[24]
Cyclizine	$PNEC_{DW+F}$ (child)	1.00E+06	ng/L	[22]
Cyclophosphamide	$AWQC_{DW}$	4.80E+02	ng/L	[21]
Dehydronifedipine[a]	$PNEC_{DW+F}$	1.10E+06	ng/L	[24]
Diazepam	DWEL	4.80E+00	μg/L	[30]
Diclofenac	DWEL	4.80E+01	μg/L	[30]
Digoxigenin[a]	$PNEC_{DW+F}$	1.00E+03	ng/L	[24]
Digoxin	$PNEC_{DW+F}$ (child)	1.00E+03		[22]
Digoxin	$PNEC_{DW+F}$	1.00E+03	ng/L	[24]
Diltiazem	$PNEC_{DW+F}$	2.00E+05	ng/L	[24]
Doxycycline	$PNEC_{DW+F}$	4.30E+05	ng/L	[24]
Duloxetine	PNEC (adults)	2.83E+01	μg/L	[18]
	PNEC (children)	1.91E+01	μg/L	[18]

(continued)

Table 3 (continued)

COPCs	Environmental criterial value type	Values	Unit	References
Dutasteride	PNEC$_{DW+F}$ (child)	1.00E+01	ng/L	[22]
Enalapril	DWEL	6.90E+00	μg/L	[30]
Enalaprilat[a]	PNEC$_{DW+F}$	1.00E+06	ng/L	[24]
Erythromycin-H$_2$O	PNEC$_{DW+F}$	5.70E+05	ng/L	[24]
Fluoxetine	PNEC$_{DW+F}$	4.10E+04	ng/L	[24]
	DWEL	3.00E+01	μg/L	[30]
Fluticasone	PNEC$_{DW+F}$ (child)	1.20E+03	ng/L	[22]
Gemfibrozil	PNEC$_{DW+F}$	7.90E+05	ng/L	[24]
	DWEL	3.90E+01	μg/L	[30]
Halofantrine	PNEC$_{DW+F}$ (child)	2.90E+06	ng/L	[22]
Hydrochlorothiazide	PNEC$_{DW+F}$ (child)	8.70E+04	ng/L	[22]
Ibuprofen	PNEC$_{DW+F}$	1.60E+06	ng/L	[24]
Indomethacin	AWQC$_{DW}$	7.80E+05	ng/L	[21]
Lamivudine	PNEC$_{DW+F}$ (child)	2.30E+05	ng/L	[22]
Lamotrigine	PNEC$_{DW+F}$ (child)	1.70E+05	ng/L	[22]
Lincomycin	PNEC$_{DW+F}$	3.60E+05	ng/L	[24]
Melphalan	PNEC$_{DW+F}$ (child)	3.00E+01	ng/L	[22]
Meprobamate	DWEL$_{DW\text{-toxicity}}$	2.63E+05	ng/L	[27]
	DWEL	1.80E+02	μg/L	[30]
Mercaptopurine	PNEC$_{DW+F}$ (child)	3.00E+01	ng/L	[22]
Metaprolol	PGV	5.00E+01	μg/L	[29]
Metformin	PNEC$_{DW+F}$ (child)	1.10E+06		[22]
	PNEC$_{DW+F}$	8.90E+05	ng/L	[24]
Nabumetone	PNEC$_{DW+F}$ (child)	5.20E+06	ng/L	[22]
Naproxen	DWEL	1.40E+03	μg/L	[30]
Naratriptan	PNEC$_{DW+F}$ (child)	1.70E+04	ng/L	[22]
Norfloxacin	PNEC$_{DW+F}$	2.70E+06	ng/L	[24]
Olanzapine	PNEC (adults)	4.40E+01	μg/L	[18]
	PNEC (children)	3.59E+01	μg/L	[18]
Ondansetron	PNEC$_{DW+F}$ (child)	3.40E+04	ng/L	[22]
Oxytetracycline	PNEC$_{DW+F}$	4.30E+05	ng/L	[24]
Paroxetine metabolite[a]	PNEC$_{DW+F}$	4.10E+04	ng/L	[24]
Phenazone	PGV	1.25E+02	μg/L	[29]
Phenytoin	DWEL$_{DW\text{-toxicity}}$	2.03E+06	ng/L	[27]
	DWEL	5.80E+00	μg/L	[30]
Proguanil	PNEC$_{DW+F}$ (child)	1.40E+06	ng/L	[22]
Ranitidine	PNEC$_{DW+F}$ (child)	1.50E+05	ng/L	[22]
	PNEC$_{DW+F}$	1.60E+05	ng/L	[24]
Risperidone	DWEL	4.10E-01	μg/L	[30]
Ropinirole	PNEC$_{DW+F}$ (child)	1.70E+03	ng/L	[22]
Rosiglitazone	PNEC$_{DW+F}$ (child)	1.00E+04	ng/L	[22]
Salmeterol	PNEC$_{DW+F}$ (child)	7.20E+02	ng/L	[22]
Simvastatin	DWEL	1.60E+01	μg/L	[30]
Sulfamethoxazole	PGV	4.40E+02	μg/L	[29]
	PNEC$_{DW+F}$	1.90E+06	ng/L	[24]
	DWEL	8.40E+03	μg/L	[30]

(continued)

Table 3 (continued)

COPCs	Environmental criterial value type	Values	Unit	References
Sulfathiozole	$PNEC_{DW+F}$	7.20E + 05	ng/L	[24]
Sumatriptan	$PNEC_{DW+F}$ (child)	3.40E + 05	ng/L	[22]
Tetracycline	$PNEC_{DW+F}$	4.30E + 05	ng/L	[24]
Topotecan	$PNEC_{DW+F}$ (child)	3.00E + 01	ng/L	[22]
Triamterene	$PNEC_{DW+F}$ (child)	1.00E + 06	ng/L	[22]
Triclosan	DWEL	3.60E + 02	μg/L	[30]
Trimethoprim	$PNEC_{DW+F}$ (child)	1.30E + 05	ng/L	[22]
	$PNEC_{DW+F}$	6.00E + 04	ng/L	[24]
	DWEL	3.00E + 03	μg/L	[30]
Triprolidine	$PNEC_{DW+F}$ (child)	6.90E + 04	ng/L	[22]
Vinorelbine	$PNEC_{DW+F}$ (child)	3.00E + 01	ng/L	[22]
Warfarin	$PNEC_{DW+F}$	2.30E + 03	ng/L	[24]
Zanamivir	$PNEC_{DW+F}$ (child)	4.50E + 04	ng/L	[22]
Zidovudine	$PNEC_{DW+F}$ (child)	2.00E + 05	ng/L	[22]

[a]A metabolite of a parent pharmaceutical compound

of allergic reactions to phenoxymethylpenicillin was judged to be "possible for very sensitive persons" [25]. The author notes that assumptions on biodegradability of the substance play a very important role in the qualitative conclusion and also that several of the exposure assumptions taken during the exercise were conservative such that risk is unlikely.

The author also assessed human health risks to a Dutch population arising from the presence of 17α-EE2 and CPA [25]. Christensen concluded that the reasonable worst case exposure for EE2 was 0.085 μg/day (i.e., 6.32×10^{-7} mg EE2/kg body weight/day) and suggested that no significant risk exists in this context, as prepubescent boys produce 6 μg/day endogenously and adult males produce 45–48 μg. The author notes that the possibility of carcinogenic outcomes from CPA cannot be ruled out, as it is thought to proceed through a nonthreshold MOA. These exposures were apportioned among several potential routes of exposure, including ingestion of fish tissue, crops, drinking water, dairy products, meat, and inhalation of ambient air.

Another example of potential risk was illustrated by Kummerer and Al-Ahmad [18]. They concluded that the potential for excess cancers in children could not be fully excluded as a result of exposures to CPA through drinking water, primarily as a result of the use of additional UFs. The authors noted that neither Schulman et al. nor Webb et al., both of whom assessed risk from CPA, accounted for the possibility of increased risk to children [20, 31]. Another cytotoxic chemotherapy drug, 5-FU, was judged as unlikely to pose a risk to human health, as the MOS between intake values and therapeutic doses was 10^8–10^{10} and between 300 and 30,000 for a TTC value.

The other HHRAs published to date indicate a *de minimus* level of risk with regard to most APIs in the environment of developed counties [17–30]. The MOS and HQs are noted in Table 4. As might be expected, MOS were higher for adults than for children, when such comparisons are made [17, 26].

Schwab et al. performed a complex comparison between three separate PNECs (DW, fish, combined) and MECs from USGS data and PECs generated via PhATE modeling [23]. Their hazard quotients (HQs) ranged from 0.33 for ciprofloxacin ($PNEC_{DW+F}$ to PhATE PEC) to 9.1×10^{-8} for oxytetracycline ($PNEC_F$ to PhATE PEC). Aside from ciprofloxacin, HQs for metformin (0.11), ranitidine (0.1), and warfarin (0.11) were also within an order of magnitude of a conclusion of potential human health risks. The authors concluded that the "presence of low levels of APIs in surface waters and drinking water poses no appreciable risk to human health."

Webb et al., using approaches they had previously published, compared daily intake of APIs in drinking water (based on a monitored value in German surface water) to recommended daily therapeutic dosages [31]. The MOS between the calculated intake and therapeutic dosage values ranged from 1×10^4 (salbutamol) to 1×10^9 (ioxitalamic acid and iothalamic acid, X-ray contrast media).

Cunningham et al. determined that the MOS between their calculated PNECs and PECs for CBZ and two metabolites ranged between 340 and 6,560, and as such there was "no appreciable risk to human health from environmental exposures from drinking water and fish consumption" [22]. In an earlier study, they calculated MOS for North America and the EU ranging from 14 (amoxicillin, NA) to 1.79×10^{10} (halofantrine, EU). Bercu et al. calculated MOS between 147 and 642 for the three neuroactive APIs in their study [17]. Crane et al. pursued a tiered risk characterization, in which they targeted APIs with an MOS of less than 1,000 for further study; only nine of the 364 compounds under study fell within that MOS range and three of these (d-9-THC, cocaine, and LSD) are illegal drugs [30].

Microbial Resistance

A potential indirect risk from environmental APIs (and vAPIs) is the development of microbial strains that are resistant to antibiotics, as a result of ongoing selection in WWTP sludge and other compartments [33, 144, 145]. Bacterial strains resistant to multiple antibiotics have been detected in biofilms formed in surface water and in drinking water distribution systems [146]. Experiments have also shown that bacteria found in hospital wastewater effluent are resistant to different types of antibiotics than other effluents [146]. A greater prevalence of *Acetinobacter* bacteria were observed downstream from hospital and API manufacturing wastewater outflows than in the upstream area [147]. Similar results were seen for *Escherichia coli* in wastewater and for Enterobacteriaceae and *Aeromonas* species in surface water affected by urban effluents [148, 149].

The possibility remains that increased number of resistant organisms observed in the environment is related to excretion of such microbes from the gut of humans and animals being treated with antibiotics [21, 140, 150]. Per Kummerer, a strain of

Table 4 Risk characterization metrics for active pharmaceutical ingredients in risk assessment exercises published to date

Pharmaceutical	RC type		Value	Conclusion	References
17α-Ethinylestradiol	Undefined		Qualitative	Insignificant risk	[26]
	TD/DWI	MOS	1.00E+04	No risk to human health	[32]
5-Fluorouracil	TTC/DWI	MOS	3.0E+02–3.0E+04	Unlikely to pose a risk to human health	[25]
Abacavir	NA PEC/PNEC	HQ	7.60E−05	No appreciable risk to human health	[22]
	EU PEC/PNEC	HQ	2.40E−05	No appreciable risk to human health	[22]
Acetominophen	HQ$_{MEC-DW+F}$	HQ	2.10E−03	No appreciable risk to human health	[24]
	HQ$_{PEC-DW+F}$	HQ	9.70E−02	No appreciable risk to human health	[24]
Acetylsalicylate	Conc$_{max-DW}$/PGV	HQ	5.00E−03	No appreciable concern for human health	[29]
	Conc$_{max-SW}$/PGV	HQ	3.00E−03	No appreciable concern for human health	[29]
	Conc$_{max}$/AWQC		Qualitative	Not a risk to human health	[21]
	TD/DWI	MOS	1.50E+06	No risk to human health	[32]
Acyclovir/valacyclovir	NA PEC/PNEC	HQ	3.70E−04	No appreciable risk to human health	[22]
	EU PEC/PNEC	HQ	1.90E−04	No appreciable risk to human health	[22]
Albendazole	NA PEC/PNEC	HQ	4.40E−07	No appreciable risk to human health	[22]
	EU PEC/PNEC	HQ	6.00E−07	No appreciable risk to human health	[22]
Albuterol	NA PEC/PNEC	HQ	6.10E−05	No appreciable risk to human health	[22]
	EU PEC/PNEC	HQ	2.60E−04	No appreciable risk to human health	[22]
	HQ$_{MEC-DW+F}$	HQ	3.60E−04	No appreciable risk to human health	[24]
	HQ$_{PEC-DW+F}$	HQ	6.20E−03	No appreciable risk to human health	[24]
	TD/DWI	MOS	1.00E+04	No risk to human health	[32]
Alkylating chemotherapeutic agents[a]	DWE/TTC	HQ	9.0E−03–1.4E−01	Risk to healthy adults is low	[28]
	TTC/DWE	MOS	7.0E+00–1.1E+02	Risk to healthy adults is low	[28]
	DWE/TTC	HQ	1.8E−02–2.7E−01	Risk to healthy adults is low	[28]
	TTC/DWE	MOS	3.8E+00–5.4E+01	Risk to healthy adults is low	[28]

(continued)

Table 4 (continued)

Pharmaceutical	RC type		Value	Conclusion	References
Amoxycillin trihydrate	NA PEC/PNEC	HQ	6.70E−02	No appreciable risk to human health	[22]
	EU PEC/PNEC	HQ	1.40E−02	No appreciable risk to human health	[22]
Amprenavir/fosamprenavir	NA PEC/PNEC	HQ	4.70E−06	No appreciable risk to human health	[22]
	EU PEC/PNEC	HQ	1.30E−06	No appreciable risk to human health	[22]
Anthracycline antibiotics[b]	DWE/TTC	HQ	6.7E−04–4.0E−05	Risk to healthy adults is low	[28]
	TTC/DWE	MOS	1.5E+03–2.5E+04	Risk to healthy adults is low	[28]
	DWE/TTC	HQ	1.2E−03–8.0E−04	Risk to healthy adults is low	[28]
	TTC/DWE	MOS	8.3E+02–1.3E+04	Risk to healthy adults is low	[28]
Antimetabolite chemotherapeutic agents[c]	DWE/TTC	HQ	1.7E−03–2.7E−02	Risk to healthy adults is low	[28]
	TTC/DWE	MOS	3.7E+01–5.7E−02	Risk to healthy adults is low	[28]
	DWE/TTC	HQ	3.5E−03–5.0E−02	Risk to healthy adults is low	[28]
	TTC/DWE	MOS	2.0E+01–2.9E+02	Risk to healthy adults is low	[28]
Atenolol	$Conc_{max-DW}$/DWEL	HQ	2.50E−04	Not relevant to human health	[30]
	TD/DWI	MOS	5.00E+06	No risk to human health	[32]
Atomoxetine	Adults	MOS	2.86E+02	No appreciable risk to human health	[18]
	Children	MOS	2.14E+02	No appreciable risk to human health	[18]
Atorvastatin	$Conc_{max-DW}$/DWEL	HQ	1.50E−04	Not relevant to human health	[30]
Atovaquone	NA PEC/PNEC	HQ	2.10E−06	No appreciable risk to human health	[22]
	EU PEC/PNEC	HQ	2.10E−05	No appreciable risk to human health	[22]
Beclomethasone	NA PEC/PNEC	HQ	4.10E−05	No appreciable risk to human health	[22]
	EU PEC/PNEC	HQ	4.80E−04	No appreciable risk to human health	[22]
Benzylpenicillin		MOS	6.00E+06	No risk to human health	[32]
Betamethasone	NA PEC/PNEC	HQ	1.20E−03	No appreciable risk to human health	[22]
	EU PEC/PNEC	HQ	2.70E−03	No appreciable risk to human health	[22]
Betaxolol	TD/DWI	MOS	1.00E+06	No risk to human health	[32]
Bezafibrate	TD/DWI	MOS	7.41E+06	No risk to human health	[32]
Bisoprolol	TD/DWI	MOS	2.50E+05	No risk to human health	[32]
Bupropion	NA PEC/PNEC	HQ	2.40E−04	No appreciable risk to human health	[22]
	EU PEC/PNEC	HQ	4.60E−06	No appreciable risk to human health	[22]

Compound	Expression	Type	Value	Conclusion	Ref
Carazolol	TD/DWI	MOS	1.50E+06	No risk to human health	[32]
Carbamazepine	TD/DWI	MOS	6.67E+06	No risk to human health	[32]
	TD/DWI	MOS	4.00E+06	No risk to human health	[32]
	PNEC/MEC$_{90th}$	MOS	1.50E+03	No appreciable risk to human health	[23]
	PNEC/PEC$_{90th}$	MOS	6.80E+02	No appreciable risk to human health	[23]
	Average, DW	HQ	1.30E−05	No risk to human health	[27]
	Conc$_{max-sw}$/PGV	HQ	3.00E−02	No appreciable concern for human health	[29]
	Conc$_{max-DW}$/PGV	HQ	2.00E−01	No appreciable concern for human health	[29]
Carvedilol	Conc$_{max-DW}$/DWEL	HQ	1.80E−03	Not relevant to human health	[30]
	NA PEC/PNEC	HQ	1.00E−04	No appreciable risk to human health	[22]
	EU PEC/PNEC	HQ	9.90E−04	No appreciable risk to human health	[22]
Cefazolin	NA PEC/PNEC	HQ	6.50E−03	No appreciable risk to human health	[22]
	EU PEC/PNEC	HQ	3.50E−03	No appreciable risk to human health	[22]
Ceftazidime	NA PEC/PNEC	HQ	6.70E−04	No appreciable risk to human health	[22]
	EU PEC/PNEC	HQ	3.20E−04	No appreciable risk to human health	[22]
Cefuroxime	NA PEC/PNEC	HQ	3.60E−04	No appreciable risk to human health	[22]
	EU PEC/PNEC	HQ	2.30E−03	No appreciable risk to human health	[22]
Celiprolol	TD/DWI	MOS	2.00E+07	No risk to human health	[32]
Chloramphenicol	TD/DWI	MOS	7.50E+07	No risk to human health	[32]
Chlorotetracycline	TD/DWI	MOS	2.50E+07	No risk to human health	[32]
Cimetidine	NA PEC/PNEC	HQ	7.40E−04	No appreciable risk to human health	[22]
	EU PEC/PNEC	HQ	3.10E−04	No appreciable risk to human health	[22]
	HQ$_{MEC-DW+F}$	HQ	1.40E−03	No appreciable risk to human health	[24]
	HQ$_{PEC-DW+F}$	HQ	2.20E−02	No appreciable risk to human health	[24]
Ciprofloxacin	HQ$_{MEC-DW+F}$	HQ	1.30E−03	No appreciable risk to human health	[24]
	HQ$_{PEC-DW+F}$	HQ	3.30E−01	No appreciable risk to human health	[24]
Clarithromycin	TD/DWI	MOS	1.25E+07	No risk to human health	[32]

(continued)

Table 4 (continued)

Pharmaceutical	RC type	Value	Conclusion	References
Clavulanic acid	NA PEC/PNEC	2.50E−04	No appreciable risk to human health	[22]
	EU PEC/PNEC	3.20E−04	No appreciable risk to human health	[22]
Clenbuterol	TD/DWI	1.00E+03	No risk to human health	[32]
Clofibrate	$Conc_{max}$/AWQC	Qualitative	Not a risk to human health	[21]
Clofibrate	TD/DWI	1.25E+07	No risk to human health	[32]
Clofibric acid[d]	$Conc_{max-SW}$/PGV	5.00E−03	No appreciable concern for human health	[29]
	$Conc_{max-DW}$/PGV	3.00E−03	No appreciable concern for human health	[29]
	TD/DWI	3.57E+06	No risk to human health	[32]
Cloxacillin	TD/DWI	1.00E+07	No risk to human health	[32]
Codeine	$HQ_{MEC-DW+F}$	3.50E−02	No appreciable risk to human health	[24]
	$HQ_{PEC-DW+F}$	8.40E−02	No appreciable risk to human health	[24]
Cyclizine	NA PEC/PNEC	2.90E−07	No appreciable risk to human health	[22]
	EU PEC/PNEC	1.60E−06	No appreciable risk to human health	[22]
Cyclophosphamide	Undefined	Qualitative	Unlikely to contribute a significant risk	[26]
	RR1.5	2.90E−05	Unlikely to pose a risk to human health	[19]
	RR1.5	2.10E−05	Unlikely to pose a risk to human health	[19]
	$Conc_{max}$/AWQC	Qualitative	Not a risk to human health	[21]
	TD/DWI	5.00E+04	No risk to human health	[32]
	TD/DWI	1.00E+04	No risk to human health	[32]
Dehydrato-erythromycin	TD/DWI	2.50E+07	No risk to human health	[32]
Dehydronifedipine[d]	$HQ_{MEC-DW+F}$	2.60E−05	No appreciable risk to human health	[24]
	$HQ_{PEC-DW+F}$	2.00E−03	No appreciable risk to human health	[24]
Diazepam	$Conc_{max-DW}$/DWEL	5.20E−05	Not relevant to human health	[30]
	TD/DWI	1.50E+05	No risk to human health	[32]
Diclofenac	HQ	5.20E−06	Not relevant to human health	[30]
	TD/DWI	2.08E+06	No risk to human health	[32]

Dicloxacillin	TD/DWI	MOS	5.00E+06	No risk to human health	[32]
Digoxin	NA PEC/PNEC	HQ	2.00E−03	No appreciable risk to human health	[22]
	EU PEC/PNEC	HQ	6.40E−04	No appreciable risk to human health	[22]
	$HQ_{MEC-DW+F}$	HQ	N/A	No appreciable risk to human health	[24]
	$HQ_{PEC-DW+F}$	HQ	1.30E−02	No appreciable risk to human health	[24]
Digoxigenin[d]	$HQ_{MEC-DW+F}$	HQ	3.90E−03	No appreciable risk to human health	[24]
	$HQ_{PEC-DW+F}$	HQ	1.30E−02	No appreciable risk to human health	[24]
Diltiazem	$HQ_{MEC-DW+F}$	HQ	2.40E−04	No appreciable risk to human health	[24]
	$HQ_{PEC-DW+F}$	HQ	6.00E−02	No appreciable risk to human health	[24]
Doxycycline	$HQ_{MEC-DW+F}$	HQ	1.20E−04	No appreciable risk to human health	[24]
	$HQ_{PEC-DW+F}$	HQ	4.30E−03	No appreciable risk to human health	[24]
	TD/DWI	MOS	2.50E+05	No risk to human health	[32]
Duloxetine	Adults	MOS	2.18E+02	No appreciable risk to human health	[18]
	Children	MOS	1.47E+02	No appreciable risk to human health	[18]
Dutasteride	NA PEC/PNEC	MOS	2.90E−03	No appreciable risk to human health	[22]
	EU PEC/PNEC	MOS	2.30E−03	No appreciable risk to human health	[22]
Enalapril	$Conc_{max-DW}/DWEL$	HQ	3.60E−05	Not relevant to human health	[30]
Enalaprilat[d]	$HQ_{MEC-DW+F}$	HQ	4.60E−05	No appreciable risk to human health	[24]
	$HQ_{PEC-DW+F}$	HQ	6.30E−05	No appreciable risk to human health	[24]
Erythromycin-H_2O	$HQ_{MEC-DW+F}$	HQ	3.00E−03	No appreciable risk to human health	[24]
	$HQ_{PEC-DW+F}$	HQ	1.30E−02	No appreciable risk to human health	[24]
Etofibrate	TD/DWI	MOS	7.50E+06	No risk to human health	[32]
Fenofibrate	TD/DWI	MOS	2.50E+06	No risk to human health	[32]
	TD/DWI	MOS	1.19E+06	No risk to human health	[32]
Fenoprofen	TD/DWI	MOS	9.00E+07	No risk to human health	[32]
Fenoterol	TD/DWI	MOS	5.00E+04	No risk to human health	[32]
Fluoxetine	$HQ_{MEC-DW+F}$	HQ	2.90E−04	No appreciable risk to human health	[24]
	$HQ_{PEC-DW+F}$	HQ	3.10E−02	No appreciable risk to human health	[24]
	$Conc_{max-DW}/DWEL$	HQ	1.70E−05	Not relevant to human health	[30]

(continued)

Table 4 (continued)

Pharmaceutical	RC type		Value	Conclusion	References
Fluticasone	NA PEC/PNEC	HQ	1.70E−03	No appreciable risk to human health	[22]
	EU PEC/PNEC	HQ	2.00E−04	No appreciable risk to human health	[22]
Gemfibrozil	$HQ_{MEC-DW+F}$	HQ	1.00E−03	No appreciable risk to human health	[24]
	$HQ_{PEC-DW+F}$	HQ	2.20E−02	No appreciable risk to human health	[24]
	$Conc_{max-DW}$/DWEL	HQ	5.40E−05	Not relevant to human health	[30]
	TD/DWI		1.20E+08	No risk to human health	[32]
Halofantrine	NA PEC/PNEC		N/A (no sales)	No appreciable risk to human health	[22]
	EU PEC/PNEC	HQ	5.60E−11	No appreciable risk to human health	[22]
Hydrochlorothiazide	NA PEC/PNEC	HQ	2.60E−02	No appreciable risk to human health	[22]
	EU PEC/PNEC	HQ	1.70E−02	No appreciable risk to human health	[22]
Ibuprofen	$HQ_{MEC-DW+F}$	HQ	6.40E−04	No appreciable risk to human health	[24]
	$HQ_{PEC-DW+F}$	HQ	8.30E−02	No appreciable risk to human health	[24]
	TD/DWI	MOS	2.00E+08	No risk to human health	[32]
Ifosfamide	TD/DWI	MOS	1.08E+08	No risk to human health	[32]
	TD/DWI	MOS	2.16E+07	No risk to human health	[32]
	RR1.5	HQ	5.60E−05	Unlikely to pose a risk to human health	[19]
	RR1.5	HQ	1.10E−03	Unlikely to pose a risk to human health	[19]
Indometacine	TD/DWI	MOS	5.00E+06	No risk to human health	[32]
	$Conc_{max}$/AWQC	MOS	Qualitative	Not a risk to human health	[21]
Ketoprofen	TD/DWI	MOS	1.00E+07	No risk to human health	[32]
Lamivudine	NA PEC/PNEC	HQ	1.80E−03	No appreciable risk to human health	[22]
	EU PEC/PNEC	HQ	5.00E−04	No appreciable risk to human health	[22]
Lamotrigine	NA PEC/PNEC	HQ	4.10E−05	No appreciable risk to human health	[22]
	EU PEC/PNEC	HQ	1.70E−04	No appreciable risk to human health	[22]
Lincomycin	$HQ_{MEC-DW+F}$	HQ	2.00E−03	No appreciable risk to human health	[24]
	$HQ_{PEC-DW+F}$	HQ	5.90E−05	No appreciable risk to human health	[24]
Melphalan	NA PEC/PNEC	HQ	2.60E−04	No appreciable risk to human health	[22]
	EU PEC/PNEC	HQ	2.00E−04	No appreciable risk to human health	[22]

Compound	Parameter	Metric	Value	Description	Ref
Meprobamate	Average, DW	HQ	2.48E−05	No risk to human health	[27]
	$Conc_{max\text{-}DW}$/DWEL	HQ	2.40E−04	Not relevant to human health	[30]
Mercaptopurine	NA PEC/PNEC	HQ	2.40E−02	No appreciable risk to human health	[22]
	EU PEC/PNEC	HQ	3.80E−02	No appreciable risk to human health	[22]
Metaprolol	$Conc_{max\text{-}DW}$/PGV	HQ	4.00E−02	No appreciable concern for human health	[29]
	$Conc_{max\text{-}SW}$/PGV	HQ	4.00E−03	No appreciable concern for human health	[29]
Metformin	NA PEC/PNEC	HQ	1.60E−02	No appreciable risk to human health	[22]
	EU PEC/PNEC	HQ	2.00E−02	No appreciable risk to human health	[22]
	$HQ_{MEC\text{-}DW+F}$	HQ	1.70E−04	No appreciable risk to human health	[24]
	$HQ_{PEC\text{-}DW+F}$	HQ	1.10E−01	No appreciable risk to human health	[24]
Methicillin	TD/DWI	MOS	2.00E+07	No risk to human health	[32]
Metropolol	TD/DWI	MOS	2.50E+06	No risk to human health	[32]
Nabumetone	NA PEC/PNEC	HQ	2.40E−04	No appreciable risk to human health	[22]
	EU PEC/PNEC	HQ	4.90E−06	No appreciable risk to human health	[22]
Nadolol	TD/DWI	MOS	4.00E+06	No risk to human health	[32]
Nafcillin	TD/DWI	MOS	1.00E+07	No risk to human health	[32]
Naproxen	$Conc_{max\text{-}DW}$/DWEL	HQ	3.60E−07	Not relevant to human health	[30]
Naratriptan	NA PEC/PNEC	HQ	8.70E−06	No appreciable risk to human health	[22]
	EU PEC/PNEC	HQ	9.00E−06	No appreciable risk to human health	[22]
Norfloxacin	$HQ_{MEC\text{-}DW+F}$	HQ	4.40E−05	No appreciable risk to human health	[24]
	$HQ_{PEC\text{-}DW+F}$	HQ	5.90E−05	No appreciable risk to human health	[24]
Olanzapine	Adults	MOS	6.42E+02	No appreciable risk to human health	[18]
	Children	MOS	5.24E+02	No appreciable risk to human health	[18]
Ondansetron	NA PEC/PNEC	HQ	1.40E−03	No appreciable risk to human health	[22]
	EU PEC/PNEC	HQ	7.80E−04	No appreciable risk to human health	[22]
Oxacillin	TD/DWI	MOS	1.00E+07	No risk to human health	[32]

(continued)

Table 4 (continued)

Pharmaceutical	RC type	Value	Conclusion	References
Oxytetracycline	$HQ_{MEC-DW+F}$	7.90E−04	No appreciable risk to human health	[24]
	$HQ_{PEC-DW+F}$	4.50E−06	No appreciable risk to human health	[24]
	TD/DWI	2.50E+07	No risk to human health	[32]
Pentoxifylline	TD/DWI	6.00E+07	No risk to human health	[32]
	TD/DWI	3.00E+07	No risk to human health	[32]
Phenazon	TD/DWI	1.50E+06	No risk to human health	[32]
	$Conc_{max-DW}$/PGV	2.00E−04	No appreciable concern for human health	[29]
	$Conc_{max-SW}$/PGV	9.00E−04	No appreciable concern for human health	[29]
	TD/DWI	1.50E+06	No risk to human health	[32]
Phenoxymethylpenicillin	Undefined	Qualitative	Risk is possible for very sensitive persons	[26]
	TD/DWI	1.00E+07	No risk to human health	[32]
Phenytoin	Average, DW	1.84E−05	No risk to human health	[27]
	$Conc_{max-DW}$/DWEL	2.60E−03	Not relevant to human health	[30]
Proguanil	NA PEC/PNEC	6.60E−06	No appreciable risk to human health	[22]
	EU PEC/PNEC	1.00E−05	No appreciable risk to human health	[22]
Propranolol	TD/DWI	3.00E+06	No risk to human health	[32]
Ranitidine	NA PEC/PNEC	5.30E−03	No appreciable risk to human health	[22]
	EU PEC/PNEC	1.90E−02	No appreciable risk to human health	[22]
	$HQ_{MEC-DW+F}$	3.10E−03	No appreciable risk to human health	[24]
	$HQ_{PEC-DW+F}$	2.70E−02	No appreciable risk to human health	[24]
Risperidone	$Conc_{max-DW}$/DWEL	8.40E−04	Not relevant to human health	[30]
Ropinirole	NA PEC/PNEC	1.20E−04	No appreciable risk to human health	[22]
	EU PEC/PNEC	7.50E−05	No appreciable risk to human health	[22]
Rosiglitazone	NA PEC/PNEC	5.00E−04	No appreciable risk to human health	[22]
	EU PEC/PNEC	1.60E−04	No appreciable risk to human health	[22]

Compound	Metric	Method	Value	Risk	Ref
Roxithromycin	TD/DWI	MOS	3.75E+06	No risk to human health	[32]
Salicylic acid	TD/DWI	MOS	1.50E+08	No risk to human health	[32]
Salmeterol	NA PEC/PNEC	HQ	2.80E−06	No appreciable risk to human health	[22]
	EU PEC/PNEC	HQ	1.70E−05	No appreciable risk to human health	[22]
Simvastatin	$\text{Conc}_{max-DW}/\text{DWEL}$	HQ	1.50E−05	Not relevant to human health	[30]
Sotalol	TD/DWI	MOS	8.00E+06	No risk to human health	[32]
Sulfamethazine	TD/DWI	MOS	5.00E+07	No risk to human health	[32]
Sulfamethoxazole	$\text{Conc}_{max-DW}/\text{PGV}$	HQ	7.00E−04	No appreciable concern for human health	[29]
	$\text{Conc}_{max-SW}/\text{PGV}$	HQ	3.00E−04	No appreciable concern for human health	[29]
	$\text{HQ}_{MEC-DW+F}$	HQ	1.00E−03	No appreciable risk to human health	[24]
	$\text{HQ}_{PEC-DW+F}$	HQ	9.70E−03	No appreciable risk to human health	[24]
	$\text{Conc}_{max-DW}/\text{DWEL}$	HQ	3.60E−07	Not relevant to human health	[30]
	TD/DWI	HQ	2.00E+07	No risk to human health	[32]
Sulfathiozole	$\text{HQ}_{MEC-DW+F}$	HQ	3.50E−05	No appreciable risk to human health	[24]
	$\text{HQ}_{PEC-DW+F}$	HQ	3.90E−05	No appreciable risk to human health	[24]
Sumatriptan	NA PEC/PNEC	HQ	3.80E−06	No appreciable risk to human health	[22]
	EU PEC/PNEC	HQ	8.10E−07	No appreciable risk to human health	[22]
Terbutalin	TD/DWI	MOS	1.25E+04	No risk to human health	[32]
Tetracycline	$\text{HQ}_{MEC-DW+F}$	HQ	2.60E−04	No appreciable risk to human health	[24]
	$\text{HQ}_{PEC-DW+F}$	HQ	1.50E−02	No appreciable risk to human health	[24]
	TD/DWI	HQ	2.50E+07	No risk to human health	[32]
Timolol	TD/DWI	MOS	2.00E+06	No risk to human health	[32]
Topotecan	NA PEC/PNEC	HQ	1.60E−04	No appreciable risk to human health	[22]
	EU PEC/PNEC	HQ	3.00E−04	No appreciable risk to human health	[22]
Triamterene	NA PEC/PNEC	HQ	1.50E−04	No appreciable risk to human health	[22]
	EU PEC/PNEC	HQ	7.30E−05	No appreciable risk to human health	[22]
Triclosan	$\text{Conc}_{max-DW}/\text{DWEL}$	HQ	3.30E−06	Not relevant to human health	[30]

(continued)

Table 4 (continued)

Pharmaceutical	RC type		Value	Conclusion	References
Trimethoprim	NA PEC/PNEC	HQ	4.20E−03	No appreciable risk to human health	[22]
	EU PEC/PNEC	HQ	1.30E−03	No appreciable risk to human health	[22]
	$HQ_{MEC-DW+F}$	HQ	1.20E−02	No appreciable risk to human health	[24]
	$HQ_{PEC-DW+F}$	HQ	6.20E−02	No appreciable risk to human health	[24]
	$Conc_{max-DW}$/DWEL	HQ	8.30E−08	Not relevant to human health	[30]
	TD/DWI	MOS	5.00E+06	No risk to human health	[32]
Triprolidine	NA PEC/PNEC	HQ	4.20E−05	No appreciable risk to human health	[22]
	EU PEC/PNEC	HQ	2.40E−05	No appreciable risk to human health	[22]
Vinorelbine	NA PEC/PNEC	HQ	6.70E−04	No appreciable risk to human health	[22]
Vinorelbine	EU PEC/PNEC	HQ	1.30E−03	No appreciable risk to human health	[22]
Warfarin	$HQ_{MEC-DW+F}$	HQ	2.20E−04	No appreciable risk to human health	[24]
	$HQ_{PEC-DW+F}$	HQ	1.10E−01	No appreciable risk to human health	[24]
Zanamivir	NA PEC/PNEC	HQ	8.90E−07	No appreciable risk to human health	[22]
	EU PEC/PNEC	HQ	2.90E−06	No appreciable risk to human health	[22]
Zidovudine	NA PEC/PNEC	HQ	1.50E−03	No appreciable risk to human health	[22]
	EU PEC/PNEC	HQ	4.30E−04	No appreciable risk to human health	[22]

TD/DWI therapeutic dose/drinking water intake; TTC/DWI threshold of toxicologic concern/drinking water intake; $Conc_{max-SW}$/PGV max. concentration in surface water/provisional guideline value; $Conc_{max-DW}$/PGV max. concentration in drinking water/provisional guideline value; $Conc_{max}$/AWQC max. concentration/ambient water quality criteria; DWE/TTC drinking water exposure/threshold of toxicologic concern; TTC/DWE threshold of toxicologic concern/drinking water exposure; $Conc_{max-DW}$/DWEL max. concentration in drinking water/drinking water equivalent levels

[a]Oxaliplatin, temozolomide, cisplatin, carboplatin, cyclophosphamide
[b]Epirubicin, doxorubicin
[c]Gemcitabine, fludarabine, capecitabine
[d]Metabolite

bacteria known to be resistant to seven different antibiotics was released into a laboratory-simulated WWTP [150]. Within 2 weeks, the resistant bacterium could not be detected in the system, suggesting that continuous input of resistant microbes may be more important than selection events occurring inside a WWTP. Indeed, it has been suggested that levels of antibiotics found in WWTP or aquatic compartments may be too high to generate resistance [150]. Other studies have indicated that natural environments are more likely to be the reservoir of genetic material that confer resistance [2].

Kim and Aga propose a framework of risk assessment for the development of antibiotic-resistant microorganisms and their eventual likely impact on human health [144]. This conceptual model could provide a calculated probability of an increase in the annual rate of antibiotic-resistant infections through ingestion in drinking water or through other exposure pathways. Interestingly, Webb et al. note that the ADI developed for trimethoprim in their HHRA exercise takes into account the possibility of selection for resistant bacteria in natural gut flora, though it certainly does not consider the indirect events in WWTP [31].

Uncertainty

There are two main sources of uncertainty in HHRA: variability and lack of knowledge. Studies conducted to date on APIs in the environment and the potential for adverse human health outcomes are fraught with issues related to these two sources [35]. Several parameters associated with exposure and potential toxicity of APIs can show high degrees of variability. Ironically, the uncertainties associated with HHRA for APIs in the environment are slightly fewer than for ecological or environmental risk assessment (ERA), as in most cases a MOA for the COPC is better understood than it would be for a nonhuman target.

The limited amount of information on the presence of these materials in the environment gives rise to uncertainty regarding exposure scenarios, and selection of potential hazards for further study. Perhaps more importantly, there is also a dearth of toxicity and dose–response information, which induces most risk assessors to use therapeutic doses as points of departure in assessment of dose–response relationships. In particular, the ability of risk assessors to evaluate potential risks arising from chronic, low-dose exposures is hindered by a lack of data. Further, no suitable mechanism exists at this time for quantification of the importance of drug–drug interactions as well, though several observations indicate that exposure to mixtures can cause more complex responses [129].

Uncertainty in Substance Selection/Hazard Assessment

Several rationales were offered for API selection in the HHRAs completed to date. One suggested procedure was to consider whether the APIs were among the top prescribed or to choose substances for which MECs were available [23]. The work of Benotti et al. demonstrates that prescription information is a poor predictor of exposure in drinking water, as several of the APIs that persisted through treatment were not in the top 200 prescribed drugs [130]. Though the work of Crane et al. analyzed risks from 396 compounds, the vast majority of APIs have not been assessed [30]. It is to be anticipated that future efforts will include prioritization of APIs to be analyzed in the environment [24, 26, 27]. Obviously, it is sensible to focus on compounds that are expected to exert toxicological effects at very low concentrations, including EDCs and genotoxic APIs such as cancer chemotherapeutics.

Mixtures

Experts in the field of APIs in the environment seem to agree that there is a high level of uncertainty regarding the potential significance of API mixture effects [7, 129]. Part of this uncertainty stems from a lack of an accepted and validated risk assessment methodology for exposure events of this type. For antibiotics in particular, if the possibility of additivity or synergism are not taken into account, it is likely that real-world risk will be underestimated [140]. Some authors have suggested the use of toxicogenomic or metabolomics approaches to understand mixture toxicology and ecotoxicology, and certainly new effort in this area is expected [151]. Bioassays, including the yeast estrogen screening (YES) assay, have been used to assess the importance of mixtures of EDCs, and other classes of chemicals with similar toxicological mechanisms [152–154]; this general technique may prove useful for pharmaceuticals in the future. There is also a lack of information on drug–chemical interactions, which may be important [5, 155–157]. Indeed, one author has recommended a UF of 100,000 be applied to all APIs to account for mixtures, sensitive subpopulations, and the more standard issues of differences in dose duration [158].

One of the published HHRAs to date attempts to deal quantitatively with the potential significance of mixture effects [30]. The authors chose to combine potential concentrations of non-steroidal anti-inflammatory drugs (NSAIDs) into a grouped parameter. Total NSAIDs in their exercise thus had the lowest MOS [77]. They adopted a similar approach for statins, though the margin of safety in that case was over 1,000. Rowney et al. adopted a similar approach to alkylating (oxaliplating, temozolomide, cisplating, carboplatin, CPA) and antimetabolite (gemcitabine, fludarabine, capecitabine) chemotherapeutics, as well as anthracycline antibiotics (epirubicin, doxorubicin) [27]. Kumar and Xagoraraki considered mixture effects of their analytes of interest and consulted the RxList internet drug index and HSDB to determine whether any interaction between these compounds had been observed in previous studies [26]; however, this approach would not consider the presence of other APIs in a real-life exposure scenario for human or environmental receptors.

Metabolites, Conjugates, and Degradation By-Products

The large number of potential targets for HHRA is further complicated by the generation of metabolites and conjugates through human and biotic environmental processes, and the potential generation of degradation by-products [18, 159]. In some exercises, conjugates and metabolites were treated as parent compounds for the purposes of risk assessment [17]. Deconjugation is an important possibility to account for, as it can result in WWTP effluent concentrations being higher than influent values [91]. Generally, the importance of metabolites is not considered in risk assessments, probably because it adds another layer of complexity to an already complex issue. Cunningham et al. did assess potential risks from CBZ-diol and CBZ-n-glucuronide, the two major metabolites of CBZ, and found them unlikely to pose a risk to human health [21]. However, the authors did note a lack of toxicology data for these two substances and were forced to use essentially proxy data. The same was true for Snyder's analysis of hydroxylated metabolites of atorvastatin and simstatin [29]. Also, conjugates were assumed to be returned to parent compound in the environment. The lack of data on toxicological properties of metabolites and conjugates of APIs would be expected to be the rule.

Uncertainty in Toxicity Values

In several of the HHRAs conducted to date, UFs were used to account for the more obvious sources of variability: interspecies and interindividual differences, database quality, etc. Rationales for application of various UFs in HHRA was again well articulated by Schwab et al. [23]. In one instance, the authors mined clinical trial data to derive chemical-specific adjustment factors (CSAFs) that ranged from 10 to 35 for variability in human responses [17]. Based on this information, it is worthwhile to reexamine the default factor of 10 used in many other studies. In another exercise, total UFs integrated into an ADI estimate ranged from 9 to 1,000 across a relatively small set of APIs [23].

Most of the data that are available to serve as a POD for establishment of ADIs and/or PNECs is derived from acute toxicity endpoints. Safety values using such values might be underestimating potential risks from chronic exposures [160]. The use of conservative UFs to correct for these concerns may provide safety values that are indeed protective, but the default UFs employed in several HHRAs have generally been applied to industrial chemicals, as opposed to APIs that are designed to generate an effect in a human target [21–23]. Therapeutic dose values have also been used for PODs, and it is unclear as to whether this is appropriate [23, 31]. Furthermore, there is a deficit of information on reproductive effects and mechanisms of action for many of the APIs under examination [6, 15, 128, 161–163]. There is also not enough information regarding the bioaccumulative potential of APIs, as they were often estimated based on $\log K_{ow}$ values [17, 135].

TTCs are derived from a class of compounds as opposed to a specific compound of interest and as such as associated with significant uncertainty when used as a toxicity value [24, 25]. Though the value is expected to be protective as it is theoretically based on the most potent substance in the most sensitive system, the possibility of incomplete data cannot be discounted, among other potential uncertainties.

Theoretically, the use of a UF to account for interindividual differences should be protective of sensitive subpopulations. These populations include the elderly (who may be compromised in terms of their capacity for detoxification), children (who receive comparatively larger doses), and pregnant women [129]. This factor may also be adequate to account for individuals who are exposed to additional quantities of medications that they are consuming [20]. However, more study would be useful in determining whether the default factor is sufficiently protective. Variations in standard practice among nations could lead to different conclusions from very similar estimated intakes [18, 20, 31]. It would also be helpful to understand whether the current UFs are protective of mixture effects; if data suggest that mixture effects play an important role in the toxicity of a substance, then it is possible an additional UF may be needed.

Uncertainty in Exposure

Several of the available HHRAs considered potential human exposures to API through ingestion of drinking water, surface water, and/or fish [18, 21–23]. However, none incorporated the possibility of exposures through other media or other routes. Though it is likely that these routes do not contribute significantly to overall exposure, their exclusion is a source of uncertainty for the conclusions of the HHRA. There has also been no exploration of potential exposures through unintentional water reuse [33].

One of the primary uncertainties in exposure relates to lack of knowledge of how APIs move through the environment, including degradation and partitioning to various environmental media [34, 160]. Though the literature has expanded greatly in recent years on the topics of fate and behavior, as well as removal of APIs in WWTP and drinking water processing, the data exist only for a limited number of substances and thus our ability to accurately model the occurrence of these APIs in the environment is also limited.

There is of course uncertainty associated with MECs that would be used in HHRAs, related to sampling bias and/or analytical methodology [56, 120, 132, 155]. The exploration of new techniques capable of analyzing the concentrations of multiple APIs at once is an additional source, as well as the variable usage of the available techniques between studies that are used to determine an MEC for exposure assessment. The seasonal and diurnal variability of API concentrations also imparts uncertainty to any single value MEC used as an exposure value. As noted by Daughton et al., the expansion of the universe of published MECs has paradoxically increased uncertainty surrounding concentrations of any single API in the environment [81]. Cunningham et al. rightly point out that MECs represent a snapshot, or at best a series of snapshots, and suggest that the use of models may enable the assessor to take into account more variables [22].

Modeling of PECs also brings a certain level of uncertainty, as so little data exist with which to validate the calculations. Some issues have been raised with the accuracy of the most basic calculation typically used to create PECs, as employed by Kummerer and Al-Ahmad [18, 49]. Generation of PECs through more complex models such as PhATE, iSTREEM or GREAT-ER would include uncertainty through required input parameters, including usage rates based on prescription numbers [37], variability in excretion of metabolites or conjugates [18, 56], fate and behavior of API in multiple compartments [164], flow conditions and surface water usage patterns [23], the removal of APIs in WWTP [76, 165, 166], and relative source contributions [4, 56, 124, 133, 167]. The accuracy of computer models will always be dependent upon the quality of the data entered and the assumptions associated with the model [133]. In that context, it is difficult to provide the model with data on metabolism, excretion, or fate and behavior that can accurately portray the inherent variability of these parameters. For example, it has been shown that ibuprofen can degrade in the environment at half-lives ranging from less than 1 to 50 days [160].

Efforts to understand detailed usage patterns may be seen as an intrusion on medical record-keeping, and thus such parameters may be difficult to compile. Regardless, information on prescription volume may be uninformative as to ultimate exposure concentrations that would be used in an HHRA [130]. Further, the usage of illicit substances is even less well understood than legal APIs. The quality of data on sales and production of APIs is somewhat unclear, so this is also a source of uncertainty in the modeling of PECs [36, 84]. Schwab et al. note that PECs generated by PhATE rely on the per capita consumption of the APIs under study and as such may underestimate exposure in some areas [23].

Also, it is worth nothing that in North America, PhATE does not cover urban areas, whose water supplies are not thought to be significantly impacted by WWTP effluent. It would be informative to have information from many of the populous areas not covered by the PhATE model, whose water supplies are not derived from sources impacted by WWTP.

Little effort has been devoted to quantitative or semiquantitative evaluations of uncertainty. Boeije et al. performed a Monte Carlo analysis to better understand the uncertainty surrounding the removal of APIs in WWTP and surface water [165]. Also, MCA was used to examine the uncertainty associated with degradation of ibuprofen and naproxen as a result of ozonation [122]. These techniques have been applied to risk assessments for ecological receptors [168, 169]. Variables that could be assessed via probabilistic distributions have been identified above, including temporal flow conditions [76], mixture effects [157], and obviously environmental API concentrations [155, 170]. To an extent, distributions for WWTP removal and degradation are included in modeling tools for exposure concentrations [133, 164]. Recently, a HHRA published by Kumar and Xagoraraki performed Monte Carlo simulations in an attempt to assess uncertainty in ADI and exposure parameters water ingestion rate, API concentration, and body weight. Ultimately, they concluded that variability in the ADI contributed more than 95% of variability to the risk estimates in all risk scenarios [26]. This finding may serve to highlight the need to develop more robust measures of dose–response relationships for APIs in the environment.

Conclusion

Current practice in HHRA for APIs in the environment centers on a number of uncertainties. Firstly, the scientific community is ill equipped at this point to generate a reliable estimate of exposure. Most of the efforts conducted to date employ modeling of exposure concentrations using PhATE or GREAT-ER, as few MECs are available for most APIs in drinking water. Also, the relative importance of exposures through surface water and soil is very poorly understood. In the setting of safety values, there is inconsistency in the choice of PODs, as some practitioners begin with therapeutic doses and some with NOELs or LOELs. It is also clear that risk assessors are not well served by considering "pharmaceuticals" as a broad class.

Regardless, the assessments conducted to date all indicate a negligible degree of risk associated with human exposures to APIs through drinking water and/or ingestion of fish tissue. As illustrated above, these conclusions are associated with a significant degree of uncertainty as to the potential hazards of long-term, low-dose exposure to APIs, lack of understanding of exposure in most developing countries, and also with regard to the potential additive or synergistic effects of API mixtures. Almost no effort has been expended on quantification of the uncertainties surrounding risk conclusions made so far through probabilistic techniques, though undoubtedly this is a next step in the natural progression of HHRA practice in this area. Due to the conservative nature of most of the assumptions taken in HHRAs performed so far, it is unlikely we will see any evidence of significant risk from APIs unless further study uncovers contamination of surface and drinking water on a large scale.

The areas in which a critical need exists for further research include:

1. Realistic monitoring of the occurrence of APIs in multiple environmental compartments

It has been stated that most monitoring of APIs has been undertaken in areas where contamination is expected. This is somewhat comforting, as it suggests that the use of such data in HHRA will provide a conservative risk estimate. However, the overall lack of comprehensive data is a significant area of uncertainty in such exercises. Also, this shortcoming makes it difficult for regulatory agencies and members of the public to prioritize the need for further API research and fully understand potential exposures.

2. Fate and behavior of APIs

The ability of APIs to move through the environment has already been demonstrated for a number of selected compounds. With the vast number of substances at issue, it is understandable that relatively little information is available on the mechanisms through which APIs partition to different compartments, are degraded by biotic and abiotic processes, and persist in the environment. Again, more information on more pharmaceutical substances would be useful in understanding potential exposures and in prioritizing APIs for further study or regulatory action.

3. Efficacy of various technologies for the removal of APIs from waste- and drinking water

As explored more in Chap. 9 of this volume, the removal of APIs varies among technologies employed in WWTP and drinking water treatment. It is also clear that removal is not complete and that complete information is not available for the suite of APIs that enter the environment. From an exposure perspective, this information is invaluable; any computer model designed to predict exposure concentrations would greatly benefit from further study in this area.

4. Robust exposure assessment, including further validation of current computer models

Many potential exposure pathways exist for APIs in the environment, as described in Fig. 2. To date, only a few studies have examined the sources and transport media involved in these pathways, and thus robust exposure information is generally not available. A prime example is the use of surface water data for drinking water exposure calculations; this choice would be expected to be conservative, but if we are to make rational decisions on action or inaction, complete data on real-world exposures should not be optional. The importance of intentional and unintentional water reuse should also be evaluated in the context of exposure.

5. Assessment of potential for low-dose, chronic adverse human health outcomes

Due to the "pseudopersistent" nature of APIs in the environment, the possibility of low-dose, chronic effects cannot be discounted. Most of the toxicological research that has taken place on these compounds does not test for this possibility, and thus hazard identification and setting of safety values may be an incomplete process. It is quite possible that the conservative assumptions used in current practice for HHRA in this arena are adequate to account for this possibility, but there is ample need for understanding the potential for nontherapeutic effects of APIs.

6. Mixture effects, including anthropogenic and naturally occurring complex chemicals

There are some indications that mixtures of APIs may cause changes in the function of therapeutic API intake in humans. This possibility again highlights the needs for understanding low-dose effects and in the context of the complex mixture of APIs that might be expected to occur. Virtually no effort has been expended thus far on the interaction between APIs and nonpharmaceutical complex substances (anthropogenic and naturally occurring) that form the cocktail in which we live our lives. Further, the importance of environmental (i.e., through drinking water, surface water, or ingestion of fish) exposures to an API which shares a MOA with a compound being taken therapeutically has not yet been considered.

7. Development of antibiotic resistance in response to constant influx of antibiotics

Currently, the evidence on resistance selection in bacteria is mixed, but relatively little work has been done in this area. The rise of new strains of multiresistant pathogens is certainly of interest to the public health community and thus more study of this scenario is required.

Though currently there seems to be little cause for concern to human health with regard to APIs in the environment of developed countries, copious research should be devoted to deepening our knowledge, especially in the area of EDCs and genotoxic APIs. Further characterization of the environmental prevalence of APIs in various compartments will also be an area of ongoing study. A more comprehensive evaluation of human exposure to environmental APIs, perhaps through biomonitoring, would be useful.

References

1. NCHS (2008) Health, united states, 2008 with special feature on the health of young adults. NCHS, Hyattsville
2. Zuccato E, Castiglioni S, Fanelli R, Reitano G, Bagnati R, Chiabrando C, Pomati F, Rossetti C, Calamari D (2006) Pharmaceuticals in the environment in Italy: causes, occurrence, effects and control. Environ Sci Pollut Res Int 13:15–21
3. Kumar A, Chang B, Xagoraraki I (2010) Human health risk assessment of pharmaceuticals in water: issues and challenges ahead. Int J Environ Res Public Health 7:3929–3953
4. Fent K, Weston AA, Caminada D (2006) Ecotoxicology of human pharmaceuticals. Aquat Toxicol 76:122–159
5. Carlsson C, Johansson AK, Alvan G, Bergman K, Kuhler T (2006) Are pharmaceuticals potent environmental pollutants? Part I: environmental risk assessments of selected active pharmaceutical ingredients. Sci Total Environ 364:67–87
6. Stuer-Lauridsen F, Birkved M, Hansen LP, Lutzhoft HCH, Halling-Sorensen B (2000) Environmental risk assessment of human pharmaceuticals in Denmark after normal therapeutic use. Chemosphere 40:783–793
7. Doerr-MacEwen NA, Haight ME (2006) Expert stakeholders' views on the management of human pharmaceuticals in the environment. Environ Manage 38:853–866
8. Kolpin DW, Furlong ET, Meyer MT, Thurman EM, Zaugg SD, Barber LB, Buxton HT (2002) Pharmaceuticals, hormones, and other organic wastewater contaminants in U.S. Streams, 1999–2000: a national reconnaissance. Environ Sci Technol 36:1202–1211
9. Colborn T, vom Saal FS FS, Soto AM (1993) Developmental effects of endocrine-disrupting chemicals in wildlife and humans. Environ Health Perspect 101:378–384
10. Safe S (2004) Endocrine disruptors and human health: is there a problem. Toxicology 205:3–10
11. Kime DE, Nash JP (1999) Gamete viability as an indicator of reproductive endocrine disruption in fish. Sci Total Environ 233:123–129
12. Jobling S, Beresford N, Nolan M, Rodgers-Gray T, Brighty GC, Sumpter JP, Tyler CR (2002) Altered sexual maturation and gamete production in wild roach (*Rutilus rutilus*) living in rivers that receive treated sewage effluents. Biol Reprod 66:272–281
13. Oaks JL, Gilbert M, Virani MZ, Watson RT, Meteyer CU, Rideout BA, Shivaprasad HL, Ahmed S, Chaudhry MJ, Arshad M, Mahmood S, Ali A, Khan AA (2004) Diclofenac residues as the cause of vulture population decline in pakistan. Nature 427:630–633
14. Jones OAH, Voulvoulis N, Lester JN (2002) Aquatic environmental assessment of the top 25 English prescription pharmaceuticals. Water Res 36:5013–5022
15. Ferrari B, Mons R, Vollat B, Fraysse B, Paxeus N, Lo Giudice R, Pollio A, Garric J (2004) Environmental risk assessment of six human pharmaceuticals: Are the current environmental

risk assessment procedures sufficient for the protection of the aquatic environment? Environ Toxicol Chem 23:1344–1354

16. Workgroup OOEC (2008) White paper: aquatic life criteria for contaminants of emerging concern, Part I: general challenges and recommendations. United States Environmental Protection Agency, Washington, DC

17. Bercu JP, Parke NJ, Fiori JM, Meyerhoff RD (2008) Human health risk assessments for three neuropharmaceutical compounds in surface waters. Regul Toxicol Pharm 50:420–427

18. Kummerer K, Al-Ahmad A (2010) Estimation of the cancer risk to humans resulting from the presence of cyclophosphamide and ifosfamide in surface water. Environ Sci Pollut Res Int 17:486–496

19. Emmanuel E, Pierre MG, Perrodin Y (2009) Groundwater contamination by microbiological and chemical substances released from hospital wastewater: health risk assessment for drinking water consumers. Environ Int 35:718–726

20. Schulman LJ, Sargent EV, Naumann BD, Faria EC, Dolan DG, Wargo JP (2002) A human health risk assessment of pharmaceuticals in the aquatic environment. Hum Ecol Risk Assess 8:657–680

21. Cunningham VL, Binks SP, Olson MJ (2009) Human health risk assessment from the presence of human pharmaceuticals in the aquatic environment. Regul Toxicol Pharmacol 53:39–45

22. Cunningham VL, Perino C, D'Aco VJ, Hartmann A, Bechter R (2010) Human health risk assessment of carbamazepine in surface waters of north America and Europe. Regul Toxicol Pharmacol 56:343–351

23. Schwab BW, Hayes EP, Fiori JM, Mastrocco FJ, Roden NM, Cragin D, Meyerhoff RD, D'Aco VJ, Anderson PD (2005) Human pharmaceuticals in US surface waters: a human health risk assessment. Regul Toxicol Pharmacol 42:296–312

24. Johnson AC, Jurgens MD, Williams RJ, Kummerer K, Kortenkamp A, Sumpter JP (2008) Do cytotoxic chemotherapy drugs discharged into rivers pose a risk to the environment and human health? An overview and UK case study. J Hydrol 348:167–175

25. Christensen FM (1998) Pharmaceuticals in the environment—a human risk? Regul Toxicol Pharmacol 28:212–221

26. Kumar A, Xagoraraki I (2010) Human health risk assessment of pharmaceuticals in water: an uncertainty analysis for meprobamate, carbamazepine, and phenytoin. Regul Toxicol Pharmacol 57:146–156

27. Rowney NC, Johnson AC, Williams RJ (2009) Cytotoxic drugs in drinking water: a prediction and risk assessment exercise for the Thames catchment in the United Kingdom. Environ Toxicol Chem 28:2733–2743

28. Schriks M, Heringa MB, van der Kooi MM, de Voogt P, van Wezel AP (2010) Toxicological relevance of emerging contaminants for drinking water quality. Water Res 44:461–476

29. Snyder SA (2008) Occurrence, treatment, and toxicological relevance of EDCS and pharmaceuticals in water. Ozone Sci Eng 30:65–69

30. Crane C, Maycock D, Crane D, Fawell J, Goslan E (2007) Desk based review of current knowledge on pharmaceuticals in drinking water and estimation of potential levels. Drinking Water Inspectorate http://dwi.defra.gov.uk/research/completed-research/reports/dwi70-2-213.pdf

31. Webb S, Ternes T, Gibert M, Olejniczak K (2003) Indirect human exposure to pharmaceuticals via drinking water. Toxicol Lett 142:157–167

32. Boxall AB, Kolpin DW, Halling-Sorensen B, Tolls J (2003) Are veterinary medicines causing environmental risks? Environ Sci Technol 37:286A–294A

33. Daughton CG, Ternes TA (1999) Pharmaceuticals and personal care products in the environment: agents of subtle change? Environ Health Perspect 107(suppl 6):907–938

34. Kummerer K (2009) The presence of pharmaceuticals in the environment due to human use—present knowledge and future challenges. J Environ Manage 90:2354–2366

35. Hursthouse A, Kowalczyk G (2009) Transport and dynamics of toxic pollutants in the natural environment and their effect on human health: research gaps and challenge. Environ Geochem Health 31:165–187

36. Focazio MJ, Kolpin DW, Barnes KK, Furlong ET, Meyer MT, Zaugg SD, Barber LB, Thurman ME (2008) A national reconnaissance for pharmaceuticals and other organic wastewater contaminants in the united states—II untreated drinking water sources. Sci Total Environ 402:201–216
37. Erickson BE (2002) Analyzing the ignored environmental contaminants. Environ Sci Technol 36:140A–145A
38. Loraine GA, Pettigrove ME (2006) Seasonal variations in concentrations of pharmaceuticals and personal care products in drinking water and reclaimed wastewater in Southern California. Environ Sci Technol 40:687–695
39. Zuccato E, Castiglioni S (2009) Illicit drugs in the environment. Philos Transact A Math Phys Eng Sci 367:3965–3978
40. Bartelt-Hunt SL, Snow DD, Damon T, Shockley J, Hoagland K (2009) The occurrence of illicit and therapeutic pharmaceuticals in wastewater effluent and surface waters in Nebraska. Environ Pollut 157:786–791
41. Petrovic M, de Alda MJ, Diaz-Cruz S, Postigo C, Radjenovic J, Gros M, Barcelo D (2009) Fate and removal of pharmaceuticals and illicit drugs in conventional and membrane bioreactor wastewater treatment plants and by riverbank filtration. Philos Transact A Math Phys Eng Sci 367:3979–4003
42. Zuccato E, Castiglioni S, Fanelli R (2005) Identification of the pharmaceuticals for human use contaminating the Italian aquatic environment. J Hazard Mater 122:205–209
43. Kasprzyk-Hordern B, Dinsdale RM, Guwy AJ (2009) Illicit drugs and pharmaceuticals in the environment—forensic applications of environmental data. Part 1: estimation of the usage of drugs in local communities. Environ Pollut 157:1773–1777
44. Kasprzyk-Hordern B, Dinsdale RM, Guwy AJ (2009) Illicit drugs and pharmaceuticals in the environment—forensic applications of environmental data, Part 2: pharmaceuticals as chemical markers of faecal water contamination. Environ Pollut 157:1778–1786
45. Nikolaou A, Meric S, Fatta D (2007) Occurrence patterns of pharmaceuticals in water and wastewater environments. Anal Bioanal Chem 387:1225–1234
46. Zuccato E, Calamari D, Natangelo M, Fanelli R (2000) Presence of therapeutic drugs in the environment. Lancet 355:1789–1790
47. Ternes TA (1998) Occurrence of drugs in German sewage treatment plants and rivers. Water Res 32:3245–3260
48. Hirsch R, Ternes T, Haberer K, Kratz KL (1999) Occurrence of antibiotics in the aquatic environment. Sci Total Environ 225:109–118
49. Coetsier CM, Spinelli S, Lin L, Roig B, Touraud E (2009) Discharge of pharmaceutical products (PPS) through a conventional biological sewage treatment plant: MECS vs PECS? Environ Int 35:787–792
50. Comoretto L, Chiron S (2005) Comparing pharmaceutical and pesticide loads into a small Mediterranean river. Sci Total Environ 349:201–210
51. Andreozzi R, Raffaele M, Nicklas P (2003) Pharmaceuticals in STP effluents and their solar photodegradation in aquatic environment. Chemosphere 50:1319–1330
52. Ferrari B, Paxeus N, Lo Giudice R, Pollio A, Garric J (2003) Ecotoxicological impact of pharmaceuticals found in treated wastewaters: study of carbamazepine, clofibric acid, and diclofenac. Ecotoxicol Environ Saf 55:359–370
53. Holm JV, Rugge K, Bjerg PL, Christensen TH (1995) Occurrence and distribution of pharmaceutical organic compounds in the groundwater downgradient of a landfill (Grindsted, Denmark). Environ Sci Technol 29:1415–1420
54. Braga O, Smythe GA, Schafer AI, Feitz AJ (2005) Fate of steroid estrogens in Australian inland and coastal wastewater treatment plants. Environ Sci Technol 39:3351–3358
55. Falconer IR, Chapman HF, Moore MR, Ranmuthugala G (2006) Endocrine-disrupting compounds: a review of their challenge to sustainable and safe water supply and water reuse. Environ Toxicol 21:181–191
56. Ort C, Lawrence MG, Reungoat J, Eaglesham G, Carter S, Keller J (2010) Determining the fraction of pharmaceutical residues in wastewater originating from a hospital. Water Res 44:605–615

57. Fick J, Soderstrom H, Lindberg RH, Phan C, Tysklind M, Larsson DG (2009) Contamination of surface, ground, and drinking water from pharmaceutical production. Environ Toxicol Chem 28:2522–2527
58. Larsson DG, de Pedro C, Paxeus N (2007) Effluent from drug manufactures contains extremely high levels of pharmaceuticals. J Hazard Mater 148:751–755
59. Stumpf M, Ternes TA, Wilken RD, Rodrigues SV, Baumann W (1999) Polar drug residues in sewage and natural waters in the state of Rio de Janeiro, Brazil. Sci Total Environ 225:135–141
60. Han GH, Hur HG, Kim SD (2006) Ecotoxicological risk of pharmaceuticals from wastewater treatment plants in Korea: occurrence and toxicity to daphnia magna. Environ Toxicol Chem 25:265–271
61. Takahashi A, Higashitani T, Yakou Y, Saitou M, Tamamoto H, Tanaka H (2003) Evaluating bioaccumulation of suspected endocrine disruptors into periphytons and benthos in the Tama river. Water Sci Technol 47:71–76
62. Nakada N, Tanishima T, Shinohara H, Kiri K, Takada H (2006) Pharmaceutical chemicals and endocrine disrupters in municipal wastewater in Tokyo and their removal during activated sludge treatment. Water Res 40:3297–3303
63. Nakada N, Kiri K, Shinohara H, Harada A, Kuroda K, Takizawa S, Takada H (2008) Evaluation of pharmaceuticals and personal care products as water-soluble molecular markers of sewage. Environ Sci Technol 42:6347–6353
64. Cui CW, Ji SL, Ren HY (2006) Determination of steroid estrogens in wastewater treatment plant of a controceptives producing factory. Environ Monit Assess 121:409–419
65. Richardson BJ, Lam PK, Martin M (2005) Emerging chemicals of concern: pharmaceuticals and personal care products (PPCPs) in Asia, with particular reference to Southern China. Mar Pollut Bull 50:913–920
66. Managaki S, Murata A, Takada H, Tuyen BC, Chiem NH (2007) Distribution of macrolides, sulfonamides, and trimethoprim in tropical waters: ubiquitous occurrence of veterinary antibiotics in the Mekong delta. Environ Sci Technol 41:8004–8010
67. Lin AY, Tsai YT (2009) Occurrence of pharmaceuticals in Taiwan's surface waters: impact of waste streams from hospitals and pharmaceutical production facilities. Sci Total Environ 407:3793–3802
68. Bound JP, Voulvoulis N (2005) Household disposal of pharmaceuticals as a pathway for aquatic contamination in the United Kingdom. Environ Health Perspect 113:1705–1711
69. Joss A, Keller E, Alder AC, Gobel A, McArdell CS, Ternes T, Siegrist H (2005) Removal of pharmaceuticals and fragrances in biological wastewater treatment. Water Res 39:3 139–3152
70. Winker M, Faika D, Gulyas H, Otterpohl R (2008) A comparison of human pharmaceutical concentrations in raw municipal wastewater and yellowwater. Sci Total Environ 399:96–104
71. Bendz D, Paxeus NA, Ginn TR, Loge FJ (2005) Occurrence and fate of pharmaceutically active compounds in the environment, a case study: Hoje river in Sweden. J Hazard Mater 122:195–204
72. Jjemba PK (2006) Excretion and ecotoxicity of pharmaceutical and personal care products in the environment. Ecotoxicol Environ Saf 63:113–130
73. Mompelat S, Le Bot B, Thomas O (2009) Occurrence and fate of pharmaceutical products and by-products, from resource to drinking water. Environ Int 35:803–814
74. Claudel JP, Touboul P (1995) Sotalol: From "just another beta blocker" to "the prototype of Class III antidysrhythmic compound". Pacing Clin Electrophysiol 18:451–467
75. Lienert J, Gudel K, Escher BI (2007) Screening method for ecotoxicological hazard assessment of 42 pharmaceuticals considering human metabolism and excretory routes. Environ Sci Technol 41:4471–4478
76. Alder AC, Schaffner C, Majewsky M, Klasmeier J, Fenner K (2010) Fate of beta-blocker human pharmaceuticals in surface water: comparison of measured and simulated concentrations in the Glatt Valley watershed, Switzerland. Water Res 44:936–948
77. Kasprzyk-Hordern B, Dinsdale RM, Guwy AJ (2008) The occurrence of pharmaceuticals, personal care products, endocrine disruptors and illicit drugs in surface water in South Wales, UK. Water Res 42:3498–3518

78. Heberer T (2002) Occurrence, fate, and removal of pharmaceutical residues in the aquatic environment: a review of recent research data. Toxicol Lett 131:5–17
79. Park GR (2001) Drug metabolism. Br J Anaesth CEPD Rev 1:185–188
80. Daughton CG (2003) Cradle-to-cradle stewardship of drugs for minimizing their environmental disposition while promoting human health. II. Drug disposal, waste reduction, and future directions. Environ Health Perspect 111:775–785
81. Daughton CG (2009) Chemicals from the practice of healthcare: challenges and unknowns posed by residues in the environment. Environ Toxicol Chem 28:2490–2494
82. Rouen D, Dolan K, Kimber J (2001) A review of drug detection testing and an examination of urine, hair, saliva, and sweat. Vol Technical Report No. 120. National Drug and Alcohol Research Centre, Sydney, Australia
83. Daughton CG, Ruhoy IS (2009) Environmental footprint of pharmaceuticals: the significance of factors beyond direct excretion to sewers. Environ Toxicol Chem 28:2495–2521
84. Bound JP, Voulvoulis N (2006) Predicted and measured concentrations for selected pharmaceuticals in UK rivers: implications for risk assessment. Water Res 40:2885–2892
85. Glassmeyer ST, Hinchey EK, Boehme SE, Daughton CG, Ruhoy IS, Conerly O, Daniels RL, Lauer L, McCarthy M, Nettesheim TG, Sykes K, Thompson VG (2009) Disposal practices for unwanted residential medications in the United States. Environ Int 35:566–572
86. Kuspis DA, Krenzelok EP (1996) What happens to expired medications? A survey of community medication disposal. Vet Hum Toxicol 38:48–49
87. Adler NE, Koschorreck J, Rechenberg B (2008) Environmental impact assessment and control of pharmaceuticals: the role of environmental agencies. Water Sci Technol 57:91–97
88. Jones OA, Voulvoulis N, Lester JN (2003) Potential impact of pharmaceuticals on environmental health. Bull World Health Organ 81:768–769
89. Joakim Larsson DG, Fick J (2009) Transparency throughout the production chain—a way to reduce pollution from the manufacturing of pharmaceuticals? Regul Toxicol Pharmacol 53:161–163
90. Li D, Yang M, Hu J, Ren L, Zhang Y, Li K (2008) Determination and fate of oxytetracycline and related compounds in oxytetracycline production wastewater and the receiving river. Environ Toxicol Chem 27:80–86
91. Langford KH, Thomas KV (2009) Determination of pharmaceutical compounds in hospital effluents and their contribution to wastewater treatment works. Environ Int 35:766–770
92. Castiglioni S, Bagnati R, Fanelli R, Pomati F, Calamari D, Zuccato E (2006) Removal of pharmaceuticals in sewage treatment plants in Italy. Environ Sci Technol 40:357–363
93. Choi K, Kim Y, Park J, Park CK, Kim M, Kim HS, Kim P (2008) Seasonal variations of several pharmaceutical residues in surface water and sewage treatment plants of Han River, Korea. Sci Total Environ 405:120–128
94. Brun GL, Bernier M, Losier R, Doe K, Jackman P, Lee HB (2006) Pharmaceutically active compounds in Atlantic Canadian sewage treatment plant effluents and receiving waters, and potential for environmental effects as measured by acute and chronic aquatic toxicity. Environ Toxicol Chem 25:2163–2176
95. Conley JM, Symes SJ, Schorr MS, Richards SM (2008) Spatial and temporal analysis of pharmaceutical concentrations in the upper Tennessee River Basin. Chemosphere 73:1178–1187
96. Vieno NM, Tuhkanen T, Kronberg L (2005) Seasonal variation in the occurrence of pharmaceuticals in effluents from a sewage treatment plant and in the recipient water. Environ Sci Technol 39:8220–8226
97. Gros M, Petrovic M, Barcelo D (2007) Wastewater treatment plants as a pathway for aquatic contamination by pharmaceuticals in the Ebro river basin (Northeast Spain). Environ Toxicol Chem 26:1553–1562
98. Standley LJ, Rudel RA, Swartz CH, Attfield KR, Christian J, Erickson M, Brody JG (2008) Wastewater-contaminated groundwater as a source of endogenous hormones and pharmaceuticals to surface water ecosystems. Environ Toxicol Chem 27:2457–2468

99. Lapen DR, Topp E, Metcalfe CD, Li H, Edwards M, Gottschall N, Bolton P, Curnoe W, Payne M, Beck A (2008) Pharmaceutical and personal care products in tile drainage following land application of municipal biosolids. Sci Total Environ 399:50–65

100. Topp E, Monteiro SC, Beck A, Coelho BB, Boxall AB, Duenk PW, Kleywegt S, Lapen DR, Payne M, Sabourin L, Li H, Metcalfe CD (2008) Runoff of pharmaceuticals and personal care products following application of biosolids to an agricultural field. Sci Total Environ 396:52–59

101. Boxall AB, Blackwell P, Cavallo R, Kay P, Tolls J (2002) The sorption and transport of a sulphonamide antibiotic in soil systems. Toxicol Lett 131:19–28

102. Radjenovic J, Petrovic M, Barcelo D (2009) Fate and distribution of pharmaceuticals in wastewater and sewage sludge of the conventional activated sludge (CAS) and advanced membrane bioreactor (MBR) treatment. Water Res 43:831–841

103. Spongberg AL, Witter JD (2008) Pharmaceutical compounds in the wastewater process stream in Northwest Ohio. Sci Total Environ 397:148–157

104. Xia K, Bhandari A, Das K, Pillar G (2005) Occurrence and fate of pharmaceuticals and personal care products (PPCPs) in biosolids. J Environ Qual 34:91–104

105. Monteiro SC, Boxall AB (2009) Factors affecting the degradation of pharmaceuticals in agricultural soils. Environ Toxicol Chem 28:2546–2554

106. Slack RJ, Gronow JR, Voulvoulis N (2005) Household hazardous waste in municipal landfills: contaminants in leachate. Sci Total Environ 337:119–137

107. Barnes KK, Christenson SC, Kolpin DW, Focazio M, Furlong ET, Zaugg SD, Meyer MT, Barber LB (2004) Pharmaceuticals and other organic waste water contaminants within a leachate plume downgradient of a municipal landfill. Ground Water Monit Remed 24:119–126

108. Schwarzbauer J, Heim S, Brinker S, Littke R (2002) Occurrence and alteration of organic contaminants in seepage and leakage water from a waste deposit landfill. Water Res 36:2275–2287

109. Breton R, Boxall A (2003) Pharmaceuticals and personal care products in the environment: regulatory drivers and research needs. QSAR Comb Sci 22:399–409

110. Hamscher G, Sczesny S, Hoper H, Nau H (2002) Determination of persistent tetracycline residues in soil fertilized with liquid manure by high-performance liquid chromatography with electrospray ionization tandem mass spectrometry. Anal Chem 74:1509–1518

111. Boxall AB, Fogg LA, Blackwell PA, Kay P, Pemberton EJ, Croxford A (2004) Veterinary medicines in the environment. Rev Environ Contam Toxicol 180:1–91

112. Khetan SK, Collins TJ (2007) Human pharmaceuticals in the aquatic environment: a challenge to green chemistry. Chem Rev 107:2319–2364

113. Loffler D, Rombke J, Meller M, Ternes TA (2005) Environmental fate of pharmaceuticals in water/sediment systems. Environ Sci Technol 39:5209–5218

114. Latch DE, Stender BL, Packer JL, Arnold WA, McNeill K (2003) Photochemical fate of pharmaceuticals in the environment: cimetidine and ranitidine. Environ Sci Technol 37:3342–3350

115. Lin AY, Reinhard M (2005) Photodegradation of common environmental pharmaceuticals and estrogens in river water. Environ Toxicol Chem 24:1303–1309

116. Canonica S, Meunier L, von Gunten U (2008) Phototransformation of selected pharmaceuticals during UV treatment of drinking water. Water Res 42:121–128

117. Kim I, Tanaka H (2009) Photodegradation characteristics of PPCPs in water with UV treatment. Environ Int 35:793–802

118. Packer JL, Werner JJ, Latch DE, McNeill K, Arnold WA (2003) Photochemical fate of pharmaceuticals in the environment: naproxen, diclofenac, clofibric acid, and ibuprofen. Aquat Sci 65:342–351

119. Benotti MJ, Brownawell BJ (2007) Distributions of pharmaceuticals in an urban estuary during both dry- and wet-weather conditions. Environ Sci Technol 41:5795–5802

120. Zorita S, Martensson L, Mathiasson L (2009) Occurrence and removal of pharmaceuticals in a municipal sewage treatment system in the south of Sweden. Sci Total Environ 407:2760–2770

121. Baumgarten S, Schroder HF, Charwath C, Lange M, Beier S, Pinnekamp J (2007) Evaluation of advanced treatment technologies for the elimination of pharmaceutical compounds. Water Sci Technol 56:1–8

122. Stackelberg PE, Gibs J, Furlong ET, Meyer MT, Zaugg SD, Lippincott RL (2007) Efficiency of conventional drinking-water-treatment processes in removal of pharmaceuticals and other organic compounds. Sci Total Environ 377:255–272

123. Vieno NM, Harkki H, Tuhkanen T, Kronberg L (2007) Occurrence of pharmaceuticals in river water and their elimination in a pilot-scale drinking water treatment plant. Environ Sci Technol 41:5077–5084

124. Huber MM, Gobel A, Joss A, Hermann N, Loffler D, McArdell CS, Ried A, Siegrist H, Ternes TA, von Gunten U (2005) Oxidation of pharmaceuticals during ozonation of municipal wastewater effluents: a pilot study. Environ Sci Technol 39:4290–4299

125. Ternes TA, Stuber J, Herrmann N, McDowell D, Ried A, Kampmann M, Teiser B (2003) Ozonation: a tool for removal of pharmaceuticals, contrast media and musk fragrances from wastewater? Water Res 37:1976–1982

126. Paraskeva P, Graham NJ (2002) Ozonation of municipal wastewater effluents. Water Environ Res 74:569–581

127. Ternes TA (2001) Pharmaceuticals and metabolites as contaminants of the aquatic environment. In: Daughton CG, Jones-Lepp TL (eds) Pharmaceuticals and personal care products in the environment - scientific and regulatory issues, ACS Symposium series 791. American Chemical Society, Washington, DC, pp 39–54

128. Ternes TA (2001) Analytical methods for the determination of pharmaceuticals in aqueous environmental samples. Trac Trend Anal Chem 20:419–434

129. Jones OA, Lester JN, Voulvoulis N (2005) Pharmaceuticals: a threat to drinking water? Trends Biotechnol 23:163–167

130. Benotti MJ, Trenholm RA, Vanderford BJ, Holady JC, Stanford BD, Snyder SA (2009) Pharmaceuticals and endocrine disrupting compounds in U.S. Drinking water. Environ Sci Technol 43:597–603

131. Rabiet M, Togola A, Brissaud F, Seidel JL, Budzinski H, Elbaz-Poulichet F (2006) Consequences of treated water recycling as regards pharmaceuticals and drugs in surface and ground waters of a medium-sized Mediterranean catchment. Environ Sci Technol 40:5282–5288

132. Anderson PD, D'Aco VJ, Shanahan P, Chapra SC, Buzby ME, Cunningham VL, Duplessie BM, Hayes EP, Mastrocco FJ, Parke NJ, Rader JC, Samuelian JH, Schwab BW (2004) Screening analysis of human pharmaceutical compounds in U.S. Surface waters. Environ Sci Technol 38:838–849

133. Feijtel T, Boeije G, Matthies M, Young A, Morris G, Gandolfi C, Hansen B, Fox K, Holt M, Koch V, Schroder R, Cassani G, Schowanek D, Rosenblom J, Niessen H (1997) Development of a geography-referenced regional exposure assessment tool for European rivers—GREAT-ER contribution to GREAT-ER #1. Chemosphere 34:2351–2373

134. Brooks BW, Chambliss CK, Stanley JK, Ramirez A, Banks KE, Johnson RD, Lewis RJ (2005) Determination of select antidepressants in fish from an effluent-dominated stream. Environ Toxicol Chem 24:464–469

135. Brown JN, Paxeus N, Forlin L, Larsson DGJ (2007) Variations in bioconcentration of human pharmaceuticals from sewage effluents into fish blood plasma. Environ Toxicol Pharmacol 24:267–274

136. Chu S, Metcalfe CD (2007) Analysis of paroxetine, fluoxetine and norfluoxetine in fish tissues using pressurized liquid extraction, mixed mode solid phase extraction cleanup and liquid chromatography-tandem mass spectrometry. J Chromatogr A 1163:112–118

137. Nakamura Y, Yamamoto H, Sekizawa J, Kondo T, Hirai N, Tatarazako N (2008) The effects of ph on fluoxetine in Japanese Medaka (*Oryzias latipes*): acute toxicity in fish larvae and bioaccumulation in juvenile fish. Chemosphere 70:865–873

138. Ramirez AJ, Mottaleb MA, Brooks BW, Chambliss CK (2007) Analysis of pharmaceuticals in fish using liquid chromatography-tandem mass spectrometry. Anal Chem 79:3155–3163

139. Boxall AB, Johnson P, Smith EJ, Sinclair CJ, Stutt E, Levy LS (2006) Uptake of veterinary medicines from soils into plants. J Agric Food Chem 54:2288–2297

140. Jones OA, Voulvoulis N, Lester JN (2004) Potential ecological and human health risks associated with the presence of pharmaceutically active compounds in the aquatic environment. Crit Rev Toxicol 34:335–350

141. Herrman JL, Younes M (1999) Background to the adi/tdi/ptwi. Regul Toxicol Pharmacol 30:S109–S113

142. Galli CL, Marinovich M, Lotti M (2008) Is the acceptable daily intake as presently used an axiom or a dogma? Toxicol Lett 180:93–99

143. Kroes R, Galli C, Munro I, Schilter B, Tran L, Walker R, Wurtzen G (2000) Threshold of toxicological concern for chemical substances present in the diet: a practical tool for assessing the need for toxicity testing. Food Chem Toxicol 38:255–312

144. Kim S, Aga DS (2007) Potential ecological and human health impacts of antibiotics and antibiotic-resistant bacteria from wastewater treatment plants. J Toxicol Environ Health B Crit Rev 10:559–573

145. Jorgensen SE, Halling-Sorensen B (2000) Drugs in the environment. Chemosphere 40:691–699

146. Schwartz T, Kohnen W, Jansen B, Obst U (2003) Detection of antibiotic-resistant bacteria and their resistance genes in wastewater, surface water, and drinking water biofilms. FEMS Microbiol Ecol 43:325–335

147. Guardabassi L, Petersen A, Olsen JE, Dalsgaard A (1998) Antibiotic resistance in *Acinetobacter* spp. isolated from sewers receiving waste effluent from a hospital and a pharmaceutical plant. Appl Environ Microbiol 64:3499–3502

148. Reinthaler FF, Posch J, Feierl G, Wust G, Haas D, Ruckenbauer G, Mascher F, Marth E (2003) Antibiotic resistance of *E. coli* in sewage and sludge. Water Res 37:1685–1690

149. Goni-Urriza M, Capdepuy M, Arpin C, Raymond N, Caumette P, Quentin C (2000) Impact of an urban effluent on antibiotic resistance of riverine Enterobacteriaceae and *Aeromonas* spp. Appl Environ Microbiol 66:125–132

150. Kummerer K (2009) Antibiotics in the aquatic environment—a review—part II. Chemosphere 75:435–441

151. Dorne JL, Skinner L, Frampton GK, Spurgeon DJ, Ragas AM (2007) Human and environmental risk assessment of pharmaceuticals: differences, similarities, lessons from toxicology. Anal Bioanal Chem 387:1259–1268

152. Zhang Z, Feng Y, Gao P, Wang C, Ren N (2011) Occurrence and removal efficiencies of eight EDCS and estrogenicity in a STP. J Environ Monit 13:1333–1373

153. Kusk KO, Kruger T, Long M, Taxvig C, Lykkesfeldt AE, Frederiksen H, Andersson AM, Andersen HR, Hansen KM, Nellemann C, Bonefeld-Jorgensen EC (2011) Endocrine potency of wastewater: contents of endocrine disrupting chemicals and effects measured by in vivo and in vitro assays. Environ Toxicol Chem 30:413–426

154. Swart JC, Pool EJ, van Wyk JH (2011) The implementation of a battery of in vivo and in vitro bioassays to assess river water for estrogenic endocrine disrupting chemicals. Ecotoxicol Environ Saf 74:138–143

155. Lissemore L, Hao C, Yang P, Sibley PK, Mabury S, Solomon KR (2006) An exposure assessment for selected pharmaceuticals within a watershed in Southern Ontario. Chemosphere 64:717–729

156. Cleuvers M (2003) Aquatic ecotoxicity of pharmaceuticals including the assessment of combination effects. Toxicol Lett 142:185–194

157. Richards SM, Wilson CJ, Johnson DJ, Castle DM, Lam M, Mabury SA, Sibley PK, Solomon KR (2004) Effects of pharmaceutical mixtures in aquatic microcosms. Environ Toxicol Chem 23:1035–1042

158. Pomati F (2007) Pharmaceuticals in drinking water: is the cure worse than the disease? Environ Sci Technol 41:8204

159. Gurr CJ, Reinhard M (2006) Harnessing natural attenuation of pharmaceuticals and hormones in rivers. Environ Sci Technol 40:2872–2876

160. Ashton D, Hilton M, Thomas KV (2004) Investigating the environmental transport of human pharmaceuticals to streams in the united kingdom. Sci Total Environ 333:167–184

161. Henschel KP, Wenzel A, Diedrich M, Fliedner A (1997) Environmental hazard assessment of pharmaceuticals. Regul Toxicol Pharmacol 25:220–225

162. Halling-Sorensen B, Nors Nielsen S, Lanzky PF, Ingerslev F, Holten Lutzhoft HC, Jorgensen SE (1998) Occurrence, fate and effects of pharmaceutical substances in the environment—a review. Chemosphere 36:357–393

163. Bound JP, Voulvoulis N (2004) Pharmaceuticals in the aquatic environment—a comparison of risk assessment strategies. Chemosphere 56:1143–1155

164. Schowanek D, Webb S (2002) Exposure simulation for pharmaceuticals in European surface waters with greater. Toxicol Lett 131:39–50

165. Boeije GM, Wagner JO, Koormann F, Vanrolleghem PA, Schowanek DR, Feijtel TC (2000) New PEC definitions for river basins applicable to GIS-based environmental exposure assessment. Chemosphere 40:255–265

166. Sanderson H, Johnson DJ, Reitsma T, Brain RA, Wilson CJ, Solomon KR (2004) Ranking and prioritization of environmental risks of pharmaceuticals in surface waters. Regul Toxicol Pharmacol 39:158–183

167. Ruhoy IS, Daughton CG (2008) Beyond the medicine cabinet: an analysis of where and why medications accumulate. Environ Int 34:1157–1169

168. Brain RA, Sanderson H, Sibley PK, Solomon KR (2006) Probabilistic ecological hazard assessment: evaluating pharmaceutical effects on aquatic higher plants as an example. Ecotoxicol Environ Saf 64:128–135

169. Sanderson H, Johnson DJ, Wilson CJ, Brain RA, Solomon KR (2003) Probabilistic hazard assessment of environmentally occurring pharmaceuticals toxicity to fish, daphnids and algae by ecosar screening. Toxicol Lett 144:383–395

170. Cunningham VL, Buzby M, Hutchinson T, Mastrocco F, Parke N, Roden N (2006) Effects of human pharmaceuticals on aquatic life: next steps. Environ Sci Technol 40:3456–3462

171. Philips PJ, Smith SG, Kolpin DW, Zaugg SD, Buxton HT, Furlong ET, Esposito K, Stinson B (2010) Pharmaceutical formulation facilities as sources of opioids and other pharmaceuticals to wastewater treatment plant effluents. Environ Sci Technol 44(13):4910–16

172. Brooks BW, Riley TM, Taylor RD (2006) Water quality of effluent-dominated ecosystems: ecotoxicological, hydrological, and management considerations. Hydrobiologica 556:365–79

173. Brooks BW, Huggett DB, Boxall AB (2009) Pharmaceuticals and personal care products: Research needs for the next decade. Environ Toxicol Chem 28(12):2469–72

174. Monteiro SC, Boxall AB (2010) Occurrence and fate of human pharmaceuticals in the environment. Rev Environ Contam Toxicol 202:53–154

175. Daughton CG, Brooks BW (2011) Active pharmaceutical ingredients and aquatic organisms. In: Environmental Contaminants in Wildlife: Interpreting Tissue Concentrations, 2nd ed. Eds: Meador J, Beyer N. Taylor and Francis. p. 281–341

176. Valenti TV, Gould GG, Berninger JP, Connors KA, Keele NB, ProsserKN, Brooks BW (2012) Human therapeutic plasma levels of the selective serotonin reuptake inhibitor (SSRI) sertraline decrease serotonin reuptake transporter binding and shelter seeking behavior in adult male fathead minnows. Environ Sci Technol 46:2427–35

Wastewater and Drinking Water Treatment Technologies

Daniel Gerrity and Shane Snyder

Abbreviations

AOP	Advanced oxidation process
BAC	Biological activated carbon
CAS	Conventional activated sludge
CDPH	California Department of Public Health
DBP	Disinfection by-product
D_{ow}	Octanol/water distribution coefficient
EDC	Endocrine disrupting compound
EEO	Electrical energy per order of magnitude destruction
GAC	Granular activated carbon
IPR	Indirect potable reuse
K_{ow}	Octanol/water partitioning coefficient
MBR	Membrane bioreactor
MF	Microfiltration
MRL	Method reporting limit
MW	Molecular weight

D. Gerrity
Water Quality Research and Development Center, Southern Nevada Water Authority,
River Mountain Water Treatment Facility, 1299 Burkholder Boulevard, Henderson,
NV 89015, USA
e-mail: Dan.Gerrity@lvvwd.com

S. Snyder (✉)
Chemical and Environmental Engineering, University of Arizona, 1133 E. James E. Rogers Way,
P.O. Box 210011, Tucson, AZ 85721, USA
e-mail: snyders2@email.arizona.edu

B.W. Brooks and D.B. Huggett (eds.), *Human Pharmaceuticals in the Environment:
Current and Future Perspectives*, Emerging Topics in Ecotoxicology 4,
DOI 10.1007/978-1-4614-3473-3_9, © Springer Science+Business Media, LLC 2012

NF   Nanofiltration
NPDES   National Pollutant Discharge Elimination System
PAC   Powder activated carbon
PPCPs   Pharmaceuticals and personal care products
RO   Reverse osmosis
S   Solubility
SRT   Solids retention time
TOrC   Trace organic contaminant
UF   Ultrafiltration

Introduction

Although pharmaceuticals and personal care products (PPCPs) and endocrine disrupting compounds (EDCs) are often considered "emerging contaminants," researchers have been aware of their ubiquity in water for decades. As early as the 1940s, scientists were aware that certain chemicals had the ability to mimic endogenous estrogens and androgens [1, 2], and in 1965, Stumm-Zollinger and Fair of Harvard University published the first known report indicating that steroid hormones were not completely eliminated by wastewater treatment [3]. In 1977, researchers from the University of Kansas published the first known report of pharmaceutical discharge from a wastewater treatment plant (WWTP) [4].

Despite these early findings, studies related to PPCPs and EDCs in source water, drinking water, and wastewater did not become a mainstream research topic until the late 1990s and early 2000s. Potential human health effects, demonstrated impacts on aquatic ecosystems, and increased media coverage, which ultimately led to increased public awareness, were primarily responsible for the spike in scientific studies [5]. This was coupled with the development of extremely sensitive analytical methods that allowed researchers to approach parts-per-quadrillion (sub-ng L^{-1}) detection limits for a variety of trace organic contaminants (TOrCs) [6, 7]. Each of these factors increased the number and scope of scientific investigations into the presence, fate, and transport of TOrCs in natural and engineered systems.

Although there are a number of significant sources of PPCPs and EDCs in the environment, including industrial manufacturing processes and confined animal feeding operations [8], municipal wastewater is considered the primary source [9]. The occurrence of these compounds, associated by-products, and transformation products in wastewater results from their release during manufacturing, excretion after personal use, and disposal of unused quantities [10]. In 1999, Daughton and Ternes highlighted the ubiquity of pharmaceuticals, of which more than 3,000 are now available by prescription [11], due to their direct correlation to human presence: pharmaceuticals will be detected in any water supply in proximity to human populations [10]. In fact, the presence or absence of any chemical in wastewater effluent is essentially a function of analytical detection capability. In a 2008 review of TOrC occurrence in municipal wastewater effluent, Snyder et al. [8] identified pharmaceutical residues, antibiotics, steroid hormones, and fragrances as the most frequently detected compound classes, and Ternes [12] provided one of the first comprehensive evaluations of TOrC concentrations in municipal wastewater effluent and receiving waters. Fent et al. [13] also provided a comprehensive review of TOrC concentrations in wastewater effluent in addition to the modes of action and toxicological implications of those contaminants.

With respect to wastewater treatment, compound removal and transformation is highly dependent on the unit processes (e.g., secondary treatment, filtration, and disinfection) and operational variables (e.g., solids retention time [SRT] and oxidant dose) employed at a particular plant [5, 11]. Even at a single WWTP, effluent concentrations can be highly variable as they are influenced by temporal variations in influent concentrations, temperature, and dry vs. wet weather flows [12]. Once these contaminants are discharged, natural attenuation occurs through microbial degradation, dilution, adsorption to solids, photolysis, or other forms of abiotic transformation. However, these natural processes are generally insufficient to reduce TOrC concentrations to nondetect levels. Furthermore, some receiving bodies can be comprised of 50–90% wastewater effluent during dry weather conditions [10]. This ultimately leads to contamination of surface water, groundwater (i.e., after aquifer recharge or leaching from landfilled solids), and even food supplies (i.e., after plant uptake from reclaimed irrigation water) [10, 14]. Kolpin et al. [15] documented the extent of contamination (with respect to 95 TOrCs) of 139 predominantly wastewater-impacted streams in the USA. Although identified as a conservative estimate due to method limitations (i.e., method reporting limits [MRLs]), at least one TOrC was detected in 80% of the sample sites, but the concentrations were generally less than 1 μg L^{-1}. To highlight immediate impacts on drinking water supplies, Benotti et al. [11] monitored 51 TOrCs in the source water, finished drinking water, and distribution systems of 19 US utilities. Although median concentrations of the target pharmaceuticals rarely exceeded 10 ng L^{-1}, some TOrCs were detected at maximum concentrations exceeding 100 ng L^{-1}. The herbicide atrazine was even detected in systems with no known agricultural applications. Therefore, recalcitrant compounds certainly persist in drinking water supplies and ultimately contaminate finished drinking water.

Water and wastewater treatment trains are generally not designed for the removal of TOrCs. However, the interrelatedness of wastewater discharge and drinking water sources and potential effects on aquatic ecosystems now justify some consideration of TOrCs in the design process. In fact, expansion and optimization of wastewater treatment processes may be the most efficient strategy to mitigate the potential effects of these contaminants. Countless treatment processes have been evaluated for their ability to remove or destroy TOrCs. These evaluations span the continua of physicochemical treatment (e.g., media or membrane filtration), conventional oxidation (e.g., chlorine and ozone), and advanced oxidation processes (AOPs) (e.g., UV/H$_2$O$_2$) in drinking water and wastewater [16–21]. This chapter discusses the efficacy of the various treatment technologies available to water and WWTPs for TOrC removal and/or destruction. It is important to note that the TOrCs included in most studies in the literature satisfy the following four criteria: (1) high likelihood of occurrence in the environment, (2) potential toxicological relevance and significant

Table 1 Physicochemical properties of selected TOrCs[a]

Compounds	Classes	MW	S (mg L^{-1})	Log K_{OW}	pK_a
Acetaminophen	Analgesic	151.2	1.40E+4	0.46	9.38
Androstenedione	Hormone	286.4	57.8	2.75	N/A
Atrazine	Herbicide	215.7	34.7	2.61	1.7
Caffeine	Psychoactive	194.2	2.16E+4	−0.07	10.4
Carbamazepine	Anticonvulsant	236.3	18 [22]	2.45	13.9 [23]
DEET[b]	Insect repellant	191.3	9.9 [24]	2.18	0.7 (est)
Diazepam	Antianxiety	284.7	50	2.82	3.4
Diclofenac	Analgesic	296.2	2.37	4.51	4.15
Dilantin	Anticonvulsant	252.3	32	2.47	8.33
Erythromycin	Antibiotic	733.9	1.44 (est)	3.06	8.88
Estriol	Hormone	288.4	441 (est)	2.45	9.85 (est)
Estradiol	Hormone	272.4	3.6	4.01	10.4 [25]
Estrone	Hormone	270.4	30	3.13	10.4 [25]
Ethynyl estradiol	Hormone	296.4	11.3	3.67	10.4 [16]
Fluoxetine	Psychoactive	309.3	60.3 (est)	4.05	10.3 (est)
Galaxolide	Fragrance	258.4	1.75 [26]	5.9 [26]	N/A
Gemfibrozil	Antilipidemic	250.3	19 (est)	4.33 (est)	4.42
Hydrocodone	Analgesic	299.4	6,870 (est)	2.16 (est)	8.35 (est)
Ibuprofen	Analgesic	206.3	21	3.97	4.91
Iopromide	X-ray contrast	791.1	23.8 (est)	−2.05	10.2 (est)
Meprobamate	Antianxiety	218.3	4,700	0.7	10.9 (est)
Metolachlor	Pesticide	283.8	530	3.13	N/A
Musk ketone	Fragrance	294.3	0.46 [27], 1.9 [28]	4.3 [27]	N/A
Naproxen	Analgesic	230.3	15.9	3.18	4.15
Pentoxifylline	Vasodilator	278.3	7.70E+4	0.29	1.49 (est)
Progesterone	Hormone	314.5	8.81	3.87	N/A
Sulfamethoxazole	Antibiotic	253.3	610	0.89	5.5 [29]
TCEP[c]	Flame retardant	285.5	7,000	1.44	N/A
Testosterone	Hormone	288.4	23.4	3.32	N/A
Triclosan	Antimicrobial	289.5	10	4.76	7.9 [30]
Trimethoprim	Antibiotic	290.3	400	0.91	7.12

[a]Experimental values from Environmental Science Database SRC PhysProp
[b]Chemical name: *N,N*-diethyl-meta-toluamide
[c]Chemical name: Tri(chloroethyl)phosphate

public interest, (3) structural diversity resulting in a range of treatability, and (4) amenability to available analytical methods. The target pharmaceuticals often encompass several therapeutic classes, including analgesics, antibiotics, anticonvulsants, psychoactive drugs, and cholesterol-lowering medications. A subset of the target compounds discussed in this chapter in addition to their structural properties (e.g., molecular weight [MW], solubility (S), and octanol/water partitioning coefficient [K_{OW}]) are summarized in Table 1.

TOrC Occurrence in Water and Wastewater

TOrC concentrations in raw wastewater may routinely exceed 1 μg L^{-1} for a variety of compounds, particularly analgesics, some antibiotics, flame retardants, and caffeine. Fortunately, conventional biological wastewater treatment processes are particularly effective in removing compounds with high detection frequencies and concentrations—with flame retardants (e.g., TCEP), X-ray contrast media (e.g., iopromide), some psychoactive drugs (e.g., meprobamate), and herbicides (e.g., atrazine) being notable exceptions. TOrC concentrations in finished effluent are highly site specific and dependent on the unit processes and operational conditions at a particular plant. Therefore, it is difficult to identify "typical" wastewater concentrations, but concentrations from two US WWTPs are provided for context in Table 2.

Snyder et al. [18] and Benotti et al. [11] present a comprehensive reconnaissance of pharmaceuticals and EDCs in US source and drinking water, though both studies targeted systems susceptible to wastewater contamination. Snyder et al. [18] surveyed the occurrence of 36 pharmaceuticals and EDCs in the source and finished drinking water of 20 US drinking water treatment plants (Table 3). The 12 compounds that were detected in at least half of the source water samples were atrazine, caffeine, carbamazepine, DEET, gemfibrozil, ibuprofen, iopromide, meprobamate, naproxen, phenytoin, sulfamethoxazole, and TCEP. Median concentrations of detected pharmaceuticals and EDCs in source water were usually less than 10 ng L^{-1}, except for atrazine (28 ng L^{-1}), caffeine (27 ng L^{-1}), fluorene (13 ng L^{-1}), galaxolide (28 ng L^{-1}), metolachlor (15 ng L^{-1}), musk ketone (16 ng L^{-1}), and TCEP (13 ng L^{-1}), though values for fluorene, galaxolide, and musk ketone were biased by low frequencies of detection. The eight compounds that were detected in at least half of the finished drinking water samples were atrazine, caffeine, carbamazepine, DEET, ibuprofen, iopromide, meprobamate, and phenytoin. Median concentrations of detected pharmaceuticals and EDCs in finished drinking water were usually less than 10 ng L^{-1}, except for atrazine (29 ng L^{-1}), caffeine (23 ng L^{-1}), metolachlor (86 ng L^{-1}), and triclosan (43 ng L^{-1}), though values for metolachlor and triclosan were biased by low frequencies of detection.

Benotti et al. [11] surveyed the occurrence of 51 pharmaceuticals and EDCs in 19 source waters, 18 finished drinking waters, and 15 distribution systems from utilities throughout the USA (Table 4). The 11 compounds that were detected in at least half of the source water samples were atenolol, atrazine, carbamazepine, estrone, gemfibrozil, meprobamate, naproxen, phenytoin, sulfamethoxazole, TCEP, and trimethoprim. As with the previous study, median concentrations of detected pharmaceuticals and EDCs in source water usually less than 10 ng L^{-1}, except for atrazine (32 ng L^{-1}), butylbenzyl phthalate (53 ng L^{-1}), BHT (49 ng L^{-1}), diethylhexyl phthalate (150 ng L^{-1}), DEET (85 ng L^{-1}), estradiol (17 ng L^{-1}), metolachlor (17 ng L^{-1}), nonylphenol (100 ng L^{-1}), sulfamethoxazole (12 ng L^{-1}), TCEP (120 ng L^{-1}), and TCPP (180 ng L^{-1}). The values for estradiol, BHT, butylbenzyl phthalate, and diethylhexyl phthalate were biased by low frequencies of detection. Only three compounds (atrazine, meprobamate, and phenytoin) were detected in at least half of the finished drinking water samples. Median concentrations of detected pharmaceuticals and

Table 2 TOrC concentrations (ng L^{-1}) at two US wastewater treatment plants

TOrC	Wastewater treatment plant 1			Wastewater treatment plant 2		
	Primary effluent	Secondary effluent	Finished effluent	Primary effluent	Secondary effluent	Finished effluent
Acetaminophen	NA	NA	NA	170,000	<500	<500
Atenolol	1,600	730	220	2,600	430	560
Atorvastatin	98	<10	16	NA	NA	NA
Atrazine	<5	<5	<5	NA	NA	NA
Benzophenone	<1,000	<1,000	<1,000	1,000	250	440
BHA	170	<20	<20	250	16	12
Bisphenol A	550	<100	<100	430	<5.0	<5.0
Caffeine	67,000	<100	<100	120,000	<5.0	30
Carbamazepine	160	190	180	260	340	310
Cimetidine	NA	NA	NA	350	120	86
DEET	510	72	190	420	17	17
Diazepam	<5	<5	<5	NA	NA	NA
Diclofenac	120	96	63	NA	NA	NA
Diphenhydramine	NA	NA	NA	1,200	61	47
Estradiol	<1	0.52	<0.5	NA	NA	NA
Estrone	<1	6.7	<0.2	NA	NA	NA
Ethynylestradiol	<1	<1	<1	NA	NA	NA
Fluoxetine	25	33	29	25	24	10
Gemfibrozil	2,900	<5	17	290	3.6	3.6
Ibuprofen	17,000	<20	<20	30,000	<10	<10
Iopromide	<200	<200	<200	32,000	7,700	2,000
Meprobamate	1,400	470	340	280	62	61
Musk ketone	<500	<500	<500	<250	<25	<25
Naproxen	15,000	<10	13	12,000	13	27
Octylphenol	<500	<500	<500	NA	NA	NA
Phenytoin	97	120	130	NA	NA	NA
Primidone	140	140	140	<5.0	<0.50	<0.50
Progesterone	34	7.3	8.0	NA	NA	NA
Sucralose	NA	NA	NA	28,000	51,000	77,000
Sulfamethoxazole	1,900	1,500	1,500	650	480	370
TCEP	220	350	360	360	440	420
TCPP	<2,000	2,000	2,200	1,900	1,000	900
Testosterone	40	<0.5	<0.5	NA	NA	NA
Triclocarbon	NA	NA	NA	550	200	67
Triclosan	1,300	48	48	1,100	12	3.7
Trimethoprim	700	19	17	440	26	24

NA not analyzed

EDCs in finished drinking water were generally less than 10 ng L^{-1}, except for atrazine (49 ng L^{-1}), bisphenol A (25 ng L^{-1}), galaxolide (31 ng L^{-1}), nonylphenol (93 ng L^{-1}), BHT (26 ng L^{-1}), metolachlor (16 ng L^{-1}), DEET (63 ng L^{-1}), TCEP (120 ng L^{-1}), and TCPP (210 ng L^{-1}). Again, some of these median concentrations were biased by low frequencies of detection. Finally, the four compounds that were

Table 3 TOrC concentrations (ng L^{-1}) in US source and finished drinking water

Contaminant	Source ($n = 20$)			Finished ($n = 20$)		
	Max.	Med.	#	Max.	Med.	#
Acetaminophen	9.5	1.6	7	<1.0	<1.0	–
Androstenedione	1.9	1.9	1	<1.0	<1.0	–
Atrazine	570	28	17	430	29	15
Caffeine	87	27	14	83	23	12
Carbamazepine	39	3.1	18	5.7	2.8	11
DEET	28	6.9	20	30	5.1	18
Erythromycin	3.5	2.2	8	1.3	1.3	1
Estrone	1.4	1.2	2	2.3	1.7	2
Fluorene	13	13	1	<1.0	<1.0	–
Galaxolide	30	28	3	<1.0	<1.0	–
Gemfibrozil	11	4.8	13	6.5	4.2	5
Hydrocodone	1.9	1.9	1	<1.0	<1.0	–
Ibuprofen	24	4.2	16	32	3.8	13
Iopromide	46	7.6	14	31	6.5	13
Meprobamate	16	5.9	16	13	3.8	15
Metolachlor	170	15	7	160	86	4
Musk ketone	17	16	3	17	17	1
Naproxen	16	2.2	10	8	8	1
Oxybenzone	7.4	2.9	4	1.1	1.1	1
Phenytoin	13	3.2	18	6.7	2.3	14
Progesterone	1.1	1.1	1	1.1	1.1	2
Sulfamethoxazole	44	8.1	17	<1.0	<1.0	–
TCEP	66	13	15	19	5.5	7
Triclosan	30	1.9	6	43	43	1
Trimethoprim	2.3	2.2	3	1.3	1.3	1

The "#" sign represents the number of samples with reportable concentrations for that particular contaminant [31]

detected in at least half of the distribution system samples were atrazine, atenolol, meprobamate, and phenytoin. Median concentrations of detected pharmaceuticals and EDCs in the distribution systems were generally less than 10 ng L^{-1}, except for atrazine (50 ng L^{-1}), DEET (49 ng L^{-1}), metolachlor (18 ng L^{-1}), nonylphenol (97 ng L^{-1}), TCEP (150 ng L^{-1}), and TCPP (220 ng L^{-1}), though values for metolachlor and nonylphenol were biased by low frequencies of detection.

TOrC concentrations in source waters are generally a direct function of (1) the contribution of wastewater to the source, (2) TOrC occurrence in the wastewater influent, (3) unit operations and treatment efficacy at the contributing WWTPs, and (4) degree of natural attenuation after environmental discharge. Accordingly, in both of the studies presented above, TOrC occurrence in the finished drinking water was governed by (1) TOrC concentrations in the source waters and (2) removal during drinking water treatment. Due to the importance of treatment efficacy on TOrC concentrations, the following sections provide a summary of the most common technologies for drinking water and wastewater treatment. Although the treatment processes have been categorized

Table 4 TOrC concentrations (ng L^{-1}) in US source water, finished drinking water, and distribution systems

Contaminant	Source (n=19)			Finished (n=18)			Distribution (n=15)		
	Max.	Med.	#	Max.	Med.	#	Max.	Med.	#
Estradiol	17	17	1	<0.50	<0.50	–	<0.50	<0.50	–
Ethynylestradiol	1.4	1.4	1	<1.0	<1.0	–	<1.0	<1.0	–
Atenolol	36	2.3	12	18	1.2	8	0.84	0.47	8
Atorvastatin	1.4	0.80	3	<0.25	<0.25	–	<0.25	<0.25	–
Atrazine	870	32	15	870	49	15	930	50	12
BHT	49	49	1	26	26	1	<25	<25	–
Bisphenol A	14	6.1	3	25	25	1	<5.0	<5.0	–
Butylbenzyl phthalate	54	53	2	<50	<50	–	<50	<50	–
Carbamazepine	51	4.1	15	18	6.0	8	10	6.8	6
DEET	110	85	6	93	63	6	63	49	4
Diazepam	0.47	0.43	2	0.33	0.33	1	<0.25	<0.25	–
Diclofenac	1.2	1.1	4	<0.25	<0.25	–	<0.25	<0.25	–
Diethylhexyl phthalate	170	150	2	<120	<120	–	<120	<120	–
Estrone	0.90	0.30	15	<0.20	<0.20	–	<0.20	<0.20	–
Fluoxetine	3.0	0.80	3	0.82	0.71	2	0.64	0.64	1
Galaxolide	48	3	4	33	31	2	<25	<25	–
Gemfibrozil	24	2.2	11	2.1	0.48	7	1.2	0.43	4
Linuron	9.3	4.1	5	6.2	6.1	2	<0.50	<0.50	–
Meprobamate	73	8.2	16	42	5.7	14	40	5.2	11
Metolachlor	81	17	7	27	16	6	22	18	3

Naproxen	32	0.90	11	<0.50	<0.50	<0.50	–	<0.50	<0.50	–
Nonylphenol	130	100	8	100	93	110	2	110	97	2
Norfluoxetine	<0.50	<0.50	–	<0.50	<0.50	0.77	–	0.77	0.77	1
o-Hydroxy atorvastatin	1.2	0.70	3	<0.50	<0.50	<0.50	–	<0.50	<0.50	–
Phenytoin	29	5.1	14	19	6.2	16	10	16	3.6	10
p-Hydroxy atorvastatin	2.0	1.0	3	<0.50	<0.50	<0.50	–	<0.50	<0.50	–
Progesterone	3.1	2.2	4	0.57	0.57	<0.50	1	<0.50	<0.50	–
Risperidone	<2.5	<2.5	–	<2.5	<2.5	2.9	–	2.9	2.9	1
Sulfamethoxazole	110	12	17	3.0	0.39	0.32	4	0.32	0.32	1
TCEP	530	120	10	470	120	200	7	200	150	6
TCPP	720	180	8	510	210	240	5	240	220	6
Testosterone	1.2	1.1	2	<0.50	<0.50	<0.50	–	<0.50	<0.50	–
Triclosan	6.4	3.0	6	1.2	1.2	<1.0	1	<1.0	<1.0	–
Trimethoprim	11	0.80	11	<0.25	<0.25	<0.25	–	<0.25	<0.25	–

The "#" sign represents the number of samples with reportable concentrations for that particular contaminant [11]

based on their most common applications, there is certainly technology overlap between drinking water and wastewater treatment. Natural attenuation is not discussed since it is highly site specific. However, many of the treatment processes (e.g., photolysis, biological wastewater processes, and filtration) mimic natural attenuation mechanisms so there is some degree of overlap between the two concepts.

TOrC Removal During Drinking Water Treatment

Coagulation/Flocculation/Sedimentation

Coagulation involves the addition of treatment chemicals such as aluminum sulfate (alum, $Al_2(SO_4)_3$), ferric chloride ($FeCl_3$), and ferric sulfate ($Fe_2(SO_4)_3$) to promote the destabilization of small suspended particles and colloidal material [32]. During the rapid mix phase, the metal salts hydrolyze, form complexes with organic solutes, and ultimately precipitate as amorphous metal hydroxides. After the rapid mix phase, a period of slow mixing (flocculation) is often used to promote the aggregation of smaller particulates and organic matter into larger settleable flocs. These can be removed by granular media filtration either with or without prior gravity settling or dissolved air flotation [33, 34]. Conventional coagulation is generally intended for turbidity removal via destabilization of existing particles. Enhanced coagulation, which employs higher coagulant doses and/or pH reduction, is now used for the removal of dissolved organic compounds.

Westerhoff et al. [35] evaluated the efficacy of alum and ferric chloride coagulation for PPCP and EDC removal in bench-scale experiments. In four different surface waters, 34 of the 36 PPCPs and EDCs were removed by less than 15%. The two remaining compounds (DDT and benzo(a)pyrene) were removed by 31 and 70%, respectively, due to their greater hydrophobicity ($\log K_{OW} > 6.0$). Results from the bench-scale tests suggested that (1) removal efficiency and K_{OW} were linearly correlated, (2) there was no added benefit with enhanced coagulation, and (3) removal efficiencies were similar between the two coagulants. Coagulation was also deemed ineffective for PPCP and EDC removal in another study [36], which noted no significant difference in pharmaceutical concentrations before and after coagulation. A summary of these results is provided in Table 5.

Activated Carbon Adsorption

Activated carbon is a highly porous material that has typically been used for control of taste and odor problems, though its applicability is expanding due to changes in disinfection by-product (DBP) regulation and the ability for activated carbon to remove DBP precursors [37]. The two main forms of activated carbon are utilized in different ways. Powder activated carbon (PAC) is applied similar to a coagulation process and

Table 5 TOrC removal by coagulation/flocculation/sedimentation [18]

<20% removal	20–50% removal	50–80% removal	>80% removal
Acetaminophen	DDT	Benzo(a)pyrene	
Androstenedione			
Atrazine			
Caffeine			
Carbamazepine			
DEET			
Diazepam			
Diclofenac			
Erythromycin			
Estradiol			
Estriol			
Estrone			
Ethinyl estradiol			
Fluorene			
Fluoxetine			
Galaxolide			
Gemfibrozil			
Hydrocodone			
Ibuprofen			
Iopromide			
Lindane			
Meprobamate			
Metolachlor			
Musk ketone			
Naproxen			
Oxybenzone			
Pentoxifylline			
Phenytoin			
Progesterone			
Sulfamethoxazole			
TCEP			
Testosterone			
Triclosan			
Trimethoprim			

[a]10 mg alum per mg total organic carbon or equivalent dose ($[Fe^{3+}]/[Al^{3+}] = 1$) of $FeCl_3$

can be used as an additional coagulant during seasonal contaminant spikes. Granular activated carbon (GAC) requires permanent contactors configured as media filters or fixed-bed adsorbers, which can also allow for microbial growth and significant biodegradation. As with coagulation or any adsorption process, the efficacy of PAC and GAC is highly dependent on the hydrophobicity and size of the target compounds.

Many researchers have reported on the efficacy of GAC and PAC for the removal of PPCPs and EDCs [18, 35, 36, 38]. GAC and PAC treatment for trace contaminants can be hindered by the presence of high concentrations of NOM, as they compete for the same adsorption sites on the substrate. Thus, the effectiveness and life span of

Table 6 Freundlich parameters for four pharmaceuticals [36]

Contaminant	Deionized water		Groundwater	
	N	K_A	n	K_A
Bezafibrate	0.19	141	0.22	77
Carbamazepine	0.38	430	0.22	90
Clofibric acid	0.25	71	0.54	63
Diclofenac	0.19	141	0.21	36

activated carbon is highly dependent on the characteristics of the target water matrix [18]. In contrast to coagulation, the octanol–water distribution coefficient (D_{OW}), rather than K_{OW}, is a better indicator of performance for many of the compounds [18]. Removal can be correlated with K_{OW} for neutral compounds, however. In general, higher PAC concentrations lead to higher removal of most PPCPs and EDCs. Protonated bases are very susceptible to removal by PAC, because they are electrostatically attracted to negatively charged moieties on the substrate's surface. Conversely, deprotonated acids are electrostatically repelled from the surface-bound negatively charged moieties and do not adsorb. The removal of neutrally charged molecules is controlled by the hydrophobicity of a particular compound given that the mechanism for adsorption is hydrophobic exclusion from the aqueous phase: compounds with low K_{OW} values are less likely to adsorb to activated carbon [35].

Adsorption of target contaminants is often modeled with batch isotherm testing and the Freundlich isotherm model, as described below [37]:

$$q_A = K_A C_A^{1/n} \qquad\qquad 1$$

where

q_A = equilibrium adsorbent-phase concentration of contaminant (mg contaminant g^{-1} adsorbent)
K_A = Freundlich adsorption capacity parameter $((mg\ g^{-1})(L\ mg^{-1})^{1/n})$
C_A = equilibrium concentration of contaminant in solution (mg L^{-1})
n = Freundlich adsorption intensity parameter (unitless)

Empirical determination of the K_A and n parameters allows engineers to calculate the expected removals of certain compounds in addition to design criteria specific to the activated carbon reactor. Ternes et al. [36] published K_A and n values for four pharmaceuticals in deionized water and groundwater (Table 6).

With respect to general removal trends, Table 7 categorizes TOrC removal for 5 mg L^{-1} and 4–5 h of contact time with PAC [18]. In contrast to coagulation, only two of the target compounds are removed by less than 20% with PAC, and a majority of the compounds are removed by more than 50%. Activated carbon is generally superior to coagulation, but there are still compounds that are resistant to removal. Again, increased removal of target contaminants must be balanced with the additional operational costs (i.e., infrastructure, virgin material, regeneration, disposal, etc.) associated with PAC and GAC.

Table 7 TOrC removal by PAC [18]

<20% removal	20–50% removal	50–80% removal	>80% removal
Ibuprofen	DEET	Acetaminophen	Benzo(a)pyrene
Iopromide	Diclofenac	Androstenedione	Fluorene
	Erythromycin	Atrazine	Fluoxetine
	Estriol	Caffeine	Oxybenzone
	Gemfibrozil	Carbamazepine	Progesterone
	Meprobamate	DDT	Triclosan
	Metolachlor	Diazepam	
	Naproxen	Estradiol	
	Phenytoin	Estrone	
	Sulfamethoxazole	Ethinyl estradiol	
	TCEP	Galaxolide	
		Hydrocodone	
		Lindane	
		Musk ketone	
		Pentoxifylline	
		Testosterone	
		Trimethoprim	

PAC dose = 5 mg L^{-1} and contact time = 4–5 h

Ultraviolet Light (Photolysis)

Ultraviolet (UV) light has become more common in water treatment since the discovery in the late 1990s that it is highly effective for *Cryptosporidium* oocyst inactivation. Although typical disinfection doses are in the range of 20–100 mJ cm^{-2}, much higher doses (e.g., 500–1,000 mJ cm^{-2}) are usually employed for contaminant oxidation. Most UV reactors can be divided into two categories based on lamp characteristics and resulting output: (1) monochromatic/low pressure and (2) polychromatic/medium pressure. Both types of lamps contain mercury gas that emits ultraviolet light when excited by electrons. Low-pressure lamps produce a monochromatic output at 254 nm, which is extremely effective for UV disinfection, and medium-pressure bulbs produce a polychromatic output at a higher intensity that induces reactions in a broader range of contaminants. Both types of reactors are susceptible to fouling due to the lower solubility of many natural constituents (e.g., $CaCO_3$) at higher temperatures found at the surface of the bulb. High turbidity and high levels of organic matter also reduce the effectiveness of photolysis.

Photolysis modifies and destroys organic contaminants by direct bond cleavage and through reactions with inorganic constituents to form highly reactive intermediates, such as OH·. However, the extent of photolysis at typical UV disinfection doses is quite small so TOrC mitigation is not considered a synergistic benefit of UV disinfection [18]. In bench- and pilot-scale experiments, only four of 29 detected compounds were degraded by more than 20% with medium-pressure photolysis at a UV dose of 40 mJ cm^{-2} (Table 8). Medium-pressure photolysis at a UV dose of

Table 8 TOrC degradation by low-dose UV photolysis [18]

<20% degradation	20–50% degradation	50–80% degradation	>80% degradation
Androstenedione	Acetaminophen	Diclofenac	
Atrazine		Sulfamethoxazole	
Caffeine		Triclosan	
Carbamazepine			
DEET			
Diazepam			
Dilantin			
Erythromycin			
Estradiol			
Estriol			
Estrone			
Ethinyl estradiol			
Fluoxetine			
Gemfibrozil			
Hydrocodone			
Ibuprofen			
Iopromide			
Meprobamate			
Naproxen			
Oxybenzone			
Pentoxifylline			
Progesterone			
TCEP			
Testosterone			
Trimethoprim			

[a]UV dose = 40 mJ cm^{-2}

450 mJ cm^{-2} achieved significantly increased removals (Table 9), and the addition of hydrogen peroxide (H_2O_2) provided further improvements to the process [18]. The use of H_2O_2 to improve UV-based oxidation will be discussed in greater detail in relation to advanced wastewater treatment.

Structural properties of individual compounds play a role in how effectively a compound may be destroyed by photolysis. For example, aromatic compounds absorb light in the UV spectrum so compounds with aromatic centers are more susceptible to photolysis. Of the pharmaceuticals and EDCs investigated, diclofenac, sulfamethoxazole, and triclosan were most susceptible to removal by photolysis, and all have absorption spectra that overlap with the wavelength-specific peaks generated by medium-pressure lamps. Conversely, aliphatic compounds that lack conjugated double bonds and the appropriate absorption bands are very resistant to UV photolysis. Although UV photolysis may be effective at removing some pharmaceuticals and EDCs, it is generally not viable as a stand-alone treatment process as many compounds have structures that are not amenable to UV photolysis.

Table 9 TOrC destruction by high-dose UV photolysis [18]

<20% degradation	20–50% degradation	50–80% degradation	>80% degradation
Androstenedione	Carbamazepine	Atrazine	Acetaminophen
Caffeine	Gemfibrozil	Dilantin	Diclofenac
DEET	Ibuprofen	Erythromycin	Estradiol
Diazepam	Pentoxifylline	Iopromide	Estriol
Meprobamate	Progesterone		Estrone
TCEP	Testosterone		Ethinyl estradiol
	Trimethoprim		Fluoxetine
			Hydrocodone
			Naproxen
			Oxybenzone
			Sulfamethoxazole
			Triclosan

[a]UV dose = 450 mJ cm^{-2}

Free Chlorine and Chloramine

Chlorination is the most common form of disinfection due to its effectiveness against a variety of pathogens (with the exception of protozoan parasites) and the ease with which a residual can be maintained throughout a distribution system. However, many utilities are currently turning toward chloramination for residual disinfection [39] due to its greater stability in distribution systems and lower potential to form halogenated DBPs [40]. The amount of chlorine or chloramine utilized in drinking water applications is usually reported as units of concentration×time (CT). Chlorine and chloramine doses of 3 mg L^{-1} for 24 h (CT = 4,320 mg min L^{-1}) were evaluated for PPCP and EDC oxidation [18]. These results are illustrated in Tables 10 and 11, respectively.

Compounds most susceptible to removal by chlorine or chloramine often contain aromatic structures with electron-donating functional groups (e.g., hydroxyl, amine, and methoxy groups) [41, 42]. For example, steroid hormones containing phenolic groups were removed by more than 95%. Other compounds susceptible to chlorine or chloramine oxidation may contain primary amines attached to conjugated rings (e.g., trimethoprim and sulfamethoxazole), highly alkylated benzenes (e.g., gemfibrozil and hydrocodone), and polycyclic aromatic hydrocarbons (e.g., carbamazepine, benzo(a)pyrene, diclofenac, and naproxen). The most resistant compounds often lack carbon–carbon double bonds and contain carboxyl groups, ketones, heterocyclic nitrogen, or primary amides (e.g., iopromide and meprobamate) [18]. Given that some compounds are resistant to chlorine or chloramine oxidation, complete mineralization is not possible. As with any treatment technology, the potential effects of molecular (e.g., chlorinated triclosan [43]) or bulk (e.g., total organic halogens [40]) transformation products must be considered.

Table 10 TOrC oxidation by chlorination [18]

<20% degradation	20–50% degradation	50–80% degradation	>80% degradation
Androstenedione	Diazepam	Gemfibrozil	Acetaminophen
Atrazine	Galaxolide		Benzo(a)pyrene
Caffeine	Pentoxifylline		Diclofenac
Carbamazepine			Erythromycin
DDT			Estradiol
DEET			Estriol
Dilantin			Estrone
Fluorene			Ethinyl estradiol
Fluoxetine			Hydrocodone
Ibuprofen			Musk ketone
Iopromide			Naproxen
Lindane			Oxybenzone
Meprobamate			Sulfamethoxazole
Metolachlor			Triclosan
Progesterone			Trimethoprim
TCEP			
Testosterone			

[a]Chlorine concentration = 3 mg L^{-1}, contact time = 24 h, pH = 7.9–8.5

Table 11 TOrC oxidation by chloramination [18]

<20% degradation	20–50% degradation	50–80% degradation	>80% degradation
Androstenedione	Hydrocodone	Benzo(a)pyrene	Acetaminophen
Atrazine	Galaxolide	Diclofenac	Estradiol
Caffeine		Oxybenzone	Estriol
Carbamazepine			Estrone
DDT			Ethinyl estradiol
DEET			Triclosan
Diazepam			
Dilantin			
Erythromycin			
Fluorene			
Fluoxetine			
Gemfibrozil			
Ibuprofen			
Iopromide			
Lindane			
Meprobamate			
Metolachlor			
Musk ketone			
Naproxen			
Pentoxifylline			
Progesterone			
Sulfamethoxazole			
TCEP			
Testosterone			
Trimethoprim			

[a]Chloramine concentration = 3 mg L^{-1}, contact time = 24 h, pH = 8.0

Table 12 TOrC oxidation by ozonation [18]

<20% degradation	20–50% degradation	50–80% degradation	>80% degradation
TCEP	Atrazine	DEET	Acetaminophen
	Iopromide	Diazepam	Androstenedione
	Meprobamate	Dilantin	Caffeine
		Ibuprofen	Carbamazepine
			Diclofenac
			Erythromycin
			Estradiol
			Estriol
			Estrone
			Ethinyl estradiol
			Fluoxetine
			Gemfibrozil
			Hydrocodone
			Naproxen
			Oxybenzone
			Pentoxifylline
			Progesterone
			Sulfamethoxazole
			Testosterone
			Triclosan
			Trimethoprim

[a] Ozone concentration $= 2.5$ mg L^{-1} and contact time $= 24$ min

Ozone

Although relatively energy intensive, ozone is highly effective for both chemical oxidation and microbial inactivation (including *Giardia* cysts and *Cryptosporidium* oocysts). Ozone either reacts directly with organic molecules or indirectly through the formation of radical species [44]. Ozone is relatively unstable in water and wastewater (i.e., decays in minutes) so it is not possible to maintain a long-term residual. The natural decomposition of ozone into OH· is particularly relevant for wastewater applications, but H_2O_2 can also be used to drive the formation of OH· in drinking water and wastewater. For direct reactions, ozone reacts rapidly with amines, phenols, and double bonds in aliphatic compounds.

In contrast to photolysis, many PPCPs and EDCs are degraded rapidly with ozone CTs commonly used for disinfection applications (less than 20 mg min L^{-1}) [19, 45, 46]. Since molecular ozone is very effective for pharmaceutical and EDC treatment, modifying the process with H_2O_2 is not always necessary, although it does increase the reaction rate [18]. However, some recalcitrant compounds (e.g., clofibric acid and ibuprofen) may necessitate augmentation with H_2O_2 to achieve higher levels of treatment, particularly in drinking water applications where the natural ozone decomposition pathway is not as prevalent [18]. Table 12 describes the relative removals of a suite of PPCPs and EDCs by ozonation.

Table 13 Second-order ozone and OH˙ rate constants for select TOrCs [16, 25, 29, 47–55]

Compound	k''_{O_3} (M⁻¹ s⁻¹)	$k''_{OH\cdot}$ (M⁻¹ s⁻¹)
Meprobamate	<10	$(1-5)\times10^9$
Sulfamethoxazole	2.5×10^6	5.5×10^9
Trimethoprim	2.7×10^5	6.9×10^9
Carbamazepine	3×10^5	8.8×10^9
Phenytoin	~10	$(5-10)\times10^9$
Primidone	~10	$(5-10)\times10^9$
Triclosan	5.1×10^8	$(5-10)\times10^9$
Atenolol	6.3×10^5	8.0×10^9
TCEP	<10	7.4×10^8
Musk ketone	<10	$(1-5)\times10^9$
Atrazine	6	3×10^9
Gemfibrozil	~500	$(5-10)\times10^9$
Diclofenac	1×10^6	7.5×10^9
Ibuprofen	9.6	7.4×10^9
Naproxen	~1×10^5	9.6×10^9
DEET	~10	$(5-10)\times10^9$
Bisphenol A	~1×10^9	1×10^{10}

Numerous studies have developed second-order rate constants for the ozonation of PPCPs and EDCs; a subset of these rate constants is presented in Table 13. For compounds with unknown rate constants, quantitative structure activity relationships (QSARs) can be used to estimate their susceptibility to ozonation. For example, Huber et al. [16] noted that the aromatic and tertiary amine moieties found in sulfonamide and macrolide antibiotics are reactive with ozone, and all compounds within these classes should have similar reaction rates. Furthermore, the authors indicated that ketone-containing steroid hormones are likely to have rate constants that are approximately one order of magnitude less than the phenolic steroid hormones. The compounds experiencing the least amount of degradation are generally characterized by extensive branching (e.g., meprobamate and iopromide) and are sometimes designed specifically to resist oxidation (e.g., the flame retardant TCEP). As with chlorine and other oxidation processes, complete mineralization with ozone is impractical given the energy requirement and the potential to form DBPs (e.g., bromate). Thus, the potential effects of ozone transformation products must be considered.

TOrC Removal During Wastewater Treatment

Raw wastewater quality varies tremendously depending on the contributing sources (i.e., small residential communities, large urban areas, industrial discharge, etc.), and the extent of treatment ultimately depends on the intended use or effluent discharge location. For example, wastewater permitted for ocean discharge does not have the same water quality requirements as that permitted for indirect potable reuse (IPR). Conventional wastewater treatment has evolved over time but generally includes the following unit operations and processes: preliminary solids removal, primary clarification, secondary biological treatment, filtration, and disinfection [56]. Depending on the specific requirements of the discharge permit, conventional treatment may be supplemented with nutrient removal (i.e., for nitrogen or phosphorus removal) or other advanced processes to achieve a higher quality effluent. This may be required for discharge to a sensitive ecosystem (e.g., areas susceptible to algal blooms and eutrophication) or for IPR applications. Advanced treatment may include membrane treatment or AOPs, such as UV/H_2O_2 and ozone/H_2O_2.

Traditionally, wastewater treatment trains have not been designed for TOrC removal. However, the growing body of occurrence data for wastewater-derived contaminants (including PPCPs and EDCs) in surface waters [15, 57], the recognition that wastewater effluents are impacting natural waters, and the potential adverse effects on aquatic ecosystems [58] have brought these issues to the forefront. Since wastewater discharge is the primary source of PPCPs and EDCs in the environment [59], optimization of wastewater treatment processes may be the most efficient strategy to mitigate the adverse effects of these compounds. The following sections describe the general efficacy of both conventional and advanced wastewater treatment processes for PPCP and EDC mitigation.

Conventional Wastewater Treatment

Conventional wastewater treatment processes relying on physical separation, including preliminary solids removal, primary clarification, grit removal, and media filtration, provide limited reductions in TOrC concentrations. On the other hand, secondary treatment, which involves both adsorption and biological processes, can be highly effective depending on the target contaminant and operational conditions [17]. Activated sludge processes, whether in conventional activated sludge (CAS) configurations or membrane bioreactors (MBRs), may achieve high removals (up to 99%) of hormones and certain pharmaceuticals (e.g., the analgesics acetaminophen and ibuprofen), but biological treatment may be insufficient to remove the more recalcitrant compounds (e.g., the anticonvulsants phenytoin and carbamazepine) [17, 57, 60]. Limited removal efficiencies have been observed for certain antibiotics and antimicrobial compounds, such as erythromycin (10%), sulfamethoxazole (64%), and triclosan (68%) [17]. It is important to note that MBR systems contain microfiltration (MF) or ultrafiltration (UF) membranes, but it is generally the biological processes that are responsible for PPCP and EDC removal. Joss et al. [61] did not observe any relationships between structural characteristics of the compounds and efficacy of secondary treatment, but the study did identify microbial transformation, rather than sludge partitioning, as the dominant mechanism for all of the target compounds.

For the most susceptible compounds, CAS and MBRs achieve comparable removals, but some studies indicate that the longer SRTs associated with MBRs provide significant benefits with respect to the removal of recalcitrant compounds [60]. MBRs can be operated with longer SRTs due to their high microbial loads and more concentrated return activated sludge. CAS would require excessive return flows to achieve comparable SRTs. Clara et al. [62] observed a positive correlation between PPCP and EDC removal and longer SRTs. For most of the target compounds, a critical SRT of 10 days was observed, but for a small number of compounds (e.g., anticonvulsants), PPCP and EDC removal was poor regardless of SRT.

Geographic factors, such as climate, can also influence the efficacy of secondary treatment. For example, Ternes et al. [63] observed 80% to greater than 99% removals of estrogenic hormones in a Brazilian WWTP, but the removal efficiencies of those same compounds were lower (0–70%) in a German WWTP. This difference was primarily attributed to the higher water temperature of the Brazilian WWTP. Therefore, the efficacy of biological treatment is dependent on a variety of factors, including the compound of interest, process configuration, operational parameters, and geographical location. Regardless, PPCPs and EDCs are never completely removed, and they are typically detected in secondary effluent at ng L^{-1} to μg L^{-1} concentrations [64], as described earlier in Table 2.

Advanced Wastewater Treatment: Membranes

The efficacy of membranes for PPCP and EDC removal varies with membrane pore size. Low-pressure MF and UF processes are generally ineffective alternatives for TOrC removal [18] due to the fact that their pore sizes are relatively large and the MW cutoff is approximately 100,000 and 2,000 Da, respectively. Thus, PPCPs and EDCs, which are usually less than 500 Da, have the potential to easily pass through the pores. Indirect PPCP and EDC removal by MF and UF membranes is affected by physiochemical parameters. Hydrophobic compounds adsorbed onto particulates or colloids that will not pass through the membrane pores are readily rejected. High-pressure nanofiltration (NF) and reverse osmosis (RO) membranes have much tighter pores (the MW cutoff for these membranes is approximately 250 and 100 Da, respectively) so PPCPs and EDCs are generally rejected by these membranes. In fact, the concentrations of these target contaminants are generally below the MRL (often 0.25–25 ng L^{-1}) after RO and NF treatment [17, 65].

Snyder et al. [18] studied a variety of pilot and full-scale membrane processes and reported similar results, which are summarized in Table 14. They concluded that hydrophobic compounds with aliphatic substituted aromatic ring structures and high pK_a values were removed by low-pressure MF and UF membranes. This can be attributed

Table 14 TOrC removal by membrane and MBR processes [18]

	Percent removal				
Membrane size	MF	UF	UF/MBR	NF	RO
Number of systems tested	3	5	4	3	9
Acetaminophen	<20	<20	>80	20–50	>80
Androstenedione	<20	20–50	>80	50–80	>80
Atrazine	ND	<20	ND	50–80	ND
Benzo(a)pyrene	ND	>80	ND	>80	ND
Caffeine	<20	<20	>80	50–80	>80
Carbamazepine	<20	<20	20–50	50–80	>80
DDT	ND	>80	50–80	>80	ND
DEET	<20	<20	50–80	50–80	>80
Diazepam	ND	20–50	<20	50–80	>80
Diclofenac	<20	<20	<20	50–80	>80
Erythromycin	<20	20–50	20–50	>80	>80
Estradiol	<20	20–50	50–80	50–80	>80
Estriol	ND	<20	>80	50–80	>80
Estrone	<20	20–50	>80	50–80	>80
Ethinyl estradiol	ND	20–50	>80	50–80	>80
Fluorene	ND	>80	ND	>80	ND
Fluoxetine	20–50	>80	20–50	>80	>80
Galaxolide	<20	20–50	ND	50–80	>80
Gemfibrozil	<20	<20	20–50	50–80	>80
Hydrocodone	<20	<20	20–50	50–80	>80
Ibuprofen	<20	<20	50–80	50–80	>80
Iopromide	<20	<20	<20	>80	>80
Lindane	ND	20–50	ND	50–80	ND
Meprobamate	<20	<20	<20	50–80	>80
Metolachlor	ND	20–50	ND	50–80	ND
Musk ketone	<20	20–50	ND	>80	>80
Naproxen	<20	<20	>80	20–50	>80
Oxybenzone	<20	50–80	>80	>80	>80
Pentoxifylline	<20	<20	>80	50–80	>80
Phenytoin	<20	<20	<20	50–80	>80
Progesterone	ND	50–80	>80	50–80	>80
Sulfamethoxazole	<20	20–50	20–50	50–80	>80
TCEP	<20	<20	<20	50–80	>80
Testosterone	ND	20–50	>80	50–80	ND
Triclosan	20–50	>80	50–80	>80	>80
Trimethoprim	<20	<20	20–50	50–80	>80

ND not detected

to adsorption onto larger material that is readily rejected by the membrane or to electrostatic repulsion from the membrane surface. Neutrally charged or hydrophilic compounds were not removed by MF or UF membranes. Effective removal of all PPCPs and EDCs was observed following treatment with NF or RO membranes.

Advanced Wastewater Treatment: Advanced Oxidation Processes

AOPs utilize highly reactive chemical species such as free radicals to oxidize chemical contaminants in water [66]. The most common AOPs include UV/H_2O_2 and ozone/H_2O_2, but other AOP technologies, such as UV/TiO_2 (titanium dioxide) photocatalysis and nonthermal plasma (NTP), may be viable alternatives in the future [67, 68]. Although AOPs provide some level of treatment with their base mechanisms (e.g., direct photolysis of chemical contaminants from UV/H_2O_2), the dominant treatment pathway generally involves oxidation by highly reactive, nonspecific OH· [18, 19].

In general, AOPs are very effective treatment technologies for removing PPCPs and EDCs from water, though the processes are usually energy intensive. When optimized, the processes can also be very fast, given the short-lived and highly reactive nature of OH·. Huber et al. [16] reported second-order OH· rate constants for a suite of PPCPs and EDCs ranging from 3.3×10^9 to 9.8×10^9 $M^{-1}s^{-1}$ (also refer to Table 13). Snyder et al. [18] reported that treatment with ozone vs. ozone/H_2O_2 was similar in terms of overall PPCP and EDC degradation, but the AOP process yielded faster reaction rates (i.e., nearly instantaneous). In the same study, a limited number of compounds (e.g., clofibric acid and ibuprofen) were not removed by ozone alone (less than 10% removal), but were effectively removed by ozone/H_2O_2 (greater than 90% removal). It is important to note that the efficacy of ozone vs. ozone/H_2O_2 is highly dependent on the water matrix. Drinking water applications provide much greater dissolved ozone exposure, whereas ozone decomposes rapidly into OH· in wastewater applications. Therefore, in the same example presented above, the oxidation of clofibric acid and ibuprofen with molecular ozone (relative to ozone/H_2O_2) may have improved in a wastewater matrix due to rapid ozone decomposition into OH·.

For UV/H_2O_2, pharmaceutical and EDC removal was generally not attributed to direct photolysis. The addition of H_2O_2 was necessary to generate OH·, which was responsible for the oxidation of trace contaminants. Rosenfeldt and Linden [69] reported small reductions in EDC concentrations with a UV dose of 1,000 mJ cm^{-2}, but those EDCs were removed by more than 90% with the same UV dose and 15 mg L^{-1} hydrogen peroxide. The authors also calculated second-order rate constants on the order of 10^{10} $M^{-1}s^{-1}$. Snyder et al. [18] reported greater than 80% removal for 19 of 29 pharmaceuticals and EDCs following UV/H_2O_2 treatment (~375 mJ cm^{-2} and 5 mg L^{-1} hydrogen peroxide). Eight of the remaining ten compounds were between 50 and 80% removed, and only meprobamate and TCEP, which are both highly resistant to oxidation, were less than 50% removed.

Due to the highly reactive nature of OH·, scavengers such as organic matter and alkalinity reduce the efficacy of AOPs [46, 70, 71]. UV/H_2O_2 is also affected by water with high turbidity and high levels of UV absorbance, both of which reduce UV transmissivity. Therefore, it is important to understand the target water matrix

Table 15 Summary of AOP EEO values (kWh m^{-3} per log contaminant removal) for seven pharmaceuticals and EDCs

Contaminant	UV[a]	UV/H$_2$O$_2$[a,b]	UV/TiO$_2$[a,c]	NTP[d]
Atenolol	1.4	0.5	2.0	1.0
Atrazine	3.3	1.2	4.7	3.7
Carbamazepine	2.3	0.4	2.1	<0.7
Meprobamate	6.6	1.0	6.8	3.5
Phenytoin	2.1	1.0	2.2	2.0
Primidone	3.7	0.6	3.9	2.2
Trimethoprim	0.8	0.4	1.5	<0.7

[a]Benotti et al. [67]
[b]10 mg L^{-1} of H$_2$O$_2$
[c]500 mg L^{-1} of TiO$_2$
[d]Gerrity et al. [68]

when selecting the most appropriate AOP. The UV/H$_2$O$_2$ and ozone/H$_2$O$_2$ AOPs also require peroxide addition and subsequent quenching, which is a significant cost over the life of the system.

UV/TiO$_2$ photocatalysis, which generates OH˙ by irradiating a TiO$_2$ slurry or fixed film with UV light, and NTP, which generates UV light, ozone, and OH˙ with high-voltage pulses across two electrodes, are not limited by light-attenuating matrices. Additionally, these processes do not require H$_2$O$_2$ so chemical addition and quenching are not necessary; H$_2$O$_2$ may increase reaction rates, however. Benotti et al. [67] and Gerrity et al. [68] evaluated the degradation of a suite of pharmaceuticals and EDCs in surface waters with direct UV photolysis, UV/H$_2$O$_2$, UV/TiO$_2$ photocatalysis, and NTP. Table 15 provides a summary of the electrical energy per order (EEO) of magnitude destruction values for each of the processes. EEO values are a basis of comparison for many treatment options as they standardize energy consumption to the volume of water treated and the extent of treatment (i.e., kWh m^{-3} per log contaminant removal). Results from these studies indicate that of these four AOP technologies that do not use ozone, UV/H$_2$O$_2$ is the most efficient process, though UV/TiO$_2$ photocatalysis and NTP provide viable, chemical-free alternatives.

Advanced Wastewater Treatment: Indirect Potable Reuse Treatment Trains

There is an increasing global trend toward more efficient use of water resources in both urban and rural communities. In addition to innovative water management and acquisition strategies (e.g., water transfers, banking, and trading), numerous municipalities are turning to water reuse in a variety of contexts to bolster their water portfolios. Reclaimed water has the advantage of being a constant and reliable water source, and it is the only source that increases in supply relative to demand. Historically, the use of reclaimed wastewater for municipal and agricultural irrigation has been the most common and accepted application, but diminishing water supplies—primarily the result of dramatic population growth and historic drought conditions in many areas—and a greater acceptance of water reuse have led to more varied applications, including IPR.

"Unplanned" IPR can be captured colloquially in that "everyone cannot live upstream." In a more formal sense, "unplanned" IPR involves the environmental discharge of conventionally treated wastewater effluent, which is subsequently used as a drinking water source by another municipality. With respect to water quality, the discharge of wastewater effluent generally conforms to the requirements of National Pollutant Discharge Elimination System (NPDES) permits, and additional requirements are sometimes established by local entities (e.g., the California Department of Public Health [CDPH] Title 22 requirements). Many utilities are taking a proactive approach to environmental stewardship and public health by employing advanced wastewater treatment technologies (e.g., membrane filtration, biological activated carbon, and soil aquifer treatment). These additional treatment processes are common measures in many "planned" IPR systems. In a "planned" IPR system, the discharge of wastewater effluent involves some form of environmental buffer, such as soil aquifer treatment and extended storage in a reservoir, and is eventually integrated into the local potable water supply. However, "planned" IPR systems vary considerably with respect to a number of variables, including level of treatment in the WWTP, discharge mechanism, storage time in the environment, and level of treatment in the drinking water treatment plant.

The standard treatment train for "planned" IPR is generally comprised of MF or UF, RO, UV/H_2O_2, and aquifer injection (i.e., the Orange County Groundwater Replenishment District). MF and UF are included primarily as a pretreatment strategy to reduce RO fouling. As discussed earlier, the use of RO is sufficient to approach the detection limits of many TOrCs, but UV/H_2O_2 is included as an additional barrier against N-nitrosodimethylamine (NDMA), which is susceptible to UV light, and 1,4-dioxane, which is susceptible to $OH\cdot$. The CDPH Title 22 requirements for recycled water require the UV/H_2O_2 process to achieve 1.2-log destruction of NDMA and 0.5-log destruction of 1,4-dioxane. The actual operational conditions are site specific but generally require UV doses greater than 500 mJ cm^{-2} and H_2O_2 concentrations exceeding 5 mg L^{-1}. Finally, aquifer injection, which must be preceded by mineral stabilization of the RO permeate, is included as an environmental barrier primarily to increase public acceptance of the concept. Table 16 provides an example of TOrC concentrations in this type of IPR system.

Although the standard IPR treatment train is extremely effective for TOrC mitigation, the production of concentrated brines, high energy costs associated with UV oxidation and RO, and significant chemical requirements for operation and maintenance have prompted the development of alternative IPR treatment strategies. One of the most promising alternatives is comprised of filtration (i.e., media, micro-, or ultra-), ozone-based oxidation, biological activated carbon (BAC), and aquifer injection. This type of treatment train, which has already demonstrated promise in

Table 16 TOrC concentrations (ng L^{-1}) in a standard IPR treatment train

Contaminant	Secondary effluent	Microfiltration permeate	RO permeate	UV/H$_2$O$_2$ effluent
Atenolol	2,460	1,970	20	1.7
Atorvastatin	67	142	<0.25	<0.25
Carbamazepine	304	295	1.5	<0.5
Diazepam	3.8	3.4	<0.25	<0.25
Diclofenac	134	174	0.58	<0.25
Enalapril	2.8	16	<0.25	<0.25
Fluoxetine	38	32	<0.50	<0.50
Gemfibrozil	2,420	2,510	7.8	0.65
Meprobamate	339	316	1.6	0.63
Naproxen	235	245	1.0	<0.50
Phenytoin	283	258	1.3	<1.0
Risperidone	3.3	0.38	<0.25	<0.25
Sulfamethoxazole	1,300	719	2.6	<0.25
Trimethoprim	601	604	4.3	0.46

pilot- and full-scale installations in Europe and Australia [72, 73], is particularly promising for inland applications where brine disposal is an issue. Similar to the standard configuration, filtration is provided as a pretreatment step to improve the efficacy of ozonation and to reduce solids loadings on the subsequent BAC process. Ozonation is incorporated as the primary treatment mechanism for TOrC mitigation, and the BAC process is provided for the removal of oxidation by-products and recalcitrant TOrCs. Again, aquifer recharge is provided to increase public acceptance of the IPR concept. Although this alternative provides significant benefits over the standard configuration, there are certainly issues that must be considered prior to implementation, including bromate formation and pathogen regrowth in the BAC process. Bromate formation can be mitigated with the addition of H$_2$O$_2$ during the ozone process, but microbial regrowth may necessitate downstream disinfection prior to discharge. Table 17 provides an example of TOrC concentrations in a pilot-scale demonstration of UF, ozone/H$_2$O$_2$, and BAC. Ozone was dosed at 5 mg L^{-1} (mass-based ozone:total organic carbon ratio of ~1.0), and H$_2$O$_2$ was dosed at 3.5 mg L^{-1} (molar H$_2$O$_2$:ozone ratio of ~1.0) for bromate mitigation.

Advanced Wastewater Treatment: Residual Management

Advanced water and wastewater treatment technologies, such as AOPs and NF/RO membranes, are particularly effective for removing PPCPs and EDCs. However, the viability of these processes is tempered by residual management issues, including transformation products and the discharge of concentrated brine streams. With any type of oxidation process, including more conventional forms such as chlorination

Table 17 TOrC concentrations (ng L^{-1}) in an alternative IPR treatment train

Contaminant	Secondary effluent	Ultrafiltration permeate	Ozone/H_2O_2 permeate	BAC effluent
Atenolol	860	790	9.2	<1.0
Atorvastatin	20	8.1	<0.5	<0.5
Atrazine	0.83	1.1	0.39	<0.25
Benzophenone	160	130	<50	<50
Carbamazepine	300	310	<0.5	<0.5
DEET	860	920	14	<1
Diazepam	3.0	3.0	<0.25	<0.25
Diclofenac	98	79	<0.5	<0.5
Dilantin	310	110	3.0	<1.0
Fluoxetine	72	46	<0.5	<0.5
Gemfibrozil	65	60	<0.25	<0.25
Meprobamate	830	840	97	8.0
Musk ketone	50	<25	<25	<25
Naproxen	13	12	<0.5	<0.5
Phenytoin	310	110	3.0	<1.0
Primidone	230	270	11	0.66
Sulfamethoxazole	1,100	900	5.7	<0.25
TCEP	480	480	370	<10
TCPP	2,200	2,400	1,100	<100
Trimethoprim	460	240	<0.25	<0.25

and ozonation, it is impractical to achieve complete mineralization (i.e., conversion of organic molecules to water, mineral acids, and carbon dioxide). Short of complete mineralization, oxidation processes will convert target compounds into transformation products that may or may not bear toxicological significance. For example, Vanderford et al. [43] studied the chlorination of the antimicrobial compound triclosan and noted the formation of mono- and dichlorinated by-products within minutes of chlorine addition. Furthermore, Canosa et al. [74] noted that the chlorinated by-products of triclosan are more toxic than the parent compound.

Recently, researchers have begun to develop an understanding of the reaction pathways and/or transformation products that are produced following advanced treatment of waters containing PPCPs and EDCs. For example, the OH^--induced destruction of several compounds or classes of compounds, including DEET [53], fibrate pharmaceuticals [51], fluoroquinolone antibiotics [75], and beta-lactam antibiotics [76], has been documented. Research is currently underway to develop models that can predict these transformation products, their reactivity, and toxicity [77]. There is a balance that must be achieved between TOrC oxidation and the formation of transformation products. It is possible that some transformation products may carry toxicological significance, thereby requiring utilities to (1) avoid their formation or (2) implement additional treatment to remove them (e.g., BAC) or convert them into a benign form.

Membrane treatment is also affected by residual management issues. Considering that RO typically requires initial feed pressures exceeding 100 psi, it is evident that a substantial amount of energy is required to drive these processes. As the membranes foul, additional energy is required to maintain sufficient water production, and periodic chemical treatments may be required to clean the membranes. Assuming a process flow rate of 10 MGD with 90% recovery [37], the RO system would produce 9 MGD of high-quality permeate, but it would also generate 1 MGD of concentrated brine containing five to tenfold greater concentrations of PPCPs and EDCs, salts, organic matter, and other contaminants. Capital costs, operational costs, and responsible disposal of the brine stream—in addition to the loss of this precious resource—are the major limitations facing widespread use of RO membranes for wastewater treatment and water reuse. At present, disposal of brine streams is often restricted to coastal environments or to other WWTPs, limiting areas in which this technology can be implemented.

Conclusions

PPCPs and EDCs are not truly "emerging contaminants" because the water and wastewater communities have been aware of their presence in water supplies for decades. However, recent advancements in scientific and analytical methodologies in addition to increased media exposure have generated tremendous interest in this field. Recent studies have monitored the concentrations of numerous TOrCs in source water, drinking water, and wastewater to characterize the extent of contamination. Although TOrC concentrations in raw wastewater vary greatly and can often exceed 1 $\mu g\ L^{-1}$, TOrCs are often present at very low concentrations (generally less than 10 ng L^{-1}) in source water and finished drinking water. These low concentrations can be attributed to a combination of water and wastewater treatment efficacy and natural attenuation in the environment.

As mentioned earlier, the presence or absence of any chemical in water is essentially a function of analytical detection capability. Therefore, the research communities must continue to study the potential impacts of TOrCs on human health and aquatic environments. Until the scientific and regulatory communities reach consensus on the implications of PPCPs and EDCs in water, utilities will likely take a proactive approach to (1) understand the extent of contamination in their systems, (2) evaluate the efficacy of their current treatment strategy, and (3) determine whether additional measures are necessary for TOrC mitigation. As discussed in this chapter, some conventional water and wastewater treatment technologies are quite effective for the removal and/or destruction of TOrCs, and there are a number of advanced technologies that can be implemented for further TOrC reductions. However, no single treatment process is capable of 100% TOrC removal so it is important to balance the advantages and disadvantages of the various alternatives while developing a multibarrier approach to TOrC mitigation.

Acknowledgments These data were primarily collected during studies sponsored by the Water Research Foundation (formerly American Water Works Association Research Foundation (AwwaRF)) and the WateReuse Research Foundation (formerly WateReuse Foundation). The Water Research Foundation sponsored Project #2758 entitled "Evaluation of Conventional and Advanced Water Treatment Processes to Remove Endocrine Disruptors and Pharmaceutically-Active Compounds" and Project #3085 entitled "Toxicological Relevance of EDCs and Pharmaceuticals in Drinking Water." The WateReuse Research Foundation sponsored WRF-08-05 entitled "Use of Ozone in Water Reclamation for Contaminant Oxidation."

References

1. Schueler FW (1946) Sex-hormonal action and chemical constitution. Science 103:221–223
2. Sluczewski A, Roth P (1948) Effects of androgenic and estrogenic compounds on the experimental metamorphoses of amphibians. Gynecol Obstet 47:164–176
3. Stumm-Zollinger E, Fair GM (1965) Biodegradation of steroid hormones. J Water Pollut Control Fed 37:1506–1510
4. Hignite C, Azarnoff DL (1977) Drugs and drug metabolites as environmental contaminants: chlorophenoxyisobutyrate and salicylic acid in sewage water effluent. Life Sci 20:337–341
5. Snyder SA, Westerhoff P, Yoon Y, Sedlak DL (2003) Pharmaceuticals, personal care products, and endocrine disruptors in water: implications for the water industry. Environ Eng Sci 20:449–469
6. Snyder S, Vanderford B, Pearson R, Quinones O, Yoon Y (2003) Analytical methods used to measure endocrine disrupting compounds in water. Pract Period Hazard Toxic Radioact Waste Manage 7:224–234
7. Vanderford BJ, Snyder SA (2006) Analysis of pharmaceuticals in water by isotope dilution liquid chromatography/tandem mass spectrometry. Environ Sci Technol 40:7312–7320
8. Snyder SA, Vanderford BJ, Drewes J, Dickenson E, Snyder EM, Bruce GM, Pleus RC (2008) State of knowledge of endocrine disruptors and pharmaceuticals in drinking water. AWWA Research Foundation, IWA Publishing, London
9. Hollender J, Zimmermann SG, Koepke S, Krauss M, McArdell CS, Ort C, Singer H, Von Gunten U, Siegrist H (2009) Elimination of organic micropollutants in a municipal wastewater treatment plant upgraded with a full-scale post-ozonation followed by sand filtration. Environ Sci Technol 43:7862–7869
10. Daughton CG, Ternes TA (1999) Pharmaceuticals and personal care products in the environment: agents of subtle change? Environ Health Perspect 107:907–938
11. Benotti MJ, Trenholm RA, Vanderford BJ, Holady JC, Stanford BD, Snyder SA (2009) Pharmaceuticals and endocrine disrupting compounds in U.S. drinking water. Environ Sci Technol 43:597–603
12. Ternes TA (1998) Occurrence of drugs in German sewage treatment plants and rivers. Water Res 32:3245–3260
13. Fent K, Weston AA, Caminada D (2006) Ecotoxicology of human pharmaceuticals. Aquat Toxicol 76:122–159
14. Boxall ABA, Johnson P, Smith EJ, Sinclair CJ, Stutt E, Levy LS (2006) Uptake of veterinary medicines from soils into plants. J Agric Food Chem 54:2288–2297
15. Kolpin DW, Furlong ET, Meyer MT, Thurman EM, Zaugg SD, Barber LB, Buxton HT (2002) Pharmaceuticals, hormones, and other organic wastewater contaminants in U.S. streams, 1999–2000: a national reconnaissance. Environ Sci Technol 36:1202–1211
16. Huber MM, Canonica S, Park GY, Von Gunten U (2003) Oxidation of pharmaceuticals during ozonation and advanced oxidation processes. Environ Sci Technol 37:1016–1024
17. Kim SD, Cho J, Kim IS, Vanderford BJ, Snyder SA (2007) Occurrence and removal of pharmaceuticals and endocrine disruptors in South Korean surface, drinking, and waste waters. Water Res 41:1013–1021

18. Snyder SA, Wert EC, Lei H, Westerhoff P, Yoon Y (2007) Removal of EDCs and pharmaceuticals in drinking and reuse treatment processes. AWWA Research Foundation, IWA Publishing, London
19. Snyder SA, Wert EC, Rexing DJ, Zegers RE, Drury DD (2006) Ozone oxidation of endocrine disruptors and pharmaceuticals in surface water and wastewater. Ozone Sci Eng 28:445–460
20. Ternes TA, Meisenheimer M, McDowell D, Sacher F, Brauch HJ, Haist-Gulde B, Preuss G, Wilme U, Zulei-Seibert N (2002) Removal of pharmaceuticals during drinking water treatment. Environ Sci Technol 36:3855–3863
21. Westerhoff P, Yoon Y, Snyder S, Wert E (2005) Fate of endocrine-disruptor, pharmaceutical, and personal care product chemicals during simulated drinking water treatment processes. Environ Sci Technol 39:6649–6663
22. Carballa M, Omil F, Lema JM (2005) Removal of cosmetic ingredients and pharmaceuticals in sewage primary treatment. Water Res 39:4790–4796
23. Jones OAH, Voulvoulis N, Lester JN (2002) Aquatic environmental assessment of the top 25 English prescription pharmaceuticals. Water Res 36:5013–5022
24. Qiu HC, Jun HW, McCall JW (1998) Pharmacokinetics, formulation, and safety of insect repellent N, N-diethyl-3-methylbenzamide (DEET): a review. J Am Mosq Control Assoc 14:12–27
25. Deborde M, Rabouan S, Duguet J-P, Legube B (2005) Kinetics of aqueous ozone-induced oxidation of some endocrine disruptors. Environ Sci Technol 39:6086–6092
26. Noaksson E, Gustavsson B, Linderoth M, Zebuhr Y, Broman D, Balk L (2004) Gonad development and plasma steroid profiles by HRGC/HRMS during one reproductive cycle in reference and leachate-exposed female perch (Perca fluviatilis). Toxicol Appl Pharmacol 195:247–261
27. EC (2003) Risk assessment: Musk ketone. European Union risk assessment report. Final Draft June 2003
28. Simonich SL, Begley WM, Debaere G, Eckhoff WS (2000) Trace analysis of fragrance materials in wastewater and treated wastewater. Environ Sci Technol 34:959–965
29. Dodd MC, Buffle MO, Von Gunten U (2006) Oxidation of antibacterial molecules by aqueous ozone: moiety-specific reaction kinetics and application to ozone-based wastewater treatment. Environ Sci Technol 40:1969–1977
30. Loftsson T, Hreinsdottir D (2006) Determination of aqueous solubility by heating and equilibration: a technical note. AAPS PharmSciTech 7:E1–E4
31. Snyder SA, Trenholm RA, Snyder EM, Bruce GM, Pleus RC, Hemming JDC (2008) Toxicological relevance of EDCs and pharmaceuticals in drinking water. AWWA Research Foundation, IWA Publishing, London
32. Letterman RD, Amirtharajah A (1999) Coagulation and flocculation. In: Letterman RD (ed) Water quality and treatment: a handbook of community water supplies. McGraw-Hill, New York
33. Cleasby JL, Logsdon GS (1999) Granular bed and precoat filtration. In: Letterman RD (ed) Water quality and treatment: a handbook of community water supplies. McGraw-Hill, New York
34. Gregory R, Zebel TF (1999) Sedimentation and flotation. In: Letterman RD (ed) Water quality and treatment: a handbook of community water supplies. McGraw-Hill, New York
35. Westerhoff P, Mezyk SP, Cooper WJ, Minakata D (2007) Electron pulse radiolysis determination of hydroxyl radical rate constants with Suwannee river fulvic acid and other dissolved organic matter isolates. Environ Sci Technol 41:4640–4646
36. Ternes TA, Meisenheimer M, McDowell D, Sacher F, Brauch HJ, Gulde BH, Preuss G, Wilme U, Seibert NZ (2002) Removal of pharmaceuticals during drinking water treatment. Environ Sci Technol 36:3855–3863
37. Crittenden JC, Trussell RR, Hand DW, Howe KJ, Tchobanoglous G (2005) Water treatment: principles and design, 2nd edn. Wiley, New Jersey
38. Snyder SA, Adham S, Redding AM, Cannon FS, DeCarolis J, Oppenheimer J, Wert EC, Yoon Y (2007) Role of membranes and activated carbon in the removal of endocrine disruptors and pharmaceuticals. Desalination 202:156–181

39. Routt JC, Mackey E, Whitby E, Connell G, Passantine L, Noak R (2007) Tracking utility disinfection over four decades of change: disinfection survey 2007—preliminary summary. Water Quality Technology Conference, American Water Works Associon, North Carolina.
40. Hua G, Reckhow DA (2008) DBP formation during chlorination and chloramination: effect of reaction time, pH, dosage, and temperature. J Am Water Works Assoc 100:82–95
41. Reckhow DA, Singer PC, Malcolm RL (1990) Chlorination of humic materials—by-product formation and chemical interpretations. Environ Sci Technol 24:1655–1664
42. Deborde M, von Gunten U (2008) Reactions of chlorine with inorganic and organic compounds during water treatment—kinetics and mechanisms: a critical review. Water Res 42:13–51
43. Vanderford BJ, Mawhinney DB, Rosario-Ortiz FL, Snyder SA (2008) Real-time detection and identification of aqueous chlorine transformation products using QTOF MS. Anal Chem 80:4193–4199
44. Langlais RT, Reckhow DA, Brink DR (1991) Ozone in water treatment: application and engineering. Lewis Publishers, Chelsea
45. Wert EC, Rosario-Ortiz FL, Drury DD, Snyder SA (2007) Formation of oxidation byproducts from ozonation of wastewater. Water Res 41:1481–1490
46. Wert EC, Rosario-Ortiz FL, Snyder SA (2009) Effect of ozone exposure on the oxidation of trace organic contaminants in wastewater. Water Res 43:1005–1014
47. Benner J, Salhi E, Ternes T, von Gunten U (2008) Ozonation of reverse osmosis concentrate: kinetics and efficiency of beta blocker oxidation. Water Res 42:3003–3012
48. Huber MM, Goebel A, Joss A, Hermann N, Loeffler D, McArdell CS, Ried A, Siegrist H, Ternes TA, Von Gunten U (2005) Oxidation of pharmaceuticals during ozonation of municipal wastewater effluents: a pilot study. Environ Sci Technol 39:4290–4299
49. Latch DE, Packer JL, Stender BL, VanOverbeke J, Arnold WA, McNeill K (2005) Aqueous photochemistry of triclosan: formation of 2,4-dichlorophenol, 2,8-dichlorodibenzo-p-dioxin, and oligomerization products. Environ Toxicol Chem 24:517–525
50. Packer JL, Werner JJ, Latch DE, McNeill K, Arnold WA (2003) Photochemical fate of pharmaceuticals in the environment. Aquat Sci 65:342–351
51. Razavi B, Song W, Cooper WJ, Greaves J, Jeong J (2009) Free-radical-induced oxidative and reductive degradation of fibrate pharmaceuticals: kinetic studies and degradation mechanisms. J Phys Chem A 113:1287–1294
52. Rosenfeldt EJ, Linden KG, Canonica S, von Gunten U (2006) Comparison of the efficiency of {radical dot}OH radical formation during ozonation and the advanced oxidation processes O_3/H_2O_2 and UV/H_2O_2. Water Res 40:3695–3704
53. Song W, Cooper WJ, Peake BM, Mezyk SP, Nickelsen MG, O'Shea KE (2009) Free-radical induced oxidative and reductive degradation of N, N-diethyl-meta-toluamide (DEET): kinetic studies and degradation pathways. Water Res 43:635–642
54. Suarez S, Dodd MC, Omil F, von Gunten U (2007) Kinetics of triclosan oxidation by aqueous ozone and consequent loss of antibacterial activity: relevance to municipal wastewater ozonation. Water Res 41:2481–2490
55. Watts MJ, Linden KG (2009) Advanced oxidation kinetics of aqueous trialkyl phosphate flame retardants and plasticizers. Environ Sci Technol 43:2937–2942
56. Tchobanoglous G, Burton FL, Stensel HD (2004) Wastewater engineering: treatment and reuse, 4th edn. McGraw Hill, Boston
57. Snyder SA, Lei HX, Wert EC (2008) Removal of endocrine disruptors and pharmaceuticals during water treatment. In: Aga DS (ed) Fate of pharmaceuticals in the environment and in water treatment systems. CRC Press, New York
58. Snyder SA, Villeneuve DL, Snyder EM, Giesy JP (2001) Identification and quantification of estrogen receptor agonists in wastewater effluents. Environ Sci Technol 35:3620–3625
59. Glassmeyer ST, Furlong ET, Kolpin DW, Cahill JD, Zaugg SD, Werner SL, Meyer MT, Kryak DD (2005) Transport of chemical and microbial compounds from known wastewater discharges: potential for use as indicators of human fecal contamination. Environ Sci Technol 39:5157–5169

60. Radjenovic J, Petrovic M, Barcelo D (2009) Fate and distribution of pharmaceuticals in wastewater and sewage sludge of the conventional activated sludge (CAS) and advanced membrane bioreactor (MBR) treatment. Water Res 43:831–841
61. Joss A, Keller E, Alder AC, Gobel A, McArdell CS, Ternes T, Siegrist H (2005) Removal of pharmaceuticals and fragrances in biological wastewater treatment. Water Res 39:3137–3152
62. Clara M, Kreuzinger N, Strenn B, Gans O, Kroiss H (2005) The solids retention time—a suitable design parameter to evaluate the capacity of wastewater treatment plants to remove micropollutants. Water Res 39:97–106
63. Ternes TA, Stumpf M, Mueller J, Haberer K, Wilken RD, Servos M (1999) Behavior and occurrence of estrogens in municipal sewage treatment plants—I. Investigations in Germany, Canada and Brazil. Sci Total Environ 225:81–90
64. Heberer T (2002) Occurrence, fate, and removal of pharmaceutical residues in the aquatic environment: a review of recent research data. Toxicol Lett 131:5–17
65. Bellona C, Drewes JE, Oelker G, Luna J, Filteau G, Amy G (2008) Comparing nanofiltration and reverse osmosis for drinking water augmentation. J Am Water Works Assoc 100:102–116
66. Singer PC, Reckhow DA (1999) Chemical oxidation. In: Letterman RD (ed) Water quality and treatment: a handbook of community water supplies. McGraw-Hill, New York
67. Benotti MJ, Standford BD, Wert EC, Snyder SA (2009) Evaluation of a photocatalytic reactor membrane pilot system for the removal of pharmaceuticals and endocrine disrupting compounds from water. Water Res 43(6):1513–1522
68. Gerrity DW, Stanford BD, Trenholm RA, Snyder SA (2010) An evaluation of a pilot-scale nonthermal plasma advanced oxidation process for trace contaminant degradation. Water Res 44(2):493–504
69. Rosenfeldt EJ, Linden KG (2004) Degradation of endocrine disrupting chemicals bisphenol A, ethynyl estradiol, and estradiol during UV photolysis and advanced oxidation processes. Environ Sci Technol 38:5476–5483
70. Rosario-Ortiz FL, Mezyk SP, Wert EC, Doud DFR, Singh MK, Xin M, Baik S, Snyder SA (2008) Effect of ozone oxidation on the molecular and kinetic properties of effluent organic matter. J Adv Oxid Technol 11:529–535
71. Beltran FJ (2004) Ozone reaction kinetics for water and wastewater systems. CRC Press, Boca Raton
72. Reungoat J, Macova M, Escher BI, Carswell S, Mueller JF, Keller J (2010) Removal of micropollutants and reduction of biological activity in a full scale reclamation plant using ozonation and activated carbon filtration. Water Res 44:625–637
73. Stalter D, Magdeburg A, Weil M, Knacker T, Oehlmann J (2010) Toxication or detoxication? In vivo toxicity assessment of ozonation as advanced wastewater treatment with the rainbow trout. Water Res 44:439–448
74. Canosa P, Morales S, Rodriguez I, Rubi E, Cela R, Gomez M (2005) Aquatic degradation of triclosan and formation of toxic chlorophenols in presence of low concentrations of free chlorine. Anal Bioanal Chem 383:1119–1126
75. Santoke H, Song WH, Cooper WJ, Greaves J, Miller GE (2009) Free-radical-induced oxidative and reductive degradation of fluoroquinolone pharmaceuticals: kinetic studies and degradation mechanism. J Phys Chem A 113:7846–7851
76. Song WH, Chen WS, Cooper WJ, Greaves J, Miller GE (2008) Free-radical destruction of beta-lactam antibiotics in aqueous solution. J Phys Chem A 112:7411–7417
77. Lei HX, Snyder SA (2007) 3D QSPR models for the removal of trace organic contaminants by ozone and free chlorine. Water Res 41:4051–4060

Pharmaceutical Take Back Programs

Kati I. Stoddard and Duane B. Huggett

Introduction

Prior to September 2010 the US national policy for the proper disposal of pharmaceuticals was limited to published guidance provided by several federal agencies; however, on September 25, 2010 the US Drug Enforcement Administration (DEA) established the National Take Back Initiative. This program is designed to provide citizens an opportunity to safely dispose of medications they no longer need or want. In 2010 two of these DEA events were held during which 309 tons of medications were collected at thousands of take back sites across the country. Additionally, the Safe and Secure Drug Disposal Act of 2010 signed by President Obama on October 12, 2010 provided the means for the Controlled Substance Act (CSA) to be amended to allow the DEA to develop a procedure for individuals to safely dispose of their unwanted medications, including medications considered controlled under the CSA [1]. This legislation will ensure future DEA events and other take back events are legally able to dispose of controlled medications, which is critical as these medications have a high potential for diversion or abuse.

For individuals unable to participate in the national DEA events the DEA and other national agencies advise individuals to dispose of their unused, unneeded, or expired pharmaceuticals by removing them from their original containers, mixing the medications with a deterring substance, such as coffee grounds or used cat litter, placing the mixture in nondescript containers like sealable bags, and then placing the bags in the household garbage. Flushing is only recommended when the label or patient information for the prescription drug specifically calls for such a disposal method [2]. Beyond this basic medication disposal information, several

---

K.I. Stoddard • D.B. Huggett (✉)
Department of Biological Sciences, University of North Texas,
1155 Union Circle #305220, Denton, TX 76203, USA
e-mail: katistoddard@my.unt.edu; dbhuggett@unt.edu

B.W. Brooks and D.B. Huggett (eds.), *Human Pharmaceuticals in the Environment: Current and Future Perspectives*, Emerging Topics in Ecotoxicology 4, DOI 10.1007/978-1-4614-3473-3_10, © Springer Science+Business Media, LLC 2012

environmental stewardship organizations and US federal agencies, such as the Environmental Protection Agency (EPA) and the Fish and Wildlife Service (FWS), provide information to the general public on the impacts pharmaceuticals can have on the environment and provide suggestions of actions individual can take to minimize this growing environmental threat.

Despite the concerted efforts of many agencies and organizations, the message on proper disposal methods for pharmaceuticals appears to not be entirely effective at reaching target audiences. Studies have found that the most frequent methods of disposal of pharmaceuticals are the sink, toilet, and trash [3–7]. The preference among these methods appears to be significantly influenced by age with older respondents relying on the sink or toilet as a disposal method and younger respondents relying on household trash [7]. Even with strict observance to the DEA guidance for medication in household trash disposal, residues from pharmaceuticals disposed of in sanitary landfills can accumulate in leachate and escape the facilities to nearby receiving waters [8–11]. Additionally, residuals of medicines are also reaching surface waters through postconsumer excretion of pharmaceuticals and their metabolites in human and animal urine. Studies have found detectable concentrations of pharmaceuticals ranging from parts per trillion (ppt) to low end parts per billion (ppb) in both wastewater effluent [12] and drinking water [13–16]. Though there are concerted research efforts aimed at investigating the impact of pharmaceuticals on the environment, these efforts are greatly complicated by the conglomeration of pharmaceuticals and other man-made chemical products being released into the environment. Beyond considering the impact of mixtures of pharmaceuticals with other man-made chemicals and products on the environment, research is further complicated by the concept of low-dose toxicity, which stresses that levels of pharmaceuticals below what are traditionally considered harmless may have subtle or even pronounced acute and chronic toxic effects on aquatic organisms that are living in ecosystems that receive a continuous stream of these chemicals via wastewater effluent or industrial outfalls [17].

Through the dedicated efforts of toxicological and environmental research, an ever increasing volume of scientific evidence is building support for the concept that pharmaceutical residues in the environment are having a deleterious impact on the quality of the aquatic environment. Though knowledge and complete understanding of this emerging environmental threat may not be widespread among the general lay population, awareness of this issue is increasing steadily in the American and global population through news and media coverage and public health outreach campaigns. As public consciousness of the environmental implications of pharmaceuticals in water resources increases, so too has the public increased their expressed interest in policies and programs to mitigate some of the documented negative environmental impacts caused by pharmaceuticals in the environment.

Events such as the newly developed DEA Initiative are one promising type of program that have been implemented as a means to mitigate or minimize the impacts of pharmaceuticals on natural resources. Within the USA, pharmaceutical take back programs have been implemented at local, state, and, with the advent of the DEA Initiative, national levels. There is a vast array of resources available now that provide information on the various different local, national, and international programs that have been instituted with success and community support [18]. The purpose of this chapter is provide a comprehensive overview of the concept of pharmaceutical take back programs by an examining integral components of the programs such as typical objectives, methods for evaluating success, gaps or weaknesses of many take back programs, and potential or realized obstacles facing take back programs. These concepts will be further explored through the examination of a variety of successful take back programs and their results.

Objectives of Take Back Programs

Take back programs are gaining considerable popularity within communities in the USA and European nations as the public becomes increasingly aware of their individual and collective impact on the environment. However, these programs rarely are single minded in their focus on environmental protection and most often these programs strive to address several other health issues resulting from excess pharmaceuticals. The following section will present a discussion on how some take back programs have expanded their objectives beyond that of protecting the environment from contamination associated with pharmaceutical products to include objectives that address individual and public health issues.

One public health issue that many take back programs have attempted to address is the opportunity for accidental poisoning due to excess and unused pharmaceuticals being stored in the home. The following discussion will focus on the data available on poisonings and how they pertain to take back programs.

The American Association of Poison Control Centers (AAPCC) releases annual reports on the data submitted by local Poison Control Centers (PCC) to the National Poison Data System (NPDS). The NPDS is a valuable tool for researchers, policy makers, and many others as it is the only comprehensive poison surveillance system in the USA and because it collects real-time data from the 61 PCCs across the nation.

According to AAPCC the total number of human exposures to poisons in 2009 was 2,479,355 with the majority of these exposures (82.4%) being unintentional. A particularly revealing trend disclosed in the 2010 report was that the majority of fatalities reported in children 5 years old and younger were unintentional, whereas most fatalities in adults (20 years or older) were intentional. Additionally, of all the human exposures reported in 2009, 93.8% were exposures occurring at a residence. Unfortunately the data are not partitioned to provide a frequency of the types of poisons causing death in these age groups or by the location of incidence (residence or other); however, the report does list the top 25 substance categories associated with the largest number of fatalities. This listing indicates that sedatives, hypnotics, and antipsychotics rank as the number one substance category leading to poison-related fatalities, with cardiovascular drugs, opioids, acetaminophen combinations, and acetaminophen alone following sequentially [19]. As these are all medications, it seems reasonable to presume the fatalities caused by these substances likely occurred in the home. Clearly there is strong evidence to show medication home storage poses a significant poison risk for both children and adults.

The 2010 AAPCC report also revealed that among the 2,043,155 unintentional poisoning in 2009, 276,694 (11.2%) were attributed to therapeutic error and 125,742 (5.1%) were attributed to misuse. Specific therapeutic errors resulting in poisonings included unintentional double-dosing (31.4%), taking or being administered the incorrect medication (14.7%), taking or being given multiple doses within a shorter time period than recommended (9.6%), and accidental exposure to a medication belonging to someone else (9.0%). The AAPCC reports the number of poison-related fatalities in 2009 as 1,158, and although children ranging in age from newborn to under 6 years old were involved in the majority of 2009 exposures, this age group constituted only 1.8% of the poison-related fatalities. The majority of individuals reported as dying as a result of poisoning were between the ages of 20 and 59, with this age group comprising 70.9% of all poison-related fatalities [19].

The summary statistics provided in the 2010 AAPCC report are significant in our review of take back programs for several reasons. To begin with, one may speculate from the therapeutic error statistics presented in the 2010 AAPCC report that the presence of excess or expired medications in the home may be one of the leading causes of misuse of medication, ultimately resulting in avoidable deaths caused by accidental poisoning. Though the statistics presented by the 2010 AAPCC report on the frequency of death due to poisoning for age group do not indicate the type of poison, an argument can still be rationally made that the presence of excess and expired medicines in the home is a threat to all members of the household. This may be especially true for adults between 20 and 59, who may have a tendency to self-diagnose, and for young children, whose curiosity and playfulness can quickly lead to danger if medications in bottles or containers they can open are left in places accessible to them. Additionally, although the fatalities in people aged 60 to over 90 years old only account for 20.4% of the poison-related fatalities [19], concern can still be voiced for this age group as they tend to have more medicines prescribed to them, age-related memory loss is common among this group, and diminishing eyesight may hinder reading small labels on medication bottles.

In addition to reducing the occurrence and opportunity for accidental poisoning and misuse of pharmaceuticals, some take back programs have also been launched to address another public health problem resulting from the presence of excess pharmaceuticals in the home. An alarming trend that is gaining attention in the US media and which is the subject of numerous national studies and programs is the growing popularity of prescription and over-the-counter (OTC) drug abuse among teenagers and young adults. The common term for this dangerous behavior is "pharming." The nonprofit organization The Partnership for a Drug-Free America (Partnership) details in a tracking study that one in five teens (19% or 4.5 million) reports abusing prescription medication to get high and one in ten teens (10% or 2.4 million) reports abusing cough medicine for the same purpose. The abuse of OTC medications and prescription drugs is so prevalent now that the Partnership refers to the current generation of teenagers as Generation Rx [20].

The Partnership's report also reveals that the abuse of prescription and OTC medications is now as prevalent as or more prevalent than illegal drugs such as

Ecstasy, cocaine/crack, methamphetamine, and heroin [20]. This trend is also supported by the 2010 National Survey on Drug Use and Health (NSDUH) which reported nonmedical use of prescription drugs was second in US drug abuse cases only to marijuana. Comparing data from other NSDUH studies dating back to 2002, the current NSDUH highlights that although there has not been a significant overall increase in abuse of prescription pain relievers, there are other troubling signs that indicate there is a significant prescription drug abuse problem in the US. These indicators include increases in individuals dependent on pain relievers, increases in the number of people seeking substance abuse treatment, and increases in emergency room visits attributed to prescription drug abuse [21]. One possible explanation for this new social dilemma is a false perception America's teenagers and other medicine abusers have for the safety of prescribed drugs. The Partnership's study found that two out of five teens believe that prescription drugs are "safer" than illegal drugs even if they are not prescribed to them. Other prevalent misconceptions about prescription drugs held by teenagers surveyed were that there was no harm in occasionally using prescription medication without a prescription and that prescription pain medications, even those not prescribed by a doctor, where not addictive. The Partnership's study also found that teenagers believed that widespread availability and easy access to medications are a leading cause of the "pharming" problem. Three out of five teens reported that they could easily steal prescriptions from their parents' medicine cabinets. Additional views held by teenagers surveyed were that it was easy to acquire other people's medicine or that prescribed pain medicine was widely available, and that prescribed medicines, when purchased illegally, were cheap [20].

Though not specifically reported on by the Partnership, it is reasonable to presume from the youths' survey responses to questions on the accessibility of prescription and OTC medication that there is some degree of black market buying and selling of these medications. Drug abuse, whether it be with pharmaceuticals or illegal drugs, is a trend that has the potential to bring about disastrous future effects for those involved as individuals and our nation as a whole. Drug use and drug peddling are far from the foundations of a productive member of society, and continued participation in these activities threatens much more than the environment or water resources; drug dealing threatens lives. Additionally, beyond the public health issue of American teenagers abusing and potential dealing prescription medications, there is clearly an economic cost to this public health problem in the form of black market sales. Developing an effective method for estimating the economic costs of black market transactions of prescribed and OTC medications could provide very useful information necessary for estimating the additional costs and benefits of take back programs beyond those aimed at reducing environmental risks.

The presence of stored prescription and OTC medications in the home clearly has serious implications beyond the accidental misuse by adults or accidental poisoning of children as discussed previously. In addition to increased efforts to educate both parents and teenagers on the dangers of abusing prescription medications, reducing and/or eliminating teenagers' access to OTC and prescription medications, the source commonly relied upon by teenagers for their drug abuse, could clearly help reduce the problem. Though not a commonly stated objective of take back programs, pharmaceutical take back programs have the opportunity to play a critical role in solving this troubling social drug use trend by providing a safe and effective means for adults to dispose of their unused or unwanted pharmaceuticals that are usually stored in their homes. Though the campaigns and organizations designed for addressing drug abuse among the youth of this nation are not usually thought of in conjunction with pharmaceutical take back programs and environmental awareness and action programs, the potential for these philanthropic and environmental missions to unite to improve the health of our nation's population and environment is promising and could prove to be a very productive partnership.

Beyond providing a possible solution to the public health issues that arise from excess pharmaceuticals in the home, another objective take back program may also address, either purposefully or inadvertently, is improving medicine management strategies through improved knowledge and data necessary to investigate the costs and risks individuals and society are incurring from mismanagement of OTC and prescription medications. The following discussion will highlight how various elements of take back programs can address potential costs or losses in benefits society may be experiencing. Some of these losses may be hidden or unrealized by most people as they may not impact consumers directly or they may not be self-evident. Despite the seeming transparency of these costs and losses of benefits, it is possible that they may be a key factor in estimating the benefits, costs, and risks associated with pharmaceuticals in the environment and water.

One very effective means of investigating hidden costs or losses of benefits resulting from excess pharmaceuticals is to collect and analyze data gathered from pharmaceutical take back events. This has been accomplished in many individual take back programs through participation surveys. These surveys are designed to obtain data necessary to evaluate the overall effectiveness of the take back program and, in some cases, to estimate the various costs or losses of benefits associated with the accumulation of excess pharmaceuticals in homes and/or releases of these products into the environment. Survey questions usually will inquire as to demographic information, name or type (drug category or class) of medication being returned, reason for return of medication, participant satisfaction or perceptions of the take back program, and other event specific information. This type of information, along participation rates for a particular take back event, can then be analyzed to reveal significant statistics and trends in pharmaceutical use and disposal patterns and consumer behavior as it relates to pharmaceuticals. For example, the value of returned medication can be determined by researching the market value of the medication and multiplying it by the unused portion returned. Singularly, this value may only be significant to the person returning the medicine; however, when this method is used to estimate the value of all medications returned at a take back event, it provides valuable insight into the costs incurred by society by wasted pharmaceutical resources. The market value of wasted medications as well as other information that may be available from participation surveys is essential in meeting the objective of investigating costs or losses of benefits that may be burdening society as a result of poor medicine management strategies.

An additional objective that has been included in a handful of take back program is protection of patient privacy. As the majority of prescribed medicines are labeled with information specific to the patient that may be sensitive, simply throwing expired or unused medications in the trash may compromise the personal security of the patient. Just as privacy information can be stolen from mail in household trash, so too can private health information be pilfered from thrown-out medicine bottles. Pharmaceutical take back programs offer security of identity and health information for individuals because medication packaging and bottles are collected and secured from general public access at take back events.

Take back programs are gaining attention in the USA and, as has been discussed, they have far reaching objectives that cover a wide range of public health issues in addition to the goal of environmental protection through proper disposal of pharmaceuticals. However, though the objectives of preventing accidental poisoning, combating teenage medication abuse, improving medication management strategies, and protecting patient privacy are admirable and worthwhile goals, the number of programs that realize the potential to address these issues is still very limited. Yet take back programs as a whole are still in their youth and as more communities, regulators, and social and environmental activists become aware of these programs, it is very likely these individuals and groups will also come to realize that these programs cannot only provide an effective means of achieving safe disposal of medications, but they can also address a variety of public health issues.

Measuring Success

Of the many examples of current and past pharmaceutical take back programs reviewed for this work, relatively few included quantifiable and defined means of measuring the success of the program beyond calculating the amount of prescription and OTC medications collected and possibly participation at a take back event. Another useful but inconsistently reported data set is the monetary value of returned medications. Due to widely varying take back programs, unique in terms of their target audience, participation levels, frequency, duration, and several other factors, it is difficult to compare and evaluate individual programs based on data relating to the volume and value of returned medications and overall participation. However, when this information is available it can be very useful for understanding and assessing, to a limited degree, the success and the outcomes of a particular program.

As an example, Washington state operates its Unwanted Medicine Return Program Pharmaceuticals from Households: A Return Mechanism (PH:ARM) year-round and reports having collected and disposed of 35,000 pounds of pharmaceuticals since 2009 and when it began operating in October of 2006 [22] while the Bay Area Pollution Prevention Group (BAPPG) in Chicago, Illinois, reports collecting over 3 tons of expired and unused medicines between 2004 and 2007 during annual single day events [18]. Though both programs provide the duration and volume of medications collected by their programs, the subtle details such as exactly what they accepted, how many locations they established for collection, and many other unique factors of each program limit the ability to determine if one program was more successful than another and limits the ability of evaluating how successful each individual program was in its own sphere.

Though available information on the success of individual take back programs conducted throughout the USA and abroad is highly variable, a list of several take back programs that have made the details and outcomes of their programs widely available to the public has been compiled in Appendix A. This appendix provides a review of many current and past take back programs within the USA and other nations and, when available, information on the outcome of the project (e.g., volume collected, value of medications collected, and cost of program). Perhaps as the pressing need for take back programs becomes more apparent and the concept gains national momentum as a means for managing excess and unused pharmaceuticals, more programs will begin to realize the importance of identifying and quantifying measures of success for their programs. Examples of some questions take back program operators may ask themselves as they expand the scope of evaluating the success of their programs include: "Are my program's marketing and advertising methods reaching my target audience?;" "Is there an especially needful subpopulation within our area that is not participating due to lack of knowledge of the program?;" and "Are participants of our program better educated about the issues surrounding excess and unused medications following their participation in the program?" Improved measures of success would be very helpful to take back programs because if definitive measures of success are not identified it is difficult to assess where improvement can be made so that a program can be modified or expanded to be more effective and worthwhile.

Identifying Gaps or Weaknesses in Take Back Programs

Although quantifying the volume and value of medications collected does provide some indication as to the success of a program, the key motivating factors driving many take back programs can include various other reasons beyond collecting unused and excess medications. These motivating factors include issues discussed previously such as avoiding accidental poisonings and abuse of prescription drugs. While these endeavors are assuredly beneficial, the authors believe the true potential of take back programs has yet to be fully realized due to several gaps or weaknesses in the basic model of take back programs. This basic model consists of organized take back events that simply collect and dispose of medications. Beyond this, some take back programs survey participants, however it does not appear, based on an intensive literature review, that many take back programs critically analyze and report their findings from these surveys. As will be discussed, take back programs could be enhanced by expanding their program scope to address some of the gaps in the program model.

Scientific Justification

One significant gap of take back programs is that of scientific justification. The question that still remains after innumerable take back programs have been conducted is do these programs improve or mitigate the current state of the water resources and the environment in regard to pharmaceutical wastes and residues? Furthermore, are there certain classes of medications that are more harmful to the environment than others and therefore should be targeted as key classes of medications during take back events?

One promising method that may be used to estimate potential harmful impacts of pharmaceuticals to fish was developed by a team of researchers at Pfizer Global Research and Development. Their model is based on the fact that there are similar enzyme and receptor systems in both fish and mammals. Based on these similarities, the model can use existing data from toxicological and pharmacological studies on mammals to predict pharmacological responses in fish. The model compares the measured human therapeutic plasma concentration of a medication (H_TPC) to the predicted steady state plasma concentration (F_{ss}PC) in fish with the result being an effect ratio (ER). In this model there is an inverse relationship between the ER and the potential for a pharmacological response in fish, meaning the lower the ER the greater the likelihood there will be a pharmacological response in fish and thus the more likely that additional testing needs to be conducted to determine if the medication poses a threat to fish [23, 24]. As it can be time consuming and very difficult to obtain data on environmental responses to all of the human pharmaceuticals on the market, this model overcomes that hurdle by capitalizing on the vast amount of mammalian pharmacological data available and applying it in a new and innovative way. Using this model, it may be possible to determine if a certain class of pharmaceuticals poses a greater threat to fish than another class of pharmaceuticals. The information gained from application of this model could make take back programs more efficient and cost effective because it could educate take back program organizers as to which medications pose the greatest environmental risks and therefore should be the ones they target for return and proper disposal.

The presence, potential impacts, and known detrimental impacts of pharmaceuticals in aquatic environments have been researched and publicized by the professional science community as well as the mainstream media. However, despite the widespread coverage of this, to date, the authors are unaware of any research projects or other efforts designed to investigate if water quality or aquatic life in areas instituting take back programs has improved as a result of the program. One worthwhile option for filling this gap of information would be to conduct biological and chemical monitoring of streams and other receiving water bodies in areas where take back programs are in place. Biological monitoring, or biomonitoring as it is often referred to as, is a well-established environmental monitoring method which relies on the use of living organisms as sensors for water quality surveillance. Chemical monitoring utilizes proven water quality techniques and instruments to measure the chemical characteristics of the water which may be altered by pharmaceutical loads. Biological and chemical monitoring can provide the critical data necessary to determine if water quality and aquatic life have improved due to a presumable decrease in the pharmaceutical load released into the environment as a result of a take back program. Unfortunately to the knowledge of the authors, the incorporation of this type of analytical investigation before, during, and after a take back event to provide scientific justification for these programs has not been attempted or even suggested by any take back program.

Risk Perception

Risk perception is generally regarded as the intuitive assessments of risks people face based on a variety of information sources, personal experiences, and an assortment of other contributing factors. Risk perceptions are naturally developed by individuals; however, collectively groups may also cohesively form their own distinct risk perceptions, giving rise to the concept of public risk perception. Though the relationship between risk perception and behavior is very complicated, it has been shown that subjective risk perceptions may influence the actions of individuals. However despite the seemingly obvious impact risk perception is likely to have a pharmaceutical disposal behavior, the authors are unaware of any take back program incorporating risk perception into their program. To address this weakness in take back programs, this section will present some of the key concepts and areas of interest of risk perception, discuss significant research findings in this field, and discuss the applicability of these concepts and findings to pharmaceutical take back programs.

One reoccurring criticism of industry, government, scientists, and other experts with professional knowledge of risks is the wide discrepancy that exists between their objective assessments of risk and the general public's perception of risk. This disparity may be attributable in part to the complexity of factors and inputs that individuals process as they develop their personal risk perceptions. Influential factors may include such characteristics as the control individuals have over the risk, their willingness to engage or be subject to the risk, and their knowledge or understanding of the risk [25, 26]. Other inputs that influence risk perceptions include, but are not limited to, probabilities, biased news reports, confusing personal experiences, and apprehension over gambles encountered in daily life. The complex nature of these factors and inputs can confuse individuals and cause them to deny uncertainty, to over- or underestimate risks, and to assertively hold opinions relating to risks that they are unable to defend [27, 28].

Another possible cause for the disparity between public risk perception and experts' objective risk assessments is that there seem to be varying definitions of risk. Studies have shown that lay individuals vary in how they define risks. When defining risks, lay individuals may include the rich array of risk characteristics that they relied upon for the formation of their risk perceptions. In addition to those previously mentioned, these risk characteristics may include concepts such as the potential for catastrophe and the potential for impacts to future generations. This is a dramatic departure from experts' definition of risk which typically relies on a single, defined, quantifiable measure of risk such as annual fatalities [25]. Due to the complex nature of risk perception it is unlikely that there is a single explanation for the cause of the divergence in lay individuals' subjective risk perception and experts' objective assessments of risk; however, given the current knowledge of risk perception formation, it is likely that the inconsistent definitions of risk and the complexity of variables which influence risk perceptions are contributing factors to this anomaly.

Another characteristic of risk perception that has been consistently observed in risk research studies is that the lay public generally believes the current level of risk for most activities is objectionably high; indicating the perceived level of risk is beyond that of the desired level of risk and that regulatory efforts to maintain these risks are not producing acceptable results. Despite the public being dissatisfied with the level of risk management provided by regulatory mechanisms, research shows that people are willing to accept higher levels of risk for activities they may consider beneficial. Some of the characteristics shown to influence this trade-off of high risk for perceived benefit include voluntariness, familiarity, control, potential for catastrophe, knowledge of the risk, and fairness; however, no single characteristic has proved to be the sole determining factor [29, 30]. These findings are significant because government officials and other professionals involved in managing hazards could very likely increase the overall effectiveness of their risk management strategies if they addressed some of these factors in their public policies and educational campaigns.

In addition to investigating factors influencing risk perception and possible explanations for the divergence between subjective and objective risk assessments, risk research has also focused on how risk perceptions influence behavior. However despite the seemingly obvious connection between risk perception and behavior and the wide array of studies that have investigated the relationship between these two elements, for many reasons results from these studies vary widely and are difficult to compare. One possible explanation for this nonconformity of results is that most studies focus their efforts on a single risk and due to the unique characteristic of individual risks, the risk perceptions and subsequent behavioral responses to specific risks are difficult to compare across a wide spectrum of different risks. Investigation methods also vary widely which can make it difficult to compare studies and results. Another possible explanation for variation in results of these studies is that individuals usually have multiple motivations for engaging in certain actions, with risk perception being only one of many contributing factors. Regardless of the reason behind the variation in results, there are several recent risk perception studies that provide valuable insight into the complex relationship between risk perception and behavior and which may be very useful when considering the potential impact risk perception may have on pharmaceutical disposal behavior and participation in pharmaceutical take back programs.

For an example of a study investigating risk perception and behavior consider a United Kingdom (UK) survey of students which found that knowledge and risk perception had very little influence over behavior [31]. It is however important to note that this study was limited to a very selective subpopulation of university students with health or environmental backgrounds at two universities in the UK and the majority of risks respondents were questioned on were voluntary risks that have specific and direct consequences to the person engaging in the behavior (e.g., smoking, alcohol use, and illegal drugs). Similarly, a study investigating risk perception and smoking behavior in Swedish teenagers found that risk perception of lung cancer did not affect the number of cigarettes smoked [32]. This finding entirely contradicts another smoking study that found risk perception of lung cancer to be a

significant factor in the number of cigarettes smoked [33]. The Swedish study specu-
lates that limiting the study to lung cancer risks may have introduced a bias that
would explain this divergence [32]. The lack of evidence to support a link between
risk perception and behavior in the UK and Swedish studies may also be because
these studies examined risks where the individuals' choices had a direct impact on
their health and it is quite possible that human and environmental health risks
resulting from excess and unused pharmaceuticals may be very different due to the
greater separation between the behavior and consequence and due to the appeal for
social and environmental responsibility on the part of the individual [34].

To demonstrate the vastly different conclusions of some risk research studies
consider also the far different conclusion of Jakus et al. [2009] compared to the
previously mentioned survey of UK students and the Swedish smoking survey.
Jakus et al. [2009] found that for respondents living in areas with arsenic-contami-
nated drinking water, perceived risk was a statistically significant factor in the
decision of how much bottled water to purchase [35]. Another significant factor in
risk decision making may be an individual's knowledge of a particular problem.
A Swedish study investigating the concept of environmental awareness as a factor
in decision making found that individuals accounted for environmental factors
such as air pollution when making decisions about personal car use [36]. Though
the risks of air pollution and excess and unused pharmaceuticals may be different in
many ways, similar characteristics between these two risks such as delayed effects,
noncatastrophic effects, and potential impacts to future generations may make the
findings of these studies very applicable to the public health and environmental
problem of improper pharmaceutical disposal behavior. Reviewing the studies
highlighted here, it seems that one significant factor affecting the strength of the
relationship between risk perception and behavior may be the specific risk itself. As
such, take back programs aspiring to address the weakness of including risk perception in their program structure would be wise to consider studies addressing the
risks associated with the improper disposal or access to excess and unused pharmaceuticals or studies of risks with similar characteristics.

One study is particularly applicable to addressing the risk perception weakness
in the majority of take back programs. Bound et al. [2006] investigated the relationship between choice of disposal method for pharmaceuticals and risk perception
and found that there was no definite correlation between these two factors. The
authors of this study speculated that respondents may not feel that the risks posed
by pharmaceuticals in the environment are great enough or their choice of disposal
method was significant enough to make a difference, thus the individuals surveyed
did not have enough incentive to change their disposal behavior. Despite the fact
that a direct link between disposal behavior and risk perception was not found, this
survey did reveal very interesting risk perceptions about the impact of pharmaceuticals on both personal health and the environment which may provide valuable
insight for future take back programs. For example, the majority of survey participants indicated they strongly agreed or simply agreed that pharmaceuticals used
inappropriately could be potentially harmful to their personal health and more than
half agreed that improper disposal of pharmaceuticals could threaten fish or plants.

There was also a high degree of uncertainty associated with the impacts to the environment. This was speculated to be because survey participants were likely less knowledgeable about toxicology and environmental processes. Respondents also indicated they perceived a lower threat from medications that were more familiar to them, such as pain killers and antihistamines, compared to less well-known medications such as antiepileptic medication and lipid regulators. OTC medications and commonly prescribed drugs are often viewed as less potent and therefore less of a risk to the environment. This is very likely because familiarity tends to make risks either more acceptable to individuals and/or causes individuals to underestimate risks [34]. The perception that nonprescription drugs are less potent and therefore less harmful to the environment was also evident in a Canadian study that found that the percentage of respondents who believed OTC pharmaceuticals needed an appropriate disposal method was 81% while the percentage believing that unused and expired prescribed pharmaceuticals needed to be disposed of in an appropriate way was 90% [34, 37].

Though it may be difficult to draw a definitive link between risk perception in regard to excess and unused pharmaceuticals and choice of disposal method, it is important to consider what other motivations may influence disposal behavior. As there have been innumerable take back programs launched, it is clear that there are effective motivations for individuals to participate. In many cases, individuals often participate in environmental stewardship events, and other activities that are altruistic in nature, because they can see and understand the benefit of their individual efforts [34, 38]. Individuals may derive personal satisfaction from participating in such environmentally conscious programs like take back programs or they may have a desire make a small personal change in their behavior for the health of the public, themselves, their children, or the environment. Some experts suggest that perhaps the risk to the environment is not well understood by the general public or compared to other risks such as air pollution and financial instability, the risk posed by improper disposal of the pharmaceuticals is not great enough to merit a change in disposal behavior [34]. Additionally, when contemplating participating in programs such as pharmaceutical take back events, concerns about issues people recognize and understand better, such as their own health or public health, may be stronger motivators than environmental problems, which they may not understand well or acknowledge as a significant problem.

As has been illustrated here, risk perception is a complicated concept, shifting and changing with specific risks and not entirely understood yet. However the complexity of risk perception is not a viable reason for excluding it as a key element in take back programs. Due to the need to motivate local citizens to participate in a take back program, the inclusion of risk perception in these programs has the potential to make significant improvements in the success of an individual program. To bridge this gap therefore it would be advisable for take back programs to evaluate motivations of individuals participating in the program, investigate risk perceptions of their target audience, and then design their programs to account for the specific risks their target audience perceives, whether these risks be environmental, social, or a combination of many factors.

Risk Communication and Education

The well-understood purpose of risk communication is to provide individuals with the information necessary for them to reach informed decisions about risks they may encounter relating to their health, safety, or the environment [39–45]. Though a seemingly simple prospect, risk communication and education is complicated by the fact that, in addition to requiring a comprehensive knowledge of the risks the public faces, officials responsible for protecting public and environmental health and safety must also effectively communicate and educate the public while respecting the delicate relationship between risk perception, objective assessments, and the publics' response to risks. While some take back programs have made efforts to educate and communicate with their target audiences, it seems the majority of programs do not fully appreciate the critical role risk communication and education can play in the success of the program. To address this common gap in take back programs and in an effort to stress how take back programs could greatly benefit from more fully incorporating risk communication and education into their programs this section will highlight some of the leading strategies and intrinsic challenges of risk communication and education.

Effectively communicating a complex risk message requires advanced planning and a well-considered strategy. One risk communication strategy recommended by a group of experts is a straightforward four-step method largely based on what Morgan et al. [1992] described as the "mental model" approach, which is centered around the concept that individuals assess new information based on their existing knowledge and/or beliefs [46]. For example, information provided for a new topic for which individuals have no prior experience or knowledge will likely be difficult for them to understand and apply context to. Alternatively, individuals with preexisting misconceptions or erroneous information may misinterpret correct risk communication and education messages. A significant amount of research has been devoted to mental models and has revealed the important influence they have on how individuals develop skills, follow directions, and use equipment [47–53]. This concept bares striking familiarity to the risk perception concept that behaviors and actions are influenced by individuals' personal experience and prior understanding of risks.

With such a wide base of research indicating the critical role individuals' understanding and beliefs play in determining their actions, the first step for an effective risk communication and education program would logically be to determine the existing knowledge and beliefs of the target audience. The four-step method recommends accomplishing this important task by providing open-ended opportunities for individuals to express their knowledge and beliefs in regard to a particular topic. Additionally, experts stress that it is equally as important to extract both accurate and inaccurate or misguided beliefs and knowledge from their audience [46]. Focus groups, interviews, city council meetings, and a variety of other venues or methods may be employed by risk communicators and educators as they seek to understand their audience and satisfy this primary objective of risk communication and education.

Once a model of the audience's beliefs and understanding has been constructed, the next recommended step is to incorporate this information into structured surveys to evaluate how prevalent the beliefs and knowledge are within the audience. Building upon the well-structured model of the target audience's knowledge and beliefs, the final two steps in the recommend risk communication strategy are to develop and continually evaluate communication methods [46]. Methods may include distribution of educational material, broadcasting messages, and establishing an informational website. Communication methods utilized should be carefully considered in light of the objectives of the communication campaign and the characteristics of the audience. Effectively designed informational material that accounts for these important variables may be able to clarify skewed or inaccurate beliefs of lay individuals by providing the additional accurate facts necessary for them to improve and/or refine the knowledge and beliefs they currently maintain concerning the risk [46, 54].

Another model that can be applied for developing effective risk communication and education strategies is the physician–patient model of communication. In both environmental risk communication and physician–patient relationships experts provide objective information regarding facts on health, safety, uncertainties, and alternatives. Additionally, experts also are often required to convey their knowledge concerning the severity of the issue, possible alternative options, resolutions, and their advice on how to proceed forward with managing the issue. With these similarities between the two communication processes, it seems logical that some recommendations and approaches used in the physician–patient model would be appropriate for environmental risk communication [55]. One source of such recommendations is a presidential commission report on ethical problems in medicine and biomedical research issued in 1982. Though initially intended for improving health care decision making, this report provides three practical recommendations that are quite applicable for developing and improving environmental risk communication strategies. These recommendations are: approaching risk communication as a dialogue rather than an isolated occurrence, providing additional sources of information, and providing clarification on types of uncertainties associated with the risk information [55, 56].

Due to the complex nature of risks and the complications involved in educating the public about them, it should not be unexpected that experts in risk communication and education must contend with a suite of challenges as they work to provide accurate and appropriate amounts of information to the public. One ever-present challenge risk communicators and educators face is establishing and maintaining trust and credibility with the public [57]. Another challenge is accounting for the community structure and diversity of the audience. Specifically, concepts generally associated with environmental justice such as culture, economics, and life experiences of a community have been identified by experts as factors that should be kept in mind during communication efforts as these factors can lead to health and opportunity disparities between different communities [58]. Another community-based challenge of risk communication and education is overcoming the difficulty of rallying individuals to support public interests as equally as they support their individual interests. In response to this challenge, experts note that communication efforts are most effective when individuals are united as a community [58].

However, what may be the most challenging task associated with addressing community components and dynamics may be the fact that these components and dynamics are unique to each community, requiring risk communicators and educators to constantly reevaluate their message for applicability and appropriateness for the audience.

Despite the challenges associated with developing an effective risk communication and education strategy, there are a wealth of public health campaigns and environmental risk communication messages that have been launched in the past that can testify as to the effectiveness of properly planned and executed risk messages. As an example, consider the current pervasiveness of knowledge among the general public for concepts such as the importance of wearing a safety belt, the existence of global climate change, or the risks of drinking alcoholic beverages while pregnant. These risk communication successes indicate that while the task of providing critical and complex scientific information to the public concerning excess and unused pharmaceuticals in the home and the environment may initially appear to be daunting, incorporation of proven strategies and models of risk communication and education, and recognition and planning for the challenges that will inevitably confront the campaign, will very likely significantly improve the overall effectiveness of the campaign in educating the public and providing them with the information they need to make informed decisions regarding their health, safety, and the environment.

Improving Medication Management Strategies

A final powerful idea that take back programs as a whole have generally overlooked is the potential to impact healthcare costs over the long term through improving medication management strategies. As previously mentioned, data gathered from surveys completed during take back programs can yield information needed to determine the value of unused medication and the amounts and types of collected unused medications. What remains to be seen is if this information, which may not be currently available by any other means, can lead to a reduction in healthcare costs through improved use of the medical resource of pharmaceuticals.

Though the economic value of wasted pharmaceuticals is important information, unless this information is communicated to doctors, pharmacists, healthcare officials, and regulators, the value of this information is not achieving its full potential. If communicated and acted upon properly, this information could have powerful implications to improve current practices of healthcare providers and thereby possibly improve healthcare as a whole. For example, by understanding which types of medications are being wasted, it may be possible to determine if certain medications are being overprescribed, and if so, the value of these wasted resources may be used as a factor to help determine more efficient prescribing practices in an effort to reduce the pharmaceutical waste from the beginning of the chain of consumption. Improving prescribing practices has the potential to reduce the amount of a prescription drug wasted as valuable healthcare resource, which may ultimately be improperly disposed of, reduce the time a doctor spends prescribing medicine that goes unused, and/or reduce the time a patient spends visiting a doctor to receive a prescription for which a portion of the drug may go unused. As we consider the further reaching implications of reducing pharmaceuticals in the environment by implementing pharmaceutical take back programs, it is possible to also anticipate addition benefits society may gain in the form of improved medicine management strategies, which may also lead to a reduction in the time and money spent dealing with the consequences of non-use and misuse of prescription and OTC medications.

A scientific investigation into the value of wasted pharmaceutical resources and the economic and social implications resulting from this wasted resource was recently conducted in Barcelona, Spain. In this study 38 randomly selected pharmacies were surveyed as they accepted returned medicines, medical care equipment, and other items available at a pharmacy (e.g., personal care products and nutrition products). As background information, in 2002 communities and the pharmaceutical industry in Spain collaborated to develop an industry-funded program called SIGRE to facilitate the collection and disposal of unused and expired medications and medication packaging. In the Barcelona study, selected pharmacies were surveyed for seven consecutive working days with the survey period beginning on the first day a return was made. During the survey period, which spanned from February to April of 2005, 1,176 packages of medicine were returned. Due to missing information and other complications associated with determining the volume or amount of a medication remaining in a returned package, the value of the returned drugs was based on 1,119 packages and came to €8,539.90, which at the exchange rate in April of 2012 of 1€ to 1.32 US dollars equaled $11,303.80. Researchers determined that 75% of this cost or €6,463.90 ($8,549.86) was covered by the public health care system [59]. The researchers from this study note that the returns they valued during the event represent a significant unnecessary expense to the healthcare system of Spain which may be addressed in the future by improvements in prescribing, dispensing, and use of medicines in Spain [60]. Although Spain has a different national healthcare policy than the USA, which would account for differences in the value of returned medicines that would be covered under the US healthcare system, this study is still applicable to take back programs in the USA because it illustrates that medications returned equate to wasted medical resources, which ultimately are unaccounted losses in the healthcare system's budget. Additionally, the solutions of improving prescription, dispensing, and consumption practices this study puts forth to address the problem of wasting medicine resources could also be applied in the USA.

Though the concept of improving medicine management strategies through pharmaceutical take back programs is not a widely circulating concept as of yet, there is one broad-based initiative in the USA gaining support within the healthcare provider community. Practice Greenhealth™ was originally established in 2004 by an EPA grant as Hospitals for a Healthy Environment (H2E) and now is a vast network of member healthcare institutions and organizations that are committed to environmentally sustainable healthcare practices. The original purpose of H2E was to establish a national program devoted to promoting and developing environmental sustainable practices for the healthcare system which would serve to improve efficiency, health, and regulatory compliance within individual communities. As part of this program Practice Greenhealth™ has developed a comprehensive management plan to reduce pharmaceutical waste on a national level. Though not a take back program in itself, Practice Greenhealth™ recognizes that significant costs and risks are associated with disposing of pharmaceutical wastes, and as a part of the solution to this problem, Practice Greenhealth™ provides healthcare professionals with education, information, and resources on environmental management strategies that address this concern as well as other environmental concerns related to healthcare [18, 61]. With the resources, tools, and contacts within the Practice Greenhealth™ network, there seems to be the potential for a mutually beneficial relationship between local take back programs and Practice Greenhealth™ that could truly help to promote the objective of improving healthcare costs and medication management strategies.

Another example of a North American program that acknowledges the fundamental connection between take back programs and improving medicine management strategies is the Canadian-British Columbia Medication Return Program (MRP). MRP, originally named British Columbia EnviRx, was established in 1996 as a voluntary program; however, it was the pharmaceutical industry itself that eventually lobbied for mandatory product stewardship for pharmaceuticals through the establishment of the Post-Consumer Residuals Stewardship Regulation. MRP's primary objective is to protect and improve the health of the environment, economy, and consumers. MRP is supported by Canada's National Association of Pharmacy Regulatory Authorities (NAPRA) due to the program's commitment to improving consumer and child safety, reducing costs, improving therapy treatment results, and preventing detrimental impacts to the environment. One way MRP advances the cause of reducing pharmaceutical costs is through promoting the prescription and distribution of manageable medication amounts that can be completed by the patient [18]. This strategy could just as easily be incorporated into the US healthcare system to reduce pharmaceutical costs, wastes, and other problems associated with excess pharmaceuticals.

While take back programs themselves are a relatively new concept, just emerging within the last decade or so, the idea that these programs could be used to improve medicine management strategies through reducing wasted medications is an even more novel concept with few if any US take back programs mentioning this as an objective. However with the expanding interest and popularity in promoting environmentally and economically sustainable practices into more business practices and industry standards, it seems logical that the cost savings potentially available through improving medicine management strategies and medical practices will become increasingly obvious to healthcare providers, industry leaders, and regulatory leaders. Unfortunately, the degree to which healthcare costs could be reduced due to take back programs would probably remain unanswered for some time, as there will likely be considerable lag time before improved efficiency with medication management is translated into actual cost savings in the healthcare industry. Additionally, in order for these improvements in healthcare and prescription practices to be realized, an effective communication strategy must be established to educate doctors, pharmacists, and others involved in medicine on the implications of wasted pharmaceuticals and the data that are revealed through take back program surveys.

Potential Roadblocks

There is yet another facet of pharmaceutical take back programs that should be considered to complete the overview of these programs. Though these programs may provide valuable services to the community and the environment, organizers of these programs may struggle against a variety of barriers as they proceed with the planning and implementation process. The following provides brief examples of some obstacles take back programs have encountered in the past and potential solutions to these problems.

The first and possibly most obvious roadblock for these programs is lack of adequate funding. Cost is not featured as one of the program attributes in the table in Appendix A, which features a comprehensive list of example take back programs, because this information is not reported on a consistent basis and when it is reported it is usually not provided in a format that accommodates comparison between programs. For those programs that do report their costs, it seems most appropriate to consider each on a case-by-case basis due to the wide variances in program characteristics. For example, the Washington State PH:ARM state-wide continuous drop-off program reports the estimated cost of the project as a lump sum of $3.3 million, whereas the La Crosse, Wisconsin continuous drop-off program, like many other programs, reports the costs of select components of the program without an annual estimate for total program operation. Examples of program components reported include: cost of disposal of medicine waste, advertising, general staffing, security, and time and services of pharmacists and law enforcement officials. In many of the example take back programs featured in Appendix A, local pharmacies, public works departments, and businesses donated their time and services for take back programs operating within their communities. Many programs also received direct funding from federal, state, or local government programs and community organizations.

Even when programs secure adequate funding, the best of efforts can be thwarted unexpectedly by local, state, and federal laws, regulations, and ordinances. The most common legal obstacle encountered by take back programs is the Controlled Substances Act (CSA), which is administered by the DEA; however, with the enactment of the Safe and Secure Drug Disposal Act of 2010 individuals who have legal possession of controlled medications will likely encounter less hurdles as they seek to properly dispose of these medications. Medications considered to be controlled substances by the CSA include narcotics and other medications like Valium, amphetamines, Ritalin, morphine, methadone, and oxycodone. Prior to the Safe and Secure Drug Disposal Act federal law mandated that once controlled substances were prescribed to the patient the only individuals who could maintain possession of them were the patient and law enforcement officers. This restriction required take back programs wanting to include collection of controlled substances at their events to either have law enforcement officials present at collection events to take possession of controlled substances or coordinate with police stations and sheriffs' offices to allow citizens to bring controlled substances to these facilities for proper disposal. Now the law states that individuals who legally obtain controlled medications can dispose of them through agents authorized to collect and dispose of medication, so long as the disposal process is in accordance with regulations established by the US Attorney General [62]. However at the time of this writing the regulations required to be set forth by the US Attorney General had yet to be established or announced, so there is still some uncertainty as to the procedures future take back programs will need to follow to be in compliance with the CSA and the Safe and Secure Drug Disposal Act. Given the security and safety issues surrounding controlled medications, it is anticipated these rules will include some stringent requirements to prevent diversion of controlled medications being collected at take back events.

Finally, though public awareness and support may not be an initial hurdle to overcome for organizing and launching a take back program, community involvement with both individual citizens and community groups, such as businesses and public works programs, will greatly determine the success of the program. Education and risk communication play key roles in take back programs, because if citizens are either unaware of the risks or unaware of the opportunity to dispose of their excess medicines, the program will not fulfill its basic goal of ensuring proper disposal of pharmaceuticals and protecting the health of individuals and the environment. Though there are many factors which may influence the success of a particular take back program, a well-prepared and implemented education campaign which informs citizens about the risks associated with excess pharmaceuticals in the home, the impact of pharmaceuticals on the natural environment, and the details of how to participate in their local take back program will likely prove invaluable in increasing citizen participation and mustering community support.

Pharmaceutical Take Back Program Case Studies

Specific details of individual take back programs vary depending on such factors as the resources, goals, and policies of the entity organizing the program or the area in which the program will operate; however, the overall structure and operation of most take back programs are strikingly similar. In general, citizens are given the opportunity to return unused or unwanted medications to a collection center. The most common location for collection centers is a local participating pharmacy; however, hospitals, general practitioners' offices, and local police stations have also served as host locations for take back programs. Medications collected are often separated from their bottles and the pills, tablets, or other forms of the medication are placed of in a clearly marked container. The collection container may contain a deactivating liquid which renders the medications useless. This serves as a precautionary measure to ensure that in the event of a breach of security no medicinally active medications would be retrievable from the container. Filled collection containers are released to a contracted certified hazardous waste manager for disposal, which is usually accomplished via incineration. Pharmaceutical take back programs may be stand alone collections operated as 1-day events or they can be operated on a more frequent basis such as seasonally or year-round. Pharmaceutical take back programs have also been launched in conjunction with household hazardous waste collection events sponsored by local agencies or government service departments such as sanitation or environmental services.

Though there are many examples of successful take back programs, this work is not intended to provide an exhaustive review of all take back programs, but rather to highlight unique features and successes of a variety of programs. Here we provide a review of one widely publicized and documented program. This program was selected for this review due to the wide availability of information on the program and the innovative ideas integrated into the program in its efforts to promote the concept of sustainable medicine. A brief summary of additional programs in the form of a table is featured in Appendix A.

The Teleosis Institute Green Pharmacy Program (Teleosis) is one example of a pharmaceutical take back program that has experienced dramatic success. The Teleosis Institute is an organization dedicated to promoting sustainability and environmental stewardship within the healthcare industry. As a continuation of these ideals, Teleosis operated their take back program as a pilot program in Berkeley, CA, for 1 year between June 1, 2007 and June 1, 2008, partnering with a local pharmacy—Elephant Pharm. Drop-off locations for the Green Pharmacy Program were located at participating pharmacies, dentist offices, an animal hospital, the Teleosis Institute, and a healthcare facility. The Green Pharmacy Program was designed as a product stewardship model take back program in which all participants involved in pharmaceutical products, from the manufacturers, to healthcare providers, retailers, patients, and finally those who dispose of the products, were brought together as partners, equally responsible for ensuring the products were safely disposed of to reduce the impacts of pharmaceuticals on the environment [63].

With the ultimate goal of promoting and encouraging the adoption of the tenants of sustainable medicine, this program developed some very unique features. Incorporation of some of these sustainability guided program features has the potential to enhance other take back programs aimed at reducing environmental and human health risks associated with excess pharmaceuticals in the home. For example, pills collected by the Green Pharmacy Program were removed from their bottles and incinerated, while the bottles were shredded and recycled to protect patient information. The program attested that the separate disposal method had the advantages of significantly reducing both the environmental impact and cost of incineration, as fewer materials need to be incinerated. An additional advantage the program touted was that their separate disposal method was more attractive to participants because it protected their sensitive medical information [64]. The Green Pharmacy Program also incorporated a public education campaign that provided information on the take back program, proper disposal methods of pharmaceuticals, and the impact of pharmaceuticals on the environment [63].

Table 1 Top 10 Categories of pharmaceuticals returned in the Green Pharmacy Program

Category of pharmaceutical	Percent
Central nervous system (CSN)	22.62
Nutritional products	14.29
Psychotherapeutic	12.51
Gastrointestinal	8.99
Cardiovascular	8.77
Respiratory	6.00
Anti-infectives	6.00
Alternative medicines	5.69
Hormones	4.60
Immunologic	2.85

Table 2 Top 10 brand name or generic products returned in the Green Pharmacy Program

Name	Category of pharmaceutical
Acetaminophen	Analgesic, antipyretic
Aspirin	Analgesic
Tylenol	Analgesic, antipyretic
Vitamin E	Supplement
Prednisone	Corticosteroid/steroid
Ibuprofen	Nonsteroidal anti-inflammatory (NSAID)
Warfarin	Anticoagulant
Topamax	Anticonvulsant
Etodolac	NSAID
Gabapentin	Anticonvulsant

To facilitate future analysis of the program, data on all medicines collected were recorded in a national registry. This registry is known as the Unused and Expired Medicine Registry and was developed by The Community Medical Foundation for Patient Safety. Data collected was used to determine which category of medications were the most overprescribed or unused medicines, which medications were most often returned, the monetary value of the returned medicines, and to estimate the environmental impacts of the returned pharmaceuticals. Preliminary results of the Green Pharmacy Program from its inception on June 1, 2007 to December 31, 2007 indicated that a total of 690 pounds of medicines were returned through the program with an estimate of 101,359 returned pills, capsules, and tablets. The total wholesale value of these medicines was estimated to be $159,778 and total retail value was estimated to be between $228,254 and $399,445. The majority of returned medicines (60.43%) were prescriptions as opposed to OTC medicines (39.14%). Tables 1 and 2 provide the ten most frequently returned class or category of pharmaceuticals and the brand or generic name of medicines returned, respectively [63].

Although summary statistics were provided in the Teleosis' Preliminary Data Report and their *Green Pharmacy Final Report*, published in fall of 2008, no data, summary statistics, or other results are readily available from Teleosis to explain their results or findings aimed at achieving their final goal of the estimating environmental impact of returned medicines. Additionally, the Unused and Expired Medicine Registry website is still in progress and as of this writing does not have publicly available data to share with the community [65].

The Green Pharmacy Program is just one example of a successful take back program and it should be noted that the concept of pharmaceutical take back programs has been embraced by many communities and there are many exceptional and successful programs both here in the USA and abroad. Although complete summaries of all the existing and past take back programs would be informative, that extensive of a review is not necessary for understanding significant role these programs can play in potentially reducing the amount of pharmaceuticals in the environment. Rather a more efficient method of providing a comprehensive overview of the breadth of take back programs would be to examine a selection of past and existing programs instituted across the USA and in foreign nations and to consider a selection of key elements comparable between the programs. Appendix A provides a list of example programs and their comparable or distinguishable features. This appendix includes information from the previously discussed program as well as information adapted from the Illinois-Indian Sea Grant (IISG) resource guide designed to assist communities in managing programs aimed at the proper disposal of their unused pharmaceutical [18].

Though the Internet provides a vast array of information on pharmaceutical take back programs both in the USA and worldwide, this information can be regarded mainly as general in character and lacking the scientific analysis that is necessary for a deeper understanding of these programs and the possible implications they may have on the environment, society, government, the pharmaceutical industry, and other groups involved either directly or indirectly with pharmaceutical manufacturing, distribution, consumption, and disposal. One peer-reviewed article that touches on the environmental and social problems that prompted the development of take back programs and that conducts a detailed analysis of the outcomes of a pharmaceutical take back program launched in the UK is "An analysis of returned medicines in primary care", published in *Pharmacy World and Science* in 2005 by Langley et al. This study is unique in that it touches on many of the subjects previously mentioned in the "Objectives" and "Idendifying Gaps" sections of this chapter.

As a scholarly introduction into the concept of take back programs, the authors of this article provide a discussion of some of the potential impacts of unused and surplus medications, many of which were discussed at length previously in this chapter. These potential impacts include such things as the minimization for accidental poisoning and medicine misuse, prevention of detrimental effects on the environment, and accounting for costs associated with wasting a resource. The authors of this study also point out that although take back programs provide a potentially effective method for reducing or eliminating the risk of accidental poisoning or misuse and detrimental environmental impacts, these programs do not directly address the need to reduce the amount of pharmaceutical resources that are being wasted. As a budding solution to this problem, this study was launched to investigate the type and quantity of returned pharmaceuticals and the reasons given for their return.

To collect the necessary data on pharmaceutical returns and reasons for returns, two 4-week long take back programs were arranged in East Birmingham in the UK; one during August 2001 for returns to pharmacies and one in March 2002 for returns to general practitioners' (GP) offices. Returned medicines were cataloged by therapeutic category and the number of doses remaining and information concerning the person returning the medicine including age, gender, and reason for the return were collected by a pharmacist or GP. Where appropriate, additional information from patient notes or pharmacy records was acquired. The study made no efforts to advertise the take back program within the community. During the two 4-week collection events there were 114 returns totaling 340 items. The majority of returns both in number of returns (90 of 114 or 78.9%) and total items returned (298 of 340 or 87.6%) came from the pharmacy collection event as opposed to the GP collection event. A change in doctors' prescription orders was the reason given for return of the majority of medicines. Other reasons cited included, in order of frequency, clean-out of excess home medicine supplies, clean-out following a patient's death, and because the medication was expired. The value of returned medicines was estimated at £3,986 which at the exchange rate in April of 2012 of 1 GBP to 1.32 US dollars equaled $5,272.40 [59, 66].

In their discussion, the authors recognized that judging by the responses given for the return of medicines, in many cases there is unnecessary waste of medicine. The authors provided several suggestions to minimizing or eliminating this waste such as modifying prescription practices by reducing the supply of medicines provided throughout therapy, providing limited test supplies during initiation or during a change of therapy treatment, and judicious review of a patient's reaction and preference for a medicine. They also suggested establishment of more efficient use of electronic prescribing systems that are capable of tracking patterns in patients' medicine use. These systems could also be utilized to track a patient's medicine supply and prevent at-home stockpiling. The authors also expressed concern over the quantity and monetary value of the unused medicines returned during their study. In addition to the limited size and scope, this study also excluded healthcare facilities, which could very likely make a significant impact on medicine use and wastage data. These excluded facilities included such places as elderly care centers, elderly assisted living centers, and hospitals. Langley et al. [2005] advise that this data should not be used to extrapolate costs to the entire nation due to these limitations; however, they did recognize the significant results of this study in terms of the quantity of unused medicines and the considerable financial burden this waste placed on the national health care system of the UK [66].

The Green Pharmacy Program and the Langley et al. [2005] study are just two examples of take back programs that have achieved success in, not only the fundamental goal of providing a safe and proper pharmaceutical disposal option for individuals, but also in other areas such as estimating the value of returned medicines and promoting the concept of product stewardship. The inclusion of these two examples however should not distract from the fact that there are a wealth of other noteworthy take back programs in the USA and abroad, a selection of which are featured in Appendix A. The intention of this section rather was to demonstrate that the highlighted programs incorporated some innovative ideas and promoted the expansion of take back programs to address many of the social problems discussed previously in this chapter. By examining The Green Pharmacy Program and other programs featured in this section and the Appendix it should be evident that take back programs themselves have a great deal of diversity. If applied and utilized, this diversity has the potential to enrich future take back programs and provide them with information and ideas needed to expand their causes to address both public and environmental health problems associated with excess and unused pharmaceuticals.

Conclusion

Though the pervasiveness and consequences of pharmaceuticals in the natural environment is not entirely understood at this point, there is increasing evidence that there are significant impacts to the environment and public health resulting from excess and unused pharmaceuticals. One commonly understood principle in toxicology and science in general is the precautionary principle, which advocates that when in doubt one should proceed conservatively. In regard to pharmaceuticals in the environment, few would argue that adopting such an ideology would be anything but beneficial to society and the environment. Take back programs appear to be a very logical means of applying the precautionary principle to address the social and environmental problems individuals and communities are now facing as a result of excess and unused pharmaceuticals.

Provided they are properly planned, coordinated, and they can overcome potential road blocks, take back programs provide a wealth of opportunities for combating the many consequences of excess and unwanted pharmaceuticals. These consequences include not only detrimental environmental impacts, but also public health and other social issues such as accidental poisoning, abuse of pharmaceuticals, patient privacy issues, and inefficiencies in the healthcare system due to wasteful management of pharmaceutical resources. However, despite the far reaching impacts of take back programs, few programs have developed methods to evaluate the success of the program beyond tallying the amount of medications collected or the number of people participating in a program event. Additionally few programs have realized the full potential that take back programs have to offer, leaving substantial gaps in the take back program framework. In this regard, there are several elements that would greatly enhance the take back program framework including:

- Addressing the need for scientific justification through biological and chemical monitoring before, during, and after take back program events.
- Accounting for public awareness and risk perception of pharmaceuticals in the environment and in the home and using this information to promote a change in disposal behavior through risk communication and education strategies.

- Improving medication management strategies by developing a communication strategy to educate doctors and pharmacists as to the types and quantities of medications going unused and being returned to take back programs.

It is unfortunate that for many of the environmental challenges we are currently battling, such as climate change, urban sprawl, and dwindling biodiversity, we as a society only began to be aware of the issues and take decisive action after the problems were widespread and precariously endangering the balance of the environment. Rather than repeat this pattern with pharmaceuticals, why not take proactive action to prevent what is a looming environmental and social problem by promoting and establishing programs such as take back programs that can help reduce the amount of unused and excess pharmaceuticals that pose both a threat to public health and a threat to the health of the environmental. We cannot afford to continue to knowingly engage in irresponsible management of our medical resources because the impacts of such actions are bound to have serious consequences not only for the environment but also for the society and individuals.

References

1. Drug Enforcement Administration (DEA) (2011) National Take Back Initiative. http:www. deadiversion.usdoj.gov/drug_disposal/takeback/takeback.html. Accessed 20 Mar 2012
2. U.S. Food and Drug Administration (FDA) (2012) Proper disposal of prescription drugs. http://www.fda.gov/ForConsumers/ConsumerUpdates/ucm101653.htm. Accessed 28 Mar 2012
3. Bay Area Pollution Prevention Group (BAPPG) (2006) Report on the San Francisco bay area's safe medicine disposal days. http://oracwa.org/files/news/168/SFBAYSafeMeds-Report-August2006.pdf. Accessed 22 Mar 2009
4. Boivin M (1997) The cost of medication waste. Can Pharm J/Revue Pharm Can 130(4):32–39
5. Kuspis DA, Krenzelok EP (1996) What happens to expired medication? A survey of community medication disposal. Vet Hum Toxicol 38(1):48–49
6. Seehusen DA, Edwards J (2006) Patient practices and beliefs concerning disposal of medications. J Am Board Fam Med 19(6):542–547
7. Zero Waste Washington (2006) Washington Citizens for Resource Conservation (WCRC): a soundstats report. Prepared by Northwest Research Group Inc., http://www.zerowastewashington.org/images/pdfs/med-survey.pdf. Accessed 22 Mar 2012
8. Eckel WP, Ross B, Isensee R (1993) Pentobarbital found in groundwater. Groundwater 31(5): 801–804
9. Holm JV, Rugge K, Bjerg PL, Christensen TH (1995) Occurrence and distribution of pharmaceutical organic compounds in groundwater down gradient of a landfill (Grinsted Denmark). Environ Sci Technol 29(5):1415–1420
10. Ahel M, Jeličic I (2001) Phenazone analgesics in soil and groundwater below a municipal solid-waste landfill. In: Pharmaceuticals and personal care products in the environment: scientific and regulatory issues. ACS Symposium Series 791, Washington, DC, pp 100–115
11. Metzger JW (2004) Drugs in Municipal landfills and landfill teachates. In: Kümmerer K (ed), Pharmaceuticals in the environment: sources, fate, effects, and risks, 2nd edn. Springer, Berlin, pp 133–138
12. Fent K, Weston AA et al (2006) Ecotoxicology of human pharmaceuticals. Aquat Toxicol 76(2):122–159

13. Kotchen M, Kallaos J et al (2008) Pharmaceuticals in wastewater: behavior, preferences, and willingness to pay for a disposal program. J Environ Manage 90:1476–1482

14. Stackelberg PE, Furlong ET et al (2004) Persistence of pharmaceutical compounds and other organic wastewater contaminants in a conventional drinking-water-treatment plant. Sci Total Environ 329:99–113

15. Webb SF (2004) A data based perspective on the environmental risk assessment of human pharmaceuticals III-indirect human exposure. In: Kümmerer K (ed) Pharmaceuticals in the environment. Sources, fate, effects, and risk. Springer, Berlin

16. Webb SF, Ternes T et al (2003) Indirect human exposure to pharmaceuticals via drinking water. Toxicol Lett 142:157–167

17. Daughton CG, Ternes T (1999) Pharmaceuticals and Personal Care Products in the Environment: Agents of Subtle Change? Environmental Health Perspectives 107(6) 907–938

18. Boehme SE, Hinchey EK et al (2007) Disposal of unwanted medicines: a resource for action in your community. Habitats and Ecosystems. Chicago, Illinois Illinois-Indiana Sea Grant (IISG). http://www.iisgcp.org/unwantedmeds. Accessed 12 Mar 2009

19. Bronstein AC, Spyker DA et al (2010) 2009 Annual Report of the American Association of Poison Control Centers' National Poison Data System (NPDS): 27th Annual Report. Clin Toxicol 48(10):979–1179

20. The Partnership for a Drug-Free America (2006) Generation Rx: National Study Confirms Abuse of Prescription and Over-the-Counter Drugs. http://www.drugfree.org/newsroom/generation-rx-national-study-confirms-abuse-of-prescription-and-over-the-counter-drugs-%E2%80%9Cnormalized%E2%80%9D-among-teens. Accessed 21 Mar 2009

21. Substance Abuse and Mental Health Services Administration (2011) Results from the 2010 National Survey on Drug Use and Health: Summary of National Findings, NSDUH Series H-41, HHS Publication No. (SMA) 11-4658. Rockville, MD: Substance Abuse and Mental Health Services Administration

22. Grasso C (2009) The PH:ARM Pilot: pharmaceuticals from households. A return mechanism. Executive summary. www.lhwmp.org/home/HHW/documents/PHARM_2009_Exec_Summary_Web_Version.pdf. Accessed 21 Mar 2009

23. Huggett DB, Cook JC et al (2003) A theoretical model for utilizing mammalian pharmacology and safety data to prioritize potential impacts of human pharmaceuticals in fish. Hum Ecol Risk Assess 9:1789–1799

24. Huggett DB, Ericson JF et al (2004) Plasma concentrations of human pharmaceuticals as predictors of pharmacological responses in fish. In: Kummerer K (ed) Pharmaceuticals in the environment II. Springer, Heidelberg, pp 373–386

25. Slovic P (1987) Perceptions of risk. Science 236(4799):280–285

26. Slovic P, Fischhoff B et al (1985) Characterizing perceived risk. In: Kates RW, Hohenemser C, Kasperson JX (eds) Perilous progress: managing the hazards of technology. Westview, Boulder, pp 91–125

27. Henrion M, Fischhoff B (1986) Assessing uncertainty in physical constants. Am J Phys 54(9):791–798

28. Kahneman D, Slovic P et al (1982) Judgement under uncertainty: heuristics and biases. Cambridge University Press, New York

29. Fischhoff B, Slovic P et al (1978) How safe is safe enough? A psychometric study of attitudes towards technological risks and benefits. Policy Sci 9(2):127–152

30. Slovic P, Fischhoff B et al (1980) Facts and fears: understanding perceived risk. In: Schwing RC, Albers WA Jr (eds) Societal risk assessment: how safe is safe enough? Plenum Press, New York, pp 181–214

31. Cook PA, Bellis MA (2001) Knowing the risk: relationships between risk behaviour and health knowledge. Public Health 115:54–61

32. Lundborg P, Lindgren B (2004) Do they know what they are doing? Risk perceptions and smoking behaviour among Swedish teenagers. J Risk Uncertainty 28(3):261–286

33. Viscusi WK, Carvalho I et al (2000) Smoking risks in Spain: part III-determinants of smoking behavior. J Risk Uncertainty 21(2–3):213–234

34. Bound JP, Kitsou K et al (2006) Household disposal of pharmaceuticals and perception of risk to the environment. Environ Toxicol Pharmacol 21:301–307
35. Jakus PM, Shaw DW et al (2009) Risk perceptions of arsenic in tap water and consumption of bottled water. Water Resour Res 45:W05405
36. Nordlund AM, Garvill J (2003) Effects of values, problem awareness, and personal norm on willingness to reduce personal car use. J Environ Psychol 23:339–347
37. COMPAS (2002) F&DA Environmental Assessment Regulations Project Benchmark Survey. A Report to Health Canada (POR-02-13). Office of Regulatory and International Affairs http://www.hc-sc.gc.ca/ewh-semt/contaminants/person/impact/por-02-13-eng.php. Accessed 30 Mar 2012.
38. Harrison CM, Burgess J et al (1996) Rationalizing environmental responsibilities: a comparison of lay publics in the UK and the Netherlands. J Environ Change 6(3):215–234
39. Evans JS, Hawkins NC et al (1988) The value of monitoring for radon in the home: a decision analysis. J Air Pollut Control Assoc 38(11):1380–1385
40. Fischhoff B (1990) Psychology and public policy: tool or toolmaker? Am Psychol 45(5):647–653
41. Gibson M (1985) To breathe freely: risk, consent, and air. Rowman & Littlefield, Totowa
42. Gow HBF, Otway H (1990) Commuincating with the public about major accident hazards. European conference on communicating with the public about major accident hazards. Routledge, Italy
43. Morgan MG, Lave L (1990) Ethical considerations in risk communication practice and research. Risk Anal 10(3):355–358
44. Nazaroff N, Teichman K (1990) Indoor radon: exploring U.S. federal policy for controlling human exposures. Environ Sci Technol 24(6):774–782
45. Svenson O, Fischhoff B (1985) Levels of environemntal decisions. J Environ Psychol 5(1):55–67
46. Morgan MG, Fischhoff B et al (1992) Communicating risk to the public: first learn what people know and believe. Environ Sci Technol 26(11):2048–2056
47. Carroll JM, Olson JR (1987) Mental models in human-computer interaction: research issues about what the user of software knows. National Research Council, Washington, DC
48. Craik K (1943) The nature of explanation. Cambridge University Press, Cambridge
49. Gentner D, Stevens AL (1983) Mental models. Lawrence Erlbaum Associates, Hillsdale
50. Johnson-Laird P (1983) Mental models. Harvard University Press, Cambridge
51. Murphy GL, Wright JC (1984) Changes in conceptual structure with expertise: differences between real-world experts and novices. J Exp Psychol Learn Mem Cogn 10(1):144–155
52. Norman DA (1983) Some observations on mental models. In: Gentner D, Stevens A (eds) Mental models. Erlbaum, Hillsdale
53. Rouse WB, Morris NM (1986) On looking into the black box: prospects and limits in the search for mental models. Psychol Bull 100(3):349–363
54. Krug D, George B et al (1989) The effect of outlines and headings on readers' recall of text. Contemporary Educ Psychol 14(2):111–123
55. Boyle M, Holtgrove D (1989) Communicating environmental health risks using the doctor-patient model. Environ Sci Technol 23(11):1335–1337
56. President's Commission for the Study of Ethical Problems in Medicine and Biomedical and Behavioral Research (1982) Making health care decisions: report on the ethical and legal implications of informed consent in the patient-practitioner relationship. U.S. Government Printing Office, Washington, DC, pp 68–128
57. Peters RG, Covello VT, MacCallum DB (1997) The determinants of trust and credibility in environmental risk communication. An Emperical Study. Risk Anal 17(1):43–54
58. Institute of Medicine (2001) Science and risk communication: a mini-symposium sponsored by the roundtable on environmental health sciences, research, and medicine. National Academy Press, Washington, DC
59. XE (2011) Universal currency calculator. http://www.xe.com/ucc/. Accessed 03 Apr 2012

60. Coma A, Modamio P et al (2008) Returned medicines in community pharmacies in Barcelona, Spain. Pharm World Sci 30:272–277

61. Practice Greenhealth™ (2008) Pharmaceutical waste minimization. http://cms.h2e-online.org/ee/waste-reduction/waste-minimization/pharma/. Accessed 7 Aug 2009

62. OpenCongress (2011) Open Congress for the 112th United States Congress. http://www.open-congress.org/bill/111-s3397/show. Accessed 20 July 2011

63. Teleosis Institute (2007) Green Pharmacy Program: helping communities safely dispose of unused medicines: preliminary data report. Berkeley, CA, Teleosis Institue: 1–6. http://www.teleosis.org/pdf/GreenPharmacy_FullPreliminaryReport.pdf. Accessed 7 Aug 2009

64. Elephant Pharm (2008) Elephant pharm unveils groundbreaking green pharmacy program. http://www.elephantpharm.com/content/view/602/132/. Accessed 7 Aug 2009

65. CRG Medical Foundation (2006) Community Medical Foundation for Patient Safety. The Community of Competence™ and Foundation for Life. Patient Safety Registry. Available: http://www.comofcom.com/unused.html. Accessed 03 Apr 2012

66. Langley C, Marriot J et al (2005) An analysis of returned medicines in primary care. Pharm World Sci 27(4):296–299

Appendix A. Take Back Program Case Studies

B.W. Brooks and D.B. Huggett (eds.), *Human Pharmaceuticals in the Environment:*
Current and Future Perspectives, Emerging Topics in Ecotoxicology 4,
DOI 10.1007/978-1-4614-3473-3, © Springer Science+Business Media, LLC 2012

Program name or location	Category of program[a]	Operation length	Organizing body	Level of implementation	Outcome	Long-term plans
Green Pharmacy [15]	Continuous drop-off	1 year	Teleosis Institute	Regional-Northern California	Over 2,000 lbs unwanted medicines collected; educational material developed; 12 collection sites and 4 single-day events	No pharmaceutical waste in environment; sustainable medicine; develop support for TBPs nationwide
Northwest Product Stewardship Council [11]	Education Campaign	1998 to present	Coalition of government organizations in Oregon and Washington	State level	Website [11]	Foster cooperation among governments, businesses, and nonprofit groups to facilitate the incorporation of product stewardship principals into public policy
Smarxt Disposal [19]	Education Campaign	Unknown	U.S. Fish and Wildlife Service; American Pharmacists Association; Pharmaceutical Research and Manufactures of America	Nationwide	Website [19]	Raise awareness and provide practical guidance

Program	Type	Duration	Partners/Collaboration	Location	Results	Goals
Washington State Unwanted Medicine Return Program: Pharmaceuticals from Households: A Return Mechanism (PH:ARM) [2, 8]	Continuous drop-off	Pilot program operating from October 2006 to October 2008	PH:ARM Coalition[b]	Statewide in Washington	Collected and disposed of 35,000 lbs as of March 11, 2009. Statewide program projected cost $3.3 million or ~$5.60/lbs medicine collected [2, 8] Website [12]	Long-term sustainability through establishing partnerships with pharmaceutical manufacturers
Regional Excess Medication Disposal Service (RxMEDS) [9, 17]	Continuous drop-off	18 months	EPA and RxMEDS Partners[c]	St. Louis Metro Region	• Collected and disposed of 296,650 doses of returned medicines from over 892 participants over a 12 month period, consisting 9–220 collection days • Program cost of $137,849 covered by an EPA grant [10]	Increase public awareness. Reduce threat posed by improperly disposed of medicines to environment and public health. Improve pharmaceutical advertising for proper disposal of unused medications
Clark County, Washington Unwanted Medications Take Back Program [2]	Continuous drop-off	2003–2010	Collaboration between local pharmacies, state board of pharmacy, the DEA, and the county sheriff's office	Countywide, but required statewide cooperation	Exact quantity of collected medications unknown but 23 lbs of controlled medications were collected in 2006	Continue program at pharmacies and police and sheriff offices year round

(continued)

Program name or location	Category of program[a]	Operation length	Organizing body	Level of implementation	Outcome	Long-term plans
La Crosse, Wisconsin [2, 5]	Continuous drop-off	2007 to present	La Crosse County Solid Waste Department	Multiple counties	Estimated annual cost of $12,000–$15,000 No data available on quantities; however, participation reported as increasing with awareness of the program and very small quantity generators of hazardous waste expressing interest in utilizing the program as well	Information not available; presumed program will continue
(California) Bay Area Pollution Prevention Group (BAPPG) "Safe Medicines Disposal Week" [1, 2]	Single-event collection	1 week in May 2006	BAPPG	Collections held at 39 Waste Treatment Plants (WTP) from surrounding cities but regional coordination was required	3,634 lbs pharmaceutical waste collected from 1,500 participants Collection method determined not cost-effective	Collaborate with DEA to develop a cost-effective and legal collection system Develop long-term and sustainable solution for disposal of unused pharmaceuticals such as developing a continuous drop-off program at pharmacies Also considering prepaid mailers as a return method for unused medications

Unwanted Medication Disposal Drive [2]	Single-event collection	Annually 1-day collection event	Event sponsored by a variety of local public service agencies[d]	1-day event consisting of 25 collection sites in areas surrounding Chicago	Approximately 6,000 lbs OTC and prescription medications collected in 4 years; also installed permanent collection containers at 5 police stations and collected ~1,000 lbs in 1 year	Information not available; presumed program will continue
Earth Keeper Initiative [2]	Single-event collection	1 day- Earth Day 2007	Earth Keeper Initiative (an affiliation of a variety of religious groups), Cedar Tree Institute and Keweenaw Bay Indian Community	19 collections sites scattered across 15 counties in Michigan's upper peninsula	Over 1 ton of medicines collected from 2,000 participants Estimated street value of controlled substances: $500,000	Information not available; presumed program will continue
Safe Medicine Disposal for ME [2, 6, 14, 18]	Mail-back	Year-round operating since 2007	Coordination and administration of program provided by University of Maine Center on Aging	Statewide in Maine	See Tables A.1 and A.2 for preliminary findings (before program went statewide) One envelope of returned narcotics estimated to have a $7,000 street value Many envelopes contained full bottles of unused medications from mail-order or VA pharmacy services Received full bottles of antiretroviral (HIV/AIDS) drugs, which have a very high value	Continue program through funding provided by EPA's Aging Initiative

(continued)

Program name or location	Category of program[a]	Operation length	Organizing body	Level of implementation	Outcome	Long-term plans
Denton Drug Disposal Day [4]	Single-event collection and continuous drop-off	Semi-annually since April 2010	City of Denton, Texas and University of North Texas; event sponsored by a variety of local public and private entities[e]	1-day event held at area hospital in Denton, Texas	Over 2500 lbs collected during 4 separate events; has earned regional, state and national awards; established a permanent drop-off Kiosk; 1st take back event in Texas approved by the DEA and the Texas Commission on Environmental Quality Website [4]	Continue hosting semi-annual single-day events; possible expansion to multiple collection sites
National Take Back Initiative [7]	Single-event collection	Since September 2010	US DEA	Nationwide (USA)	During 3 events 499 tons of medications were collected	Continue operating single-day national collections events one or more times a year
British Columbia Medications Return Program (formerly British Columbia EnvirRx) [3, 13]	Foreign continuous drop-off	Since 1996	Administered by Post-Consumer Pharmaceutical Stewardship Association (PCPSA) Regulated by Post-Consumer Residual Stewardship Program Regulation	Nationwide (British Columbia)	From January 2007 to December 2007 52,635 lbs medicines returned Participation rate: 93% Access at 913 pharmacies	Increase public awareness and motivate citizens to dispose of excess and unused medications through the program

| Australia's Return Unwanted Medicines (RUM) Project [2, 16] | Foreign continuous drop-off | Since 1998 | The Commonwealth Department of Health | Nationwide (Australia) | From 1998 to 2002 1,675,513 lbs of pharmaceuticals were collected and destroyed; during 2005 the RUM project served 21 million citizens and collected and destroyed 696,241 lbs of pharmaceuticals | Information not available; presumed program will continue |

[a]Program categories include Continuous drop-off, Single-event collection, Mail-back, Education campaign, and Foreign continuous drop-off

[b]PH:ARM (Pharmaceuticals from Households: A Return Mechanisms) Coalition includes: Interagency Resource for Achieving Cooperation (IRAC), King County Local Hazardous Waste Management Program, Northwest Product Stewardship Council, Pacific Northwest Pollution Prevention Resource Center, Public Health Agencies in King and Seattle County, Snohomish County Solid Waste Division, Washington Citizens for Resource Conservation, and Washington State Department of Ecology

[c]Regional Excess Medication Disposal Service (RxMEDS) Partners include: Areas Resource for Community and Human Services (ARCHS), Schnuck Markets, Inc., Cintas Corporation, St. Louis College of Pharmacy, Missouri AARP, Mid-East Area Agency on Aging, St. Louis OASIS, Senior Services Plus, WK Health, Stericycle, St. Louis University, St. Louis City on Aging, MO Environmental Water Association, Metropolitan Sewer District, Living Lands and Waters, PhRMA, STL County Waste Management Program, and American Water Company

[d]Partners in Unwanted Medication Disposal Drive in Chicago, Illinois include: Cook County Sheriff's Police, Chicago Police Department, Chicago Department on Aging, Chicago Department of Public Heath, Illinois Attorney General's Office, Illinois TRIAD, and the Metropolitan Water Reclamation District of Greater Chicago

[e]Partners in Denton Drug Disposal Day Program: University of North Texas, the City of Denton Texas, Denton Independent School District, Denton Police Department, Denton County Sheriff's Dept; and Denton Regional Medical Center

Table A.1 Safe medicine disposal for ME: returned medicines data [6]

Category of medicine	Percent
Prescription	90
Controlled medicines	10
Over the counter	10
Total	73,000 pills, creams, patches, etc.

Table A.2 Safe medicine disposal for ME: top 4 category of medicines returned [6]

Medication category	Percent
Pain/anti-inflammatory	35
Heart, blood, or cholesterol medicine	34
Sleep or anti-anxiety medicine	19
Antibiotics	18

References

1. Bay Area Pollution Prevention Group (2006) Report on the San Francisco Bay Area's Safe Medicine Disposal Days. Bay Area Pollution Prevention Group, San Francisco, CA. Available: http://oracwa.org/files/news/168/SFBAYSafeMeds-Report-August2006.pdf
2. Boehme SE, Hinchey EK et al (2007) Disposal of unwanted medicines: a resource for action in your community. Habitats and Ecosystems. Chicago, Illinois Illinois-Indiana Sea Grant (IISG). http://www.iisgcp.org/unwantedmeds. Accessed 12 Mar 2009
3. Post Consumer Pharmaceutical Stewardship Association (2012) Help protect the environment: return expired medications. http://www.medicationsreturn.ca/. Accessed Jan 2009
4. City of Denton Texas (2011) Denton drug disposal day. http://www.dentondrugdisposal.com. Accessed 21 July 2011
5. City of La Crosse Wisconsin (2009) Household hazardous materials. http://www.cityoflacrosse.org/index.asp?NID=1167. Accessed Feb 2009
6. Crittenden JA, Kaye LW et al (2008) Implementing a consumer pharmaceutical mailback program: an analysis of the first year of the Safe Medicine Disposal for ME Program. 61st Annual Scientific Meeting of the Gerontological Society of America. Available: http://www.safemeddisposal.com/documents/GSAsymposiumPPT12-1-08.pdf
7. Drug Enforcement Administration (DEA) (2011) National take back initiative. http://www.deadiversion.usdoj.gov/drug_disposal/takeback/takeback_102911.html. Accessed November 2011
8. Grasso C (2009) The PH:ARM Pilot: pharmaceuticals from households. A return mechanism. Executive summary. http://www.lhwmp.org/home/HHW/documents/PHARM-2009-Exec-Summary-Web-Verson.pdf. Accessed 21 Mar 2009
9. Hayden SW, Gattas NM (2007) Regional excess medication disposal service (RXMEDS) presentation. Available: http://www.slidefinder.net/r/rxmeds2/rxmeds2/26870240
10. Kimbrough, W (2009) Prudent disposal of unwanted medications (RxMEDS) Final Report. Available: http://www.epa.gov/grants/winners/rx-meds-technical-reports508.pdf. Accessed Jan 2009
11. Northwest Product Stewardship Council (2009) http://www.productstewardship.net. Accessed Feb 2009
12. Pharmaceuticals from households: A return mechanism (PH:ARM) (2009) Washington State Unwanted Medicine Return Program. http://www.medicinereturn.com. Accessed Feb 2009

13. Vanasse, G. (2008) Medications Return Program Annual Report: January 2007 to December 2007. Ottawa, Ontario. Available: http://www.medicationsreturn.ca/ar2007.pdf.
14. University of Maine Center on Aging (2012) Safe Medicine Disposal for ME program http://www.safemeddisposal.com/index.php. Accessed Jan 2009
15. Teleosis Institute (2007) Green Pharmacy Program: helping communities safely dispose of unused medicines: preliminary data report. Berkeley, CA, Teleosis Institue: 1–6. http://www.teleosis.org/pdf/GreenPharmacy_FullPreliminaryReport.pdf. Accessed 7 Aug 2009
16. The National Return & Disposal of Unwanted Medicines Limited (2011) Returning your unwanted medicines (RUM). http://www.returnmed.com.au/. Accessed Jan 2009
17. U.S. EPA (2009) Area Resources for Community and Human Services (ARCHS). http://www.epa.gov/aging/grants/winners/archs.html. Accessed Jan 2009
18. U.S. EPA (2009) Winners of prudent disposal of unwanted medications: University of Maine Center on Aging. http://www.epa.gov/aging/grants/winners/umca.htm. Accessed Jan 2009
19. U.S. Fish and Wildlife Service, American Pharmacists Association et al (2009) Smarxt disposal: a prescription for a healthy planet. http://www.smarxtdisposal.net/index.html. Accessed Feb 2009

Index

A

Active pharmaceutical ingredient (API),
  168, 169
  adjusted daily intake and therapeutic daily
    intake values, 190–194
  adverse effect levels, 190
  bioconcentration factors, 187
  biodegradability, 197
  biosolids, 175–176
  CCL3, 170
  detection, 170–171
  dose-response assessment, 189
  drinking water, 190, 198
  environmental impacts, 168
  fate and behavior, 176–177
  fish, 186, 187
  human exposure, pathway, 178, 179
  lipid content and bioaccumulation, 187
  measured and predicted environmental
    concentrations, 179–186
  microbial resistance, 198–209
  MSW/landfills, 176
  physical–chemical properties, 65
  point of departure, 190
  predicted no-effect concentration, 190
  public databases, 87
  recreational/subsistence anglers, 187
  removal in WWTP and drinking water
    treatment, 178
  risk characterization metrics,
    198–208
  substances selection, 188–189
  surface and drinking water,
    194–197
  therapeutic doses, 189
  thresholds of toxicologic concern, 194

  uptake of, 92
  vapor pressure and density, 66
  veterinary pharmaceuticals, 176
Advanced oxidation processes,
  246–247
American Association of Poison Control
  Centers (AAPCC), 259, 260

B

Bay Area Pollution Prevention Group
  (BAPPG), 263
Bioaccumulation and effects
  antidepressants fluoxetine and sertraline,
    6–7
  aqueous effect threshold, 10, 11
  gill uptake model, 9
  hazard assessment, 9
  plasma model, 9
  toxicity, aquatic organisms, 7
  veterinary medicines, 6
  volume of distribution, 9, 10

C

Center for Biologics Evaluation and Research
  (CBER), 53, 56
Center for Drug Evaluation and Research
  (CDER), 20, 53, 56
Chlorination, 239–241
Coagulation, 234
Contaminant candidate list (CCL3), 170
Controlled Substance Act (CSA),
  257
Conventional wastewater treatment,
  243–244

B.W. Brooks and D.B. Huggett (eds.), *Human Pharmaceuticals in the Environment:*   297
Current and Future Perspectives, Emerging Topics in Ecotoxicology 4,
DOI 10.1007/978-1-4614-3473-3, © Springer Science+Business Media, LLC 2012

D
Drug regulation
    EA (*see* Environmental assessment)
    in European Union, 57
    NEPA process, 51–53
    risk assessment, 58
    in USA, 50–51

E
Ecological comparative pharmacology
    absorption and distribution, 92–93
    in aquatic plants, 86
    carrier-mediated uptake, 93–95
    drug targets
        definition, 98
        enzymes, 98, 100
        ortholog prediction, 98, 99
        protein family (Pfam) domains, 99
        receptors, 98
        type of, 99, 100
    homologous proteins, 89–90
    microarray analysis, 101–102
    mode-of-action, 100
    nontarget species, 88
    orthologous proteins, 89–90
    passive diffusion, 93–95
    pathway prediction, 101
    pharmacokinetics and pharmacodynamics,
        87–88
    phylogenetics
        one-way comparison approach, 90
        OrthoMCL, 91
        RBH-approach, 91
    plasma-protein binding
        CYP1A1 and CYP1A2, 96
        phase II metabolizing enzymes, 97
        phase I metabolizing enzymes, 96
        P450 mediated drug metabolism, 96–97
        serum albumin and orosomucoid, 95
        sex hormone-binding globulin, 95
    serotonin reuptake inhibitors (SSRIs), 86–87
    transcription-factor proteins, 102–103
Emerging contaminants, 226
Endocrine-disrupting compounds (EDC)
    AOP, 246
    biomarker exposure, 112
    chlorine and chloramine, 239
    copious effort, 168
    definition, 169
    membrane pore size, 244
    ozone, 241
    vitellogenin transcription, 126
    wastewater discharge, 243

Endocrine Disruptor Screening Program
        (EDSP), 142
Environmental analysis and exposure
    GC–MS, 5
    LC–MS, 5–6
    UPLC, 5
Environmental assessment (EA)
    acute ecotoxicity testing, 55
    CBER, 53
    CDER, 53
    CDER/CBER tiered approach, 55, 56
    chronic toxicity testing, 55–56
    depletion mechanisms, 55
    expected introductory concentration
        (EIC), 54
    extraordinary circumstances provision, 54
    IND, 54
    substance concentration, 54
Environmental Protection Agency (EPA), 22,
        50, 123, 141, 258
Environmental risk assessment (ERA)
    exposure reconstruction (*see* Exposure
        reconstruction)
    regulation, 58
    sub-lethal effects (*see* Sub-lethal effects)
Environment and its fate
    bioconcentration and transport, 79
    climate change, 80
    exposure models, 80
    organisms uptake, 75
    partitioning and persistence
        activated sludge, 70–71
        biotransformation, 73
        degradation rates, 73
        elimination rates, 69–70
        in environmental matrices, 69
        in sediments, 71
        water and air, 67
        water and sludge, 68–69
        xenobiotics, 74
    physical-chemical properties
        equilibrium process, 64–65
        ionization, 64
        nonequilibrium process, 65
        octanol-water partitioning, 66–67
        solubility and melting point, 65
        vapor pressure and density, 66
    sorption mechanisms, 79
    transformation product exposure, 79
    transport of pharmaceuticals
        aquatic exposure, 77–78
        FOCUS modeling framework, 77
        hydrology, 75
        soil contaminants, 76

surface water exposure, 76–77
terrestrial risk, 76
treatment processes, 79
Epigenetics
CYP1A phenotype, 132
DNA methylation, 129–130
noncoding RNA molecules, 131
plasticity, 130
European Centre for Ecotoxicology and
Toxicology of Chemicals
(ECETOC), 185
Exposure reconstruction
in aquatic systems, 113
batteries of molecular assays, 118
biomarkers of exposure, 112
biomonitoring data, 110
Canyon river, 114–116
chloroform, 111–112
definition, 111
epigenetics, 129–132
fractionation/biomolecular readout
approach, 116
molecular indicators, 113–115
otolith geochemistry, 132–133
PBTs, 111
phenotypic anchoring, 116
primrose pathway, 126–128
vitellogenin (see Vitellogenin)
WWTPs, 116–117

F

Fish and Wildlife Service (FWS), 258
Flocculation, 234

G

Geography-Referenced Regional Exposure
Assessment Tool for European
Rivers (GREAT-ER), 185
Green Pharmacy Final Report, 278
Green Pharmacy Program, 277–281

H

Henry's Law Constant, 67
Human health risk assessment (HHRA)
API
adjusted daily intake and therapeutic
daily intake values, 190–194
adverse effect levels, 190
bioconcentration factors, 187
biodegradability, 197
biosolids, 175–176

CCL3, 170
detection, 170–171
dose-response assessment, 189
drinking water, 190, 198
environmental impacts, 168
fate and behavior, 176–177
fish, 186, 187
human exposure, pathway, 178, 179
lipid content and bioaccumulation, 187
measured and predicted environmental
concentrations, 179–186
microbial resistance, 198–209
MSW/landfills, 176
point of departure, 190
predicted no-effect concentration, 190
recreational/subsistence anglers, 187
removal in WWTP and drinking water
treatment, 178
risk characterization metrics, 198–208
substances selection, 188–189
surface and drinking water, 194–197
therapeutic doses, 189
thresholds of toxicologic concern, 194
veterinary pharmaceuticals, 176
environment, pharmaceutical introduction
routes
excretion, 172–173
human active pharmaceutical
ingredients, 171, 172
improper disposal, 173–174
waste stream manufacturing, 174–175
seasonal variability, 175
uncertainty
exposure, 212–214
main sources, 209
substance selection/hazard assessment,
209–211
toxicity values, 211–212

I

Ibuprofen and metaprolol, 171
Indirect potable reuse (IPR) treatment trains,
247–249
Investigational New Drug (IND) applications,
54

M

Maximum expected environmental
concentration (MEEC), 143
Medication management strategies, 272–275
Membrane treatment, 251
Methamphetamine, 171

N

National Association of Chain Drug Stores,
    171
National Association of Pharmacy Regulatory
    Authorities (NAPRA), 274
National Center for Health Statistics, 168
National Environmental Policy Act of 1969
    (NEPA)
  EIS process flowchart, 51, 52
  environmental assessment, 51
  history of, 52–53
  US FDA, 52–53
National Poison Data System (NPDS), 259
National Survey on Drug Use and Health
    (NSDUH), 261

O

Otolith geochemistry, 132–133
Ozonation, 241–242

P

Persistent bioaccumulative and toxic
    chemicals (PBTs), 111
Pharamaceuticals environmental risk
    assessment (PERA)
  in Australia, 30
  in Canada, 29
  in European Union
    adsorption and sediment/water fate
      study, 27
    biodegradability, 26–27
    decision tree, 24–25
    ecotoxicity, 27
    guideline, 22
    Phase 2 Tier A, 23–24
    Phase 2 Tier B, 25–26
    Q&ADoc, 28
    registrations, 23
  existing guidelines *vs.* EU biocides, 39–41
  in Japan, 29–30
  PAS, 17
  production, 34–35
  REACH, 32–33
  Swedish environmental classification,
    31–32
  in Switzerland, 29
  in USA
    Citizen Petition, 22
    experimental evidence, 20
    guidance, 20
    water abstraction, human risk, 22
  vs. veterinary pharmacueticals (*see*
    Veterinary medicinal products)

Pharmaceutical Research and Manufacturers
    of America (PhRMA), 185
Pharmaceutical take back programs
  at-home stockpiling prevention, 280
  BAPPG, 263
  gaps/weaknesses identification
    medication management strategies,
      272–275
    motivating factors, 264
    risk communication and education,
      270–272
    risk perception, 266–269
    scientific justification, 264–265
  Green Pharmacy Final Report, 278
  Green Pharmacy Program, 277–280
  household hazardous waste collection
    events, 277
  monetary value of returned medications, 263
  objectives
    AAPCC, 259, 260
    accidental poisoning, 259, 260
    drug abuse, 261, 262
    NPDS, 259
    NSDUH, 261
    over-the-counter medications, 260–261
    participation surveys, 262
    PCC, 259
    pharming, 261
    poison-related fatalities, 260
    public health issues, 263
    therapeutic errors, 260
    value of returned medication, 262
  PH:ARM return mechanism, 263
  potential roadblocks, 275–276
  public and environmental health problems,
    281
  returned medicines, 280
  sustainable medicine, 277
  Teleosis Institute Green Pharmacy
    Program, 277–279
  Unused and Expired Medicine Registry,
    278, 279
  unwanted medications, collection centers,
    276
  unwanted medications, collection
    container, 276, 277
Pharmacologically active substances (PAS)
  analytical detections, 18
  European water framework directive,
    33–34
  formal guidelines, 18
  population density, 17
Photodegradation, 177
Poison Control Centers (PCC), 259
Practice Greenhealth™, 274

S
Safe and Secure Drug Disposal Act, 257
Sedimentation, 234
Serotonin reuptake inhibitors (SSRIs), 86
Sub-lethal effects
 ACR, 158
 causality and confidence
  biomarkers, 146–147
  general criteria/core elements, 145, 146
  hypothetical concentration–response
   curves, 145, 147
  inverted biological cascade pyramid, 145
 decision tree flowchart, 155, 156
 omics technology, 157
 in plants, 153–155
 receptor homology, 156
 risk assessment
  endocrine-sensitive endpoint, 144
  EU aquatic ecotoxicology guidance, 143
  MEEC, 143
  US Environmental Protection Agency
   (EPA), 141
  VTG/testis-ova, 144
 in vertebrates (fish) and invertebrates
  AChE inhibition, 150–151
  Daphnia magna, 152–153
  estrogen antagonists, 149–150
  Nilaparvata lugens, 153

T
Teleosis Institute Green Pharmacy Program,
  277–279
Trace organic contaminant (TOrCs)
 activated carbon adsorption, 234–237
 advanced oxidation processes, 246–247
 chlorination, 239–241
 coagulation/flocculation/sedimentation,
  234
 compounds detected, 229, 234
 concentration in raw wastewater, 229
 concentration in US source and finished
  drinking water, 229, 231
 concentration in US source water, finished
  drinking water, and distribution
  systems, 229, 231, 232
 concentration in US wastewater treatment
  plants, 229, 230
 conventional wastewater treatment,
  243–244
 indirect potable reuse treatment trains,
  247–249
 membrane and MBR processes, 244–245
 optimization, 243
 ozonation, 241–242
 physicochemical properties, 228
 residual management, 249–251
 treatment efficacy, 234
 UV photolysis, 237–238

U
Ultraperformance liquid chromatography
  (UPLC), 5
Unused and Expired Medicine Registry,
  278, 279
US Drug Enforcement Administration (DEA),
  257, 258

V
Veterinary API (vAPI), 176
Veterinary medicinal products (VMP)
 aquatic effects testing, 38
 endo/ecto-parasiticides, 37
 Phase II Tier A, 36, 37
 Phase II Tier B, 38
 terrestrial ecotoxicity, 37
 terrestrial side, 35
Vitellogenin
 atrazine, 126
 biomarker, 125
 fathead minnow, 118
  7-day deployments, 123, 124
  4-day EE2 exposure, 123, 124
  deploying indigenous fathead minnows,
   120
  experimental Lakes Area (ELA), 119
  male Pearl Dace, 121, 122
  reproductive failure, 123
  summer and fall results, 120, 121
  testicular tissues, 122, 123
 gene and protein, 128–129
 stressor, 125
 surrogate ecosystems, 125, 126

W
Wastewater and drinking water treatment
  technologies
 compound removal and transformation,
  227
 expansion and optimization, 227
 herbicide atrazine, 227
 TOrCs
  activated carbon adsorption, 234–237
  advanced oxidation processes, 246–247
  chlorination, 239–241
  coagulation/flocculation/sedimentation,
   234

Wastewater and drinking water treatment
       technologies (*cont.*)
    compounds detected, 229, 234
    concentration in raw wastewater, 229
    concentration in US source and finished
       drinking water, 229, 231
    concentration in US source water,
       finished drinking water, and
       distribution systems, 229, 231, 232
    concentration in US wastewater
       treatment plants, 229, 230
    conventional wastewater treatment,
       243–244
    IPR treatment trains, 247–249
    membrane and MBR processes, 244–245
    optimization, 243
    ozonation, 241–242
    physicochemical properties, 228
    residual management, 249–251
    treatment efficacy, 234
    UV photolysis, 237–238
Waste water treatment plants (WWTPs)
    API, 178
    deconjugation, 211
    degradation, 175
    drinking water treatment, 178
    Exposure reconstruction,
       116–117
    landfills, 176
    microbial resistance, 198–209
    semisolid waste, 172
    surface and groundwater, 174